CALCIUM IN BIOLOGY

METAL IONS IN BIOLOGY

EDITOR: **Thomas G. Spiro**
Department of Chemistry
Princeton University, Princeton, New Jersey 08540

CALCIUM IN BIOLOGY

Edited by

THOMAS G. SPIRO

Princeton University

A WILEY-INTERSCIENCE PUBLICATION

JOHN WILEY & SONS

New York • Chichester • Brisbane • Toronto • Singapore

Library of Congress Cataloging in Publication Data
Main entry under title:

Calcium in biology.

(Metal ions in biology, ISSN 0271-2911; v. 6)
"A Wiley-Interscience publication."
Includes index.
1. Calcium—Physiological effect—Addresses, essays, lectures. I. Spiro, Thomas G., 1935- II. Series.
[DNLM: 1. Calcium—Physiology. W1 ME9611AU / QV 276 C143]
QP535.C2C2633 1983 574.1'9214 83-12564
ISBN 0-471-88543-6

Printed in the United States of America

10 9 8 7 6 5 4 3 2 1

Series Preface

Metal ions are essential to life as we know it. This fact has long been recognized, and the list of essential "trace elements" has grown steadily over the years, as has the list of biological functions in which metals are known to be involved. Only recently have we begun to understand the structural chemistry operating at the biological sites where metal ions are found. This has come about largely through the application of powerful physical and chemical structure probes, particularly X-ray crystallography, to purified metalloproteins. From such studies we have learned that nature has evolved highly sophisticated ways of controlling the relatively flexible stereochemistry of metal ions. In one case after another, the structure and the reactivity of a metalloprotein active site have turned out to be different from anything previously encountered in simple compounds of the metals. Indeed, many a reasonable inference about active-site structure, based on the known properties of metal complexes in solution, have turned out to be erroneous. These surprises have inspired inorganic chemists to expand their vision of metal ion reactivity. The biological studies have spurred much fruitful synthetic and mechanistic work in inorganic chemistry, aimed at elucidating the means whereby nature achieves its stereochemical ends. The terra incognita of the biochemical functions of metal ions has become familiar territory to an increasing number of inorganic chemists and biochemists, and several of the more imposing mountains have been scaled. Vast stretches remain uncharted, and the field is alive with a sense of both accomplishment and new opportunities.

The purpose of this series is to convey some of this excitement, as well as the emerging intellectual shape of the field, to a wide audience of nonspecialists. Individual volumes will cover topics that are current and exciting—the recently scaled mountains that are still under active exploration. The chapters are not intended to be exhaustive reviews of the subject matter. Rather, they are intended to be readable accounts of the insights and directions that are emerging in active new areas of research. Volumes will appear on an occasional basis as progress in the field dictates.

Thomas G. Spiro

Princeton, New Jersey

Preface

Although the importance of calcium in biology has long been recognized, the astonishing richness and variety of biological phenomena involving calcium have become increasingly apparent in recent years. The application of more sophisticated biochemical probes reveals the growing complexity in the calcium story, but at the same time the similarities encountered in diverse systems have led to unifying concepts, for example, the role of calcium as a second messenger, and the involvement of conformationally flexible calcium-binding proteins in transmitting and modulating the calcium response.

Both the diversity and the universality of biological calcium processes are evident in the contributions to this volume. In Chapter 1, Kenneth B. Seamon and Robert H. Kretsinger analyze the calcium-modulated proteins, emphasizing the structural principles and homologies that have emerged from x-ray crystallographic studies. In Chapter 2, Milton J. Cormier discusses the multifaceted chemistry of calmodulin, the ubiquitously occurring calcium-binding protein that is directly involved in a wide variety of calcium responses. It has been appreciated for a century that calcium triggers muscle contraction, yet the molecular mechanism remains uncertain. In Chapter 3, C. C. Ashley reviews this subject and describes experiments, utilizing new probe techniques, which define the true course of calcium movement in muscle cells.

The important subject of exocytosis by secretory cells is taken up in Chapter 4 by Ronald P. Rubin, who discusses the various mechanisms by which the calcium sensitivity of the process may be explained. Bony structures are, of course, primarily composed of calcium and phosphate. E. D. Eanes and J. D. Termine give a lucid account of the current state of knowledge about biological mineralization processes in Chapter 5. Inorganic chemists have tended to overlook Ca^{2+}, because its rare gas electronic configuration renders it invisible to most of the spectroscopies that have been so useful in characterizing transition metal ions in biological systems. The substitution of lanthanide ions for Ca^{2+}, a method of obtaining spectroscopic handles on Ca^{2+} binding sites, has seen increasing application in recent years. In Chapter 6, R. Bruce Martin gives an authoritative account of the structural chemistry of Ca^{2+} and of the uses of lanthanide probes.

This volume is far from being a comprehensive treatment of calcium in biology. Many other topics might have been included. It does, however, give some of the flavor of current research in this exciting field.

THOMAS G. SPIRO

Princeton, New Jersey
September 1983

Contents

CHAPTER 1
Calcium-Modulated Proteins

KENNETH B. SEAMON

Laboratory of Bioorganic Chemistry
NIADDK NIH
Bethesda, Maryland

ROBERT H. KRETSINGER

Department of Biology
University of Virginia
Charlottesville, Virginia

CONTENTS

ABBREVIATIONS

ATPase	Adenosine triphosphatase
cyclic AMP	Cyclic adenosine monophosphate
pK_d	$-\log (K_{\text{dissociation}})$
pCa	$-\log [Ca^{2+}]$
ELC	Essential light chain of myosin
RLC	Regulatory light chain of myosin
ICBP	Intestinal calcium-binding protein
IADANS	2-(4′ Iodoacetamidoaniline)naphthalene-6-sulfonic acid
NMR	Nuclear magnetic resonance
NR%	Nucleotide replacements per 100 amino acids per 10^8 years
XANES	X-ray absorption near edge structure
EXAFS	Extended x-ray absorption fine structure
C-TNC	Cardiac troponin C
S-TNC	Skeletal troponin C

1 INTRODUCTION

1.1 Calcium as a Second Messenger

Physiologists and cell biologists have long recognized the important and unique role of calcium in biology. Ringer in 1882 first demonstrated that millimolar calcium is required to maintain contractility in eel and frog hearts (1). In 1952, Sandow suggested that the activation of contractile material in living muscle is attributed to the activation of the myosin ATPase by calcium (2). A critical link between acetylcholine-evoked secretion in the adrenal medulla and calcium was noted in 1961 by Douglas and Rubin, thus establishing the role of calcium in stimulus-secretion coupling (3). During the 1960s, Sutherland (4), Rall (5), and colleagues identified cyclic AMP as a crucial intermediary of hormonal responses and began to describe its various physiological roles. Rasmussen in 1970 reviewed the parallels and interrelationships between the functions of calcium and cyclic AMP and described calcium as a second messenger (6). There are numerous similarities between the two second messenger systems. The effects of cyclic AMP are all mediated by the intracellular cyclic AMP-dependent protein kinase, and the second messenger functions of calcium are all mediated by specific calcium-modulated proteins. We discuss the dynamics of calcium binding to these proteins and the resulting changes in conformation, which are obligatory for their functions, in this chapter.

1.2 Calcium-Modulated Proteins

In 1967, Ebashi et al. identified troponin as the component that confers calcium sensitivity to the skeletal myosin ATPase system (7). Troponin-C was subsequently identified as the calcium-binding subunit in the troponin complex, which, with troponin-I and troponin-T, regulates the calcium dependence of the skeletal and cardiac actomyosin systems. Skeletal troponin-C was the first calcium receptor protein that was not only characterized with regard to function but also purified to homogeneity.

Parvalbumin is a calcium-binding protein present in large quantities in the white muscle of fish, amphibia, and reptiles. Although its crystal structure has been determined, its function has not been established. The crystal structures of carp parvalbumin (8) and of porcine intestinal calcium-binding protein (9) have both provided a basis for understanding the structures of other homologous calcium-modulated proteins.

Although both troponin and parvalbumin were characterized before the various functions of calmodulin were fully appreciated, calmodulin is now regarded as the prototypical calcium-modulated protein. It is probably found in all cells of all metazoa and is highly conserved, as might be expected from the fact that it interacts with at least 15 different proteins. Calmodulin was first isolated as an activator of cyclic nucleotide phosphodiesterase. It is now known to activate a number of enzymes in a calcium-dependent manner, as discussed by Vanaman and Klee (10) and Cormier (Chapter 2). By contrast, troponin-C is highly specialized, interacting with only one complex, troponin-I and troponin-T, with the ultimate activation of one enzyme, myosin ATPase.

Kretsinger in 1972 suggested that troponin-C is a homolog of parvalbumin resulting from gene duplication (11). Collins et al. (12) determined the amino acid sequence of troponin-C and recognized four domains homologous with the three domains of parvalbumin. Subsequent amino acid sequence determinations established the homology of parvalbumin, troponin-C, both regulatory and essential light chains of myosin, calmodulin, intestinal calcium-binding protein, and the brain specific S-100 protein. All of these proteins are found intracellular and bind calcium under cytosolic conditions with $pK_d(Ca^{2+}) \sim 6$. We consider these two characteristics—intracellular localization and $pK_d(Ca^{2+}) \sim 6$—to define calcium-modulated proteins and to distinguish them from the various extracellular calcium-binding proteins.

1.3 Functional Characteristics of Calcium-Modulated Proteins

1.3.1 Physiological Requirements. Calcium acts as a second messenger in eliciting a physiological response only via calcium-modulated proteins. Most of these proteins are not themselves enzymes. They transmit the calcium signal only through their interactions with target enzymes and structural proteins.

Calcium-modulated proteins exist inside the cell; thus their functional characteristics are defined by the intracellular milieu. Cells normally maintain very low concentrations of the free Ca^{2+} ion ($10^{-7.5}$ *M* or pCa 7.5) against a concentration gradient of 10^4 to 10^5 with extracellular Ca^{2+} ion concentrations of pCa 3.0 to 2.5. Excitation of the cell, either mechanically, electrically, chemically, or in the retina by a photon, can result in increases in cytosolic calcium to pCa of about 5.0. These transient increases in calcium trigger physiological responses and must, therefore, be recognized by the calcium-modulated protein. Furthermore, these proteins must be able to recognize calcium fluxes in the presence of high but relatively constant concentrations of free Mg^{2+} ion ($\sim$2.0 m*M*) and K^+ ion ($\sim$0.1 *M*). This requires that the calcium-binding sites have dissociation constants for calcium of 10^{-6} *M* [$pK_d(Ca^{2+}) \sim 6$], and affinity for Mg^{2+} of at least 10^3 less and for K^+ and Na^+ at least 10^5 less than for the Ca^{2+} ion.

The calcium-modulated protein must be able to transmit the message of its bound calcium to a target enzyme or structural protein. The target protein must be able to discriminate between the apoprotein or magnesium protein and the calcium-bound form(s) of the calcium-modulated proteins. Thus the calcium-modulated protein, upon binding calcium, adopts a conformation that is different from those of the magnesium or apo forms. The existence of multiple binding sites on a single protein implies that a manifold of conformations corresponding to the different degrees of occupancy of the binding sites with calcium and with magnesium may exist.

The onset of the physiological response is triggered by an increase in the concentration of free Ca^{2+} ion. Some responses, such as contraction of striated muscle, may terminate as the concentration of Ca^{2+} ion decreases. This particular activation process, binding of calcium to troponin with its resultant conformational change, must be reversible within the time of muscle relaxation. Many calcium-modulated responses must be reversible within 0.1 sec. In other calcium-initiated responses, fertilization being an extreme example, the cascade of subsequent events may continue from a few hours to a lifetime, even though the initial pulse of calcium is reduced within a few seconds.

1.3.2 Metal-Ligand Interactions. Complex formation between metal ions and ligands consists of three fundamental steps that result in the replacement of the water molecules coordinating the ion with the binding ligands (13). The first step is the formation of an ion pair between the hydrated metal ion and the ligand:

$$M(H_2O)_n^{+a} + L^{-b} \rightleftharpoons M(H_2O)_n^{+a} L^{-b} \tag{1}$$

The formation of this ion pair is limited, in most cases, only by the diffusion rate constant. The second step involves dissociation of water from the metal ion:

$$M(H_2O)_n^{+a} L^{-b} \rightleftharpoons M(H_2O)_{n-1}^{+a} L^{-b} + H_2O \tag{2}$$

This step is generally found to be rate limiting in the formation of simple unidentate complexes. The third step is ligand substitution, resulting in the substituted complex

$$M(H_2O)_{n-1}^{+a} L^{-b} \rightleftharpoons ML^{(a-b)} + (n-1)H_2O \tag{3}$$

The rate of dehydration, corresponding to step 2, is not dependent on the ligand for most main group and transition group metals, and in many cases the bound water is so labile that the rates are close to diffusion limited (14). The rates of complex formation are characterized by three main considerations (15): (1) the rate of substitution is essentially independent of ligand, being a characteristic only of the metal ion; (2) a ligand binding more tightly than the water of hydration labilizes the remaining water molecules; and (3) the rate of substitution is inversely related to the charge and ionic radius of the metal. These criteria apply to unidentate complexes. In these situations, the ligand does not affect the rate of substitution. These conclusions generally obtain for multidentate ligands as well. The rate of complex formation between a macrocyclic antibiotic, monactin, and alkali ions is diffusion limited (15). The initial dissociation of one water molecule from the alkali ion is the rate-limiting step and close to diffusion limited. Following the substitution by one ligand from the antibiotic, the subsequent dissociations of other waters and coordination by other ligands proceed rapidly.

The rate of magnesium binding is slower than that of calcium due to the slower rate of dehydration, $10^{5.0}$ sec^{-1} for magnesium as compared to $10^{8.3}$ sec^{-1} for calcium (15). For calcium, the diffusion of the $Ca(H_2O)^{2+}_{6-7}$ ion will determine the overall reaction rate, whereas for $Mg(H_2O)^{2+}_6$, dehydration is much slower than diffusion and will determine the rate. The first ligand, almost always an oxygen atom, to bind calcium or magnesium should facilitate substitution by the other ligands such that the substitution rate by the subsequent ligands will be characterized by the above rates. The lanthanides, alkaline and alkaline earth metals, as well as cadmium, all have similar diffusion constants and dehydration rates between $10^{7.4}$ and $10^{9.5}$ sec^{-1}.

Diebler et al. proposed three criteria for the selective binding of metal ions by a biological carrier (15). These encompass many of the requirements for the binding of calcium by calcium-binding proteins. (1) The water molecules of the inner coordination sphere should all be replaced by the coordinating ligands. (2) The ligands should form a cavity of size appropriate for the desired cation such that the difference of the free energy of ligand-ligand interactions and dehydration of the metal is maximal. (3) The binding site should be flexible so as to allow a stepwise substitution of the solvent.

1.3.3 Multiple States. The binding of each Ca^{2+} ion by a calcium-modulated protein (CMP) can be described by a microscopic dissociation constant, $K_n = k_{-n}/k_n$. The binding of calcium by each domain would be independent of the state of the other domains; that is, there would be no cooperativity.

$$\mathrm{CMP} + \mathrm{Ca} \underset{k_{-1}}{\overset{k_1}{\rightleftharpoons}} \mathrm{CMPCa_1} \underset{k_{-2}}{\overset{+\mathrm{Ca}\;\; k_2}{\rightleftharpoons}} \mathrm{CMPCa_2} \underset{k_{-3}}{\overset{+\mathrm{Ca}\;\; k_3}{\rightleftharpoons}} \mathrm{CMPCa_3} \underset{k_{-4}}{\overset{+\mathrm{Ca}\;\; k_4}{\rightleftharpoons}} \mathrm{CMPCa_4}$$

1 4 6 forms 4 1

The binding of each Ca^{2+} ion induces a change in conformation of that domain and possibly of the entire protein. Although there is only one conformation of calmodulin with four calciums it is possible to have four different conformations with three calciums and six different conformations with two calciums. Without cooperatively and/or a wide range of calcium affinities, an ordered sequence of calcium binding is not expected. We need to understand the protein conformations associated with each stoichiometry of calcium binding and possibly with the various site distributions within each stoichiometry.

The rate of formation of the active conformer, CMP^*Ca_{n+1}, depends on both k_n and k_n^*, and the rate for the reverse reaction associated with calcium dissociation depends on both k_{-n} and k^*_{-n}. Identical relationships apply to magnesium binding to calcium-modulated proteins.

$$\text{CMPCa}_n \xrightleftharpoons[k_{-n}]{k_n \;(+\text{Ca})} \text{CMPCa}_{n+1} \xrightleftharpoons[k^*_{-n}]{k^*_{-n}} \text{CMP}^*\text{Ca}_{n+1}$$

$$\text{CMPMg}_n \xrightleftharpoons[k'_{-n}]{k'_n \;(+\text{Mg})} \text{CMPMg}_{n+1} \xrightleftharpoons[k^+_{-n}]{k^+_n} \text{CMP}^+\text{Mg}_{n+1}$$

There are large conformational changes associated with magnesium binding, which resemble those resulting from calcium binding. From the dissociation constants one can calculate that a significant fraction of the proteins will, under some physiological circumstances, bind both calcium and magnesium. Most people have assumed that calcium binds to apodomains only; that is, a magnesium domain would relax to the apo form before binding calcium; however, this remains to be determined.

Many calmodulin-binding proteins interact with only the CMP^*Ca_4 form of calmodulin, resulting in an activated protein or enzyme ENZ^0.

$$\text{ENZ} + \text{CMP}^*\text{Ca}_4 \rightleftharpoons \text{ENZ}^0\text{CMP}^*\text{Ca}_4$$

In contrast, apocalmodulin is an integral subunit of the hexadecameric phosphorylase kinase, $\alpha_4\beta_4\gamma_4CAM_4$. Similarly, apotroponin-C is part of troponin, which in turn is part of the entire thin filament, thick filament complex. The activation of these enzyme systems can be described:

$$\text{ENZCMP} + 4\,\text{Ca} \rightleftharpoons \text{ENZCMP*Ca}_4 \rightleftharpoons \text{ENZ}^0\text{CMP*Ca}_4$$

As written, the macroscopic dissociation constant for calcium is the product of the four individual constants $K_d = K_1 \cdot K_2 \cdot K_3 \cdot K_4$. In a system such as troponin, in which ENZ has a higher affinity for $CMPCa_4$ than for CMP, the apparent affinity of CMP for calcium is enhanced in the presence of ENZ. Since $K_d \cdot K_d^*$ (ENZ) $= K_d^0 \cdot K_d$(ENZ), if K_d(ENZ) $> K_d^*$(ENZ), then $K_d > K_d^0$.

$$\begin{array}{ccc}
\text{CMP} + 4\,\text{Ca} & \underset{K_d}{\rightleftharpoons} & \text{CMP*Ca}_4 \\
+ & & + \\
\text{ENZ} & & \text{ENZ} \\
\updownarrow \; K_d(\text{ENZ}) & & \updownarrow \; K_d^*(\text{ENZ}) \\
\text{ENZCMP} + 4\,\text{Ca} & \underset{K_d^0}{\rightleftharpoons} & \text{ENZ}^0\text{CMP*Ca}_4
\end{array}$$

1.4 The Homology Domain (EF-Hand)

The crystal structure of parvalbumin has proven to be something of a Rosetta Stone for interpreting much amino acid sequence information, as well as a wide range of chemical experiments. The recent determination of the crystal structure of the bovine intestinal calcium-binding protein has lent credence to many of the predictions and interpretations based on the "EF-hand" of parvalbumin and has revealed a range of variation initially unanticipated. The parvalbumin structure has already been described and reviewed in great detail (16). We summarize these descriptions in order to define a notation and to describe models with which experimental results can be compared.

Parvalbumin consists of three homology domains, each of which contains an α-helix about 10 residues long, a 10-residue loop, and a second length of α-helix about 10 residues long (Fig. 1). The common numbering system in Figure 2 facilitates description and comparison regardless of the type of calcium-modulated protein or the absolute number in amino acid sequence. The first α-helix begins with residue 1; for a specific domain, however, it might assume the appropriate Φ and Ψ values and hydrogen-bonding pattern at residue 0 or 2. The inner aspects of the two α-helices generally have hydrophobic residues at positions 2 and 5, 6, and 9 of the first helix and in the second helix at positions 22 and 25, 26, and 29, which face the

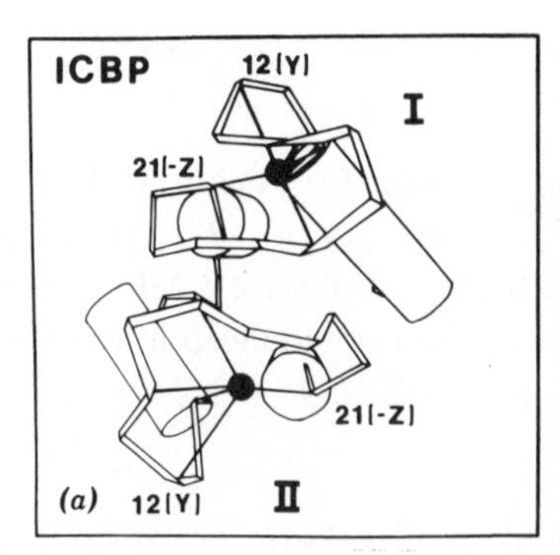

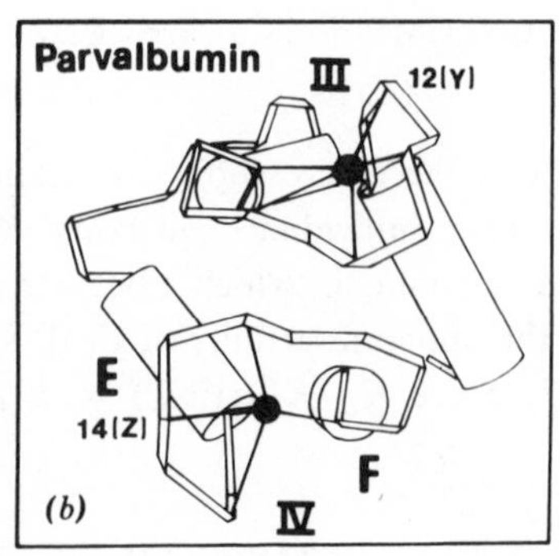

Figure 1 Pair of calmodulin (or EF-hand) domains as seen in the crystal structures of intestinal calcium-binding protein (a) and of parvalbumin (*b*). Domains I (top) and II (bottom) of ICBP and III (top) and IV (bottom) of parvalbumin are viewed down the approximate twofold axes. Cylinders represent α -helices; ● represent calcium ions. A few α-carbons are assigned domain numbers. (Figure modified from ref. 9).

hydrophobic interior. Glutamic acid frequently occurs at position 1, the start of the first helix, and at position 21, the start of the second helix.

In the standard domain the protein provides six oxygen ligands that can coordinate the Ca^{2+} or Mg^{2+} ion. The carboxylate group of aspartic or glutamic acid may coordinate calcium with either one or both oxygen atoms; we refer to it as a single "ligand." These six ligands can, approximately, be assigned to the vertices of an octahedron; the first, at postion 10, is by definition at vertex X, and the second at position 12 defines vertex Y. Figure 3 illustrates a right hand coordinate system in which the remaining ligands are at positions 14, 16, 18, and 21 and define vertices Z, $-Y$, $-X$, and $-Z$, respectively. Five of the ligands are provided by the side chains of asp, asn, glu, gln, ser, or thr. The ligand at the sixth vertex, $-Y$ at position 16, is the peptide oxygen; any side chain may be found here. The $-Z$ ligand (position 21) is usually glu or gln; the other side chains are not long enough. There is often a gly at position 15; the sharp turn here (Fig. 1) usually involves Φ, Ψ angles not accessible to other amino acids. The hydrophobic side chain of residue 17 sticks into the core of the molecule and probably stabilizes the loop.

The recognition of these characteristic residues allowed Kretsinger (11) to recognize three homology domains within parvalbumin. Although much more sophisticated algorithms (17) are now used to determine the evolutionary relationships among the proteins containing this domain, the simple scheme summarized in Figure 2 has permitted the identification of all known

Protein	Domain		0	1	2	3	4	5	6	7	8	9	10	11	12	13	14	15	16	17	18	19	20	21	22	23	24	25	26	27	28	29	30	
					n			n	n			n	X		Y		Z	G	-Y	I	-X			-Z	n			n	n			n		
Parvalbumin	II	$_{8}$D	A	D	I	A	A	A	L	E	A	C	K	A	A	D	S	—	—	F	N	H	K	A	F	—	F	A	K	V	G	L	T	S_{37}
Carp pI 4.25	III	$_{40}$A	D	D	V	K	K	A	F	A	I	I	D	Q	D	K	S	G	F	I	E	E	D	E	L	K	L	F	L	Q	N	F	K	A_{72}
	IV	$_{79}$D	G	E	T	K	T	F	L	K	A	G	D	S	D	G	D	G	K	I	G	V	D	E	F	T	A	L	V	K	A_{108}			
Troponin C	I	$_{16}$I	A	E	F	K	A	A	F	D	M	F	D	A	D	G	G	G	D	I	S	V	K	E	L	G	T	V	M	R	M	L	G	Q_{48}
Rabbit,	II	$_{52}$K	E	E	L	D	A	I	I	E	E	V	D	E	D	G	S	G	T	I	D	F	E	E	F	L	V	M	M	V	R	Q	M	K_{84}
skeletal	III	$_{92}$E	E	E	L	A	E	C	F	R	I	F	D	R	N	A	D	G	Y	I	D	A	E	E	L	A	E	I	F	R	A	S	G	E_{124}
	IV	$_{128}$D	E	E	I	E	S	L	M	K	D	G	D	K	N	N	D	G	R	I	D	F	D	E	F	L	K	M	M	E	G	V	Q_{159}	
Bovine, cardiac	I	$_{17}$K	N	E	F	K	A	A	F	D	I	F	V	^{L}G	A	E	D	G	C	I	S	T	K	E	L	G	K	V	M	R	M	L	G	Q_{49}
ICBP	I	$_{3}$P	E	E	L	K	G	I	F	E	K	Y	^{A}A	K	E	G	D	P	^{N}Q	L	S	K	E	E	L	K	L	L	L	Q	T	E	F	P_{37}
Bovine	II	$_{43}$P	S	T	L	D	E	L	F	E	E	L	D	K	N	G	D	G	E	V	S	F	E	E	F	Q	V	L	V	K	K	I	S	Q_{75}

Figure 2 Standard numbering for the homology domain with illustrative amino acid sequences. Note the deletion in parvalbumin (domain II) and the insertion in cardiac troponin C (domain I); neither binds calcium. In contrast domain I of ICBP has two inserted amino acids but still coordinates calcium with four carbonyl oxygen atoms. In parvalbumin, domain IV, and by inference in skeletal troponin C, domain I, a Glu (or an Asp in TNC) has been replaced by Gly. Water coordinates calcium at the $-X$ vertex in parvalbumin, domain IV and also by inference at the Z vertex in S-TNC, domain I. The one letter code used is: A (ala), C (cys), D (asp), E (glu), F (phe), G (gly), H (his), I (ilu), K (lys), L (leu), M (met), N (asn), P (pro), Q (gln), R (arg), S (ser), T (thr), V (val), W (trp), Y (tyr).

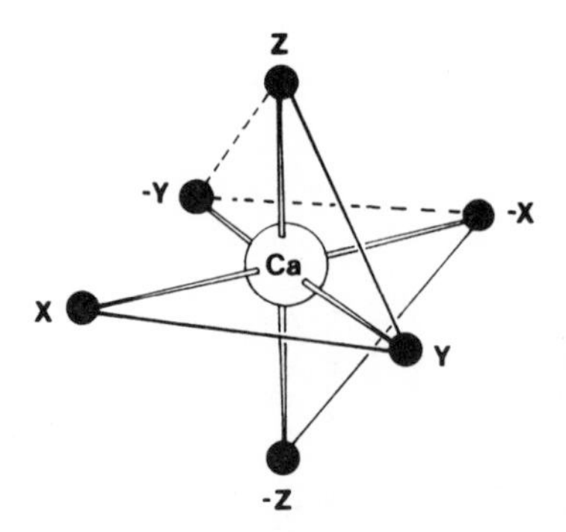

Figure 3 Octahedral coordination of the Ca^{2+} ion as occurs in the calmodulin domain. The vertices are indicated by domain number: *X* (10), *Y* (12), *Z* (14), $-Y$ (16), $-X$ (18), and $-Z$ (21). Usually, a carbonyl oxygen atom coordinates the Ca^{2+} ion at $-Y$; the other ligands come from amino acid side chains.

members of this homology family. Eight different proteins contain the number of domains and calcium-binding ability as indicated:

Protein	Domains	I	II	III	IV
Calmodulin	4	Ca	Ca	Ca	Ca
Skeletal TNC	4	Ca	Ca	Ca	Ca
Cardiac TNC	4	0	Ca	Ca	Ca
Regulatory light chain	4	Ca	0	0	0
Essential light chain	4	0	0	0	0
Parvalbumin	3		0	Ca	Ca
Intestinal CaBP	2	Ca	Ca		
S-100	2	?	Ca		

In any scheme for comparing amino acid sequences, two sorts of alignment problems occur. The numbers of residues *C*- or *N*-terminal to the domains and/or in interdomain regions vary from protein to protein. Deletions and insertions of amino acids within domains shift relative numberings. Given the simple and distinctive character of the homology domain, we use and strongly advocate adopting the following notation: Roman numerals indicate domain. Within each domain, residues 1 through 30 are assigned as already described. Interdomain residues can be designated by reference to either the preceding or the succeeding domain. Residue arg-75 of carp parvalbumin is involved in a partially buried complex of hydrogen bonds. It can equally well be called III34 or IV (-5).

The first domain of parvalbumin is now understood to have been deleted, hence the *N*-terminal domain is II. The second is III and the last, containing helices E and F of the "EF-hand," is IV. From the structure of parvalbumin it was immediately obvious that there would be minor variations on the theme previously described. Residue 98, which is 18 within the fourth domain, IV18, is glycine. It has no side chain to coordinate calcium; water is found at the -*X* vertex. In ICBP, ser-62 (II18) does not coordinate calcium directly but appears to be hydrogen bonded to a water molecule that does coordinate calcium. Similarly, in the first domain of skeletal troponin C, gly-34 is I14; water is inferred to coordinate calcium I at vertex *Z*.

The second minor variation is seen in domain III of parvalbumin. Helix D is neither an α-helix nor a 3_{10} helix but something of a bastard mixture. It still retains the characteristic hydrophobicity of the helical segments in the standard domain; residues III 22, 25, 26, and 29 are Leu, Phe, Leu, and Phe.

The first major variation is seen in parvalbumin domain II, which does not bind calcium. Regions *A* and *B* form reasonable α-helices; however, two amino acids have been deleted from the binding loop. It has been inferred from the amino acid sequences, and confirmed by direct chemical studies, that the following domains also do not bind calcium: I of cardiac TNC; II, III, and IV of the regulatory light chain, and all four of the essential light chain of myosin.

The crystal structure of bovine ICBP provides the second major variation. Although the second domain is canonical, as anticipated from the sequence, domain I coordinates calcium with four peptide oxygen atoms, in sequence ala-14 (I10 at *X*), glu-17 (I12 at *Y*), asp-19 (I14 at *Z*), and gln-22 (I16 at $-Y$). Side chains from ser-24 (I18 at $-X$) and glu-27 (I21 at $-Z$) complete the vertices. The functional and evolutionary implications of this second major variation are not understood. Note that even though two residues have been inserted over the length of the loop, it retains the same general shape and same sequence of vertices: X, Y, Z, $-Y$, $-X$, $-Z$. This conservation of topology must have some meaning since 10 distinct connections of vertices are possible for a short length of peptide chain (Fig. 3).

Domains III and IV of parvalbumin are related approximately by a twofold rotation axis. The hydrophobic inside of domain III (residues 2, 5, 6, 9, 17, 22, 25, 26, and 29) fits well with the inside of domain IV. This observation led Kretsinger and Barry (18) to postulate that homolog domains will be found in pairs—I, II and III, IV. The evolutionary relationships shown in Figure 4 are consistent with this inferred gene duplication. Domains I and II of ICBP are paired in a relationship very similar to that of domains III and

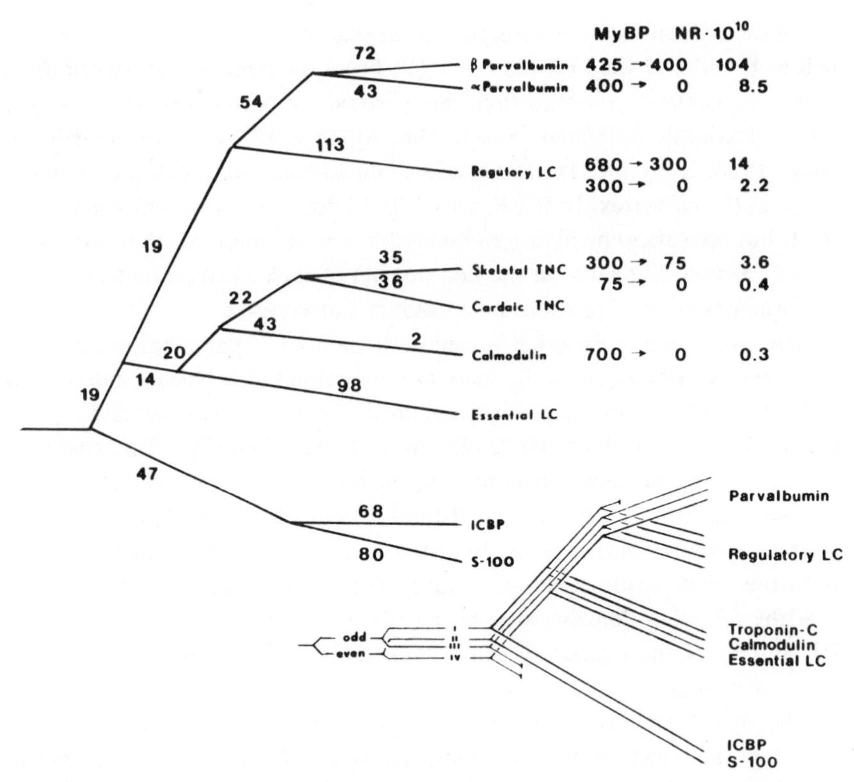

Figure 4 Evolution of the calcium-modulated proteins. The evolutionary distance of each branch is expressed in nucleotide replacements per 10^2 amino acids per 10^8 years (NR%). The rate of nucleotide replacement (NR) varies greatly, as illustrated by NR for various proteins over different time spans, millions of years before the present (MyBP). The insert shows the inferred pattern of gene duplication. (Adapted from ref. 17).

IV in parvalbumin. The Ca^{2+} ions are 12.5 Å apart in ICBP and 11.8 Å apart in parvalbumin. Nearly all chemical experiments on troponin-C and on calmodulin can be interpreted in terms of two pairs of domains. In fact, the desire to do so has created something of a self-fulfilling prophesy.

Kretsinger and Barry also suggested that in troponin-C and, by homology, in calmodulin, pair I, II matches back to back with pair III, IV thereby generating approximate point group symmetry 222 (18). Calcium-binding loops I and II are about 32 Å from loops III and IV. The surface between I, II, and III, IV at the waist of the molecule is hydrophobic. Most experiments are consistent with this "pair of pairs" idea, but few really test this most speculative part of the four domain model.

1.5 Less Well-Studied Calcium-Modulated Proteins

Our discussions focus on calmodulin, troponin-C, and parvalbumin because of their recognized importance and extensive characterizations. We briefly describe other members of this homolog family because some past studies of them contribute to our discussion. In the future, studies of these homologs, and others yet to be characterized or even discovered, will be invaluable.

1.5.1 Myosin Light Chains. All myosins characterized to date are hetero-hexamers. They consist of two nearly identical heavy chains, which have Mg-ATPase activity, two essential light chains, and two regulatory light chains. The two essential light chains have evolved from a precursor common to calmodulin and troponin-C. Although a regulatory function has yet to be assigned to the essential light chains, their presence is essential to myosin ATPase activity. Four homolog domains can be identified in their amino acid sequences; however, all four "loop regions" have lost the ability to bind calcium owing to replacements of amino acids with oxygen-containing side chains (16). Little is known of the conformation of the essential light chains either within the myosin hexamer or as isolated.

The regulatory light chains (RLC) are easily extracted from the hexamer following reaction with 5,5′-dithiobis (2-nitrobenzoic acid), hence the name DTNBLC, or by removal of divalent cations, hence EDTALC. The RLC's are phosphorylated, at different sites, by both calmodulin-dependent myosin light chain kinase and by the cyclic AMP-dependent protein kinase (19). In smooth muscle this phosphorylation is a (the) means of effecting calcium regulation of contraction. The function of this phosphorylation in striated muscle remains to be established. The RLC amino acid sequences from scallop, chicken, and rabbit muscle reveal four homolog domains. Only the first of the four domains contains a divalent cation-binding site. "Occupancy of this site with either ion allows a tight association between the regulatory light chain and the heavy chain." Conversely, "the affinity of the regulatory light chain for divalent metal ions is several orders of magnitude weaker when it is dissociated from the heavy chain" (20).

One of the two RLC's can be reversibly dissociated from the heavy chain-essential light chain complex by incubation with 10 mM EDTA at 0°C; both are dissociated at 30°C. The complete hexamer can be reassembled using RLC's from other sources. Bagshaw and Kendrick-Jones (21) exploited these properties to reassemble myosins having RLC's containing cysteine residues in different positions. These had been reacted with the spin label, *N*-(1-oxyl-2,2,6,6-tetramethyl-4-piperidinyl)iodoacetamide. The Mn^{2+} ion has several decades higher affinity for RLC than does calcium and displaces it. The cysteine of clam RLC is in domain I, and its spin label interacts strongly with the Mn^{2+} ion. The cysteines of rabbit RLC are in domains III and IV, and more than 30 Å from the Mn^{2+} ion, and do not interact. Similar results were obtained for isolated RLC. The divalent cations binding site appears to be in domain I, as indicated by the amino acid sequence.

The intact myosin hexamer from scallop has two additional calcium-specific-binding sites, which may be responsible for the calcium activation of ATPase activity. This calcium-binding capacity is not available in the isolated RLC or in the myosin remaining after the removal of one equivalent of RLC. Whether the site results from a RLC of altered conformation, or derives ligands from both light and heavy chains and is at their interface, is not known.

1.5.2 Intestinal Calcium-Binding Protein. ICBP is abundant in the small intestine and the kidney, tissues particularly active in transcellular transport of calcium (22). It is absent in rachitic animals. Its function remains unknown, but it may facilitate the diffusion of calcium. From the amino acid sequences of porcine and bovine ICBP, it was obvious that the latter half, residues 44 through 71, of the 90-amino-acid protein consists of a canonical homolog domain. The first half also contains a homolog domain; however, the optimal alignment of amino acid residues necessitates an insertion of two residues in the calcium-binding loop region.

The crystal structure determination of bovine ICBP by Szebenyi et al. (9) confirmed that the first half does indeed contain a homolog domain, as does the second half. Further, the two domains are related to one another by an approximate twofold axis, as are domains III and IV in parvalbumin. This supports the suggestion that two domains complement one another and that two pairs of domains, in similar relationship, exist in calmodulin, troponin-C, and the myosin light chains. Not surprisingly, the residue at which each helix begins and ends varies slightly from domain to domain. Similarly, the exact relationships between the two helices of a domain, and between the two domains within a pair, is different between ICBP and parvalbumin. The functional implications of these differences are not understood.

1.5.3 S-100. We know that S-100 occurs primarily in glial cells. As purified from brain tissues, it consists of two dimeric proteins—A, an $\alpha_1\beta_1$ heterodimer, and B, a β_2 homodimer. Both α and β subunits are similar to intestinal calcium binding protein (23) and have evolved from a four domain precursor by deletion of domains III and IV. From the close similarity between S-100 and ICBP, one anticipates calcium coordination in domain I similar to that seen in domain I of ICBP, and different from those of domains III and IV of parvalbumin.

The function of S-100 is unknown. Further, it is not known whether S-100 interacts with another protein or whether it might, as has been suggested for parvalbumin, serve as a calcium-binding relaxing factor or, analogous to ICBP, facilitate diffusion of calcium. The one characteristic of S-100 that is different from the other EF-hand domain proteins is that it exists as a dimer.

We know that S-100 interacts with phenothiazines and is retained, with calmodulin, on phenothiazine-Sepharose affinity columns (24). Although the evidence is somewhat contradictory, it seems that the phenothiazine site is moderately hydrophobic and relatively specific. Quite certainly the calmodulin structure consists of two pairs of homolog domains. The two pairs probably interface at their hydrophobic "backsides" opposite from the calcium-binding loops, possibly generating approximate 222 point group symmetry. In any case, this hydrophobic waist where the two halves (pairs) meet might be the binding site for phenothiazine. The S-100 monomer, either α or β, surely has a structure similar to that of ICBP, that is, a pair of domains. Monomeric ICBP does not bind phenothiazine, whereas dimeric S-100 does. The two monomers of S-100 may interact through their "backsides," as do the two pairs within monomeric calmodulin, to form a phenothiazine-binding pocket.

The S-100 binds calcium with $pK_d(Ca^{2+}) = 4.3$ at pH 7.5 (25). Calcium binding can be antagonized by monovalent cations. Calcium binding to S-100b (the β_2 dimer) causes a decrease in α-helix content with apparent $pK_d(Ca^{2+}) = 4.2$ at pH 7.5 (26). Other calcium-modulated proteins have an increased fraction of α-helix following their binding of calcium.

1.5.4 Awaiting Amino Acid Sequence Determinations. The striated muscles of various arthropods, annelids, and molluscs contain one or more acidic calcium-binding proteins in concentrations up to 2.5 g/kg wet weight (27). They are different from calmodulin, troponin-C, and parvalbumin and have a monomer molecular weight of about 20,000. Wnuk et al. (27) suggested that invertebrate sarcoplasmic calcium-binding protein may assume the function of parvalbumin, which does not occur in these invertebrates.

Oncomodulin is found in tumor cells of various origins but not in normal adult or fetal tissue (28). Its molecular weight is 11,500, and it binds two equivalents of calcium. In contrast to parvalbumin, it does activate cyclic nucleotide phosphodiesterase; this activation is inhibited by trifluoperazine. The tumor protein can also mimic calmodulin in stimulating DNA synthesis in calcium-deprived cells (29). Although the amino acid sequences of oncomodulin and of the invertebrate protein(s) have yet to be completed, it appears that they are both homologs of calmodulin.

Kretsinger (30) predicted that "calcium modulated proteins contain EF-hands and that any protein containing an EF-hand is calcium modulated." So far, this generalization seems to hold true. We emphasize, however, that in recent years many additional calcium-binding proteins from the cytosol have been identified. Some of these, such as actin, which is not a calmodulin homolog, bind divalent cations so strongly as to be in the calcium or magnesium form even in the resting cell (31). Here the calcium seems to play a structural role instead of a regulatory one. For several, such as the neutral protease of skeletal muscle (32), the reported calcium affinity is so low that they are not expected to bind calcium in the cytosol. Perhaps they function to digest the proteins of damaged cells with calcium concentrations approaching extracellular levels. There definitely are some cytosolic enzymes, such as aequorin (33) or the phospholipid-dependent *c*-kinase (34), which are well characterized and have $pK_d(Ca^{2+})$ of about 6.0. They are calcium modulated, but not via a calmodulin-like subunit. We are eager to know their amino acid sequences; possibly a homolog domain gene has been spliced into their total gene.

1.6 Evolutionary Relationships

As is so often the case in biology, evolution provides the unifying concept. Figure 4 presents a simplified version of the evolutionary pattern calculated by Goodman (17). The homolog domain is deduced to have been present in the precursor to all metazoa, if not all eukaryotes. This has subsequently been confirmed by the finding of calmodulin itself in many different tissues of various animals, plants, and fungi.

Calmodulin is most closely related to troponin-C, and these two to the essential light chains of myosin. The second major grouping includes the subfamilies of parvalbumin and the regulatory light chains of myosin. The common precursor of these five proteins contained four domains. Domain I of parvalbumin was deleted after, or at the time of, the divergence of the parvalbumins and the regulatory light chains. The two domain molecules, S-100 and ICBP, diverged from these first five at an even earlier date.

The domain pair I, II of ICBP and of S-100 is more closely related to I, II of the other proteins than to the domain pair III, IV. Domains I and III and domains II and IV are more closely related to one another than are the alternate pairings. This strongly implies that the gene coding for a precursor "odd," "even" pair of domains duplicated to form the I, II and the III, IV pairings (Fig. 4). Subsequently III, IV was deleted to form S-100 and ICBP. The earlier duplication of a single domain to form the odd, even pair is inferred. Such a single domain has yet to be identified, but it is possible that a single, unpaired domain may be encountered spliced into a large protein.

Although we have much to learn about molecular evolution, the following generalization seems valid. A single type of protein performing the same function in different organisms will evolve at a fairly constant rate (nucleotide replacements per 10^2 amino acids per 10^8 years = NR%) characteristic of that particular protein. This observation led to the term "molecular clock." The rate of evolution of parvalbumin is about 20 times faster than that of calmodulin. Because of its numerous interactions with targets, calmodulin is very highly conserved (0.3 NR%); the overwhelming majority of mutations are nonadaptive. In contrast, during those periods of "adaptive radiation" when there is frequent speciation and new functions are being found for specific proteins, the rate of evolution of all the calcium-modulated proteins is much higher ($\sim$20 NR%).

Since calmodulin is found in animals, plants, and fungi, it was probably present in the unicellular precursor of all eukaryotes. Neither calmodulin nor any of the other proteins containing homolog domains have been identified in prokaryotes. The origin of this domain and the use of calcium as a second messenger may be one of the key events in the evolution of eukaryotes.

Although these evolutionary relationships cannot reveal the subtleties of ion selectivity and conformational change, they do emphasize the universality and significance of the homolog domain. In this chapter we evaluate the various studies that relate to the dynamics of these calcium-modulated proteins.

2 METAL BINDING TO CALCIUM-MODULATED PROTEINS

2.1 Parvalbumin

Parvalbumins from both α and β subfamilies have two binding sites of equal affinity for calcium (pK_d 8.0 to 8.5) (35). These sites also bind magnesium competitively with calcium with p$K_d(Mg^{2+}) \sim 4.2$. Even though the selectivity for calcium over magnesium is 10^4, these sites are referred to as calcium-magnesium-binding sites since they bind both cations competitively at physiological concentrations. A secondary binding site has been detected by NMR relaxation studies only in parvalbumins of the β subfamily (36, 37). This site has similar affinity for calcium and for magnesium, p$K_d \sim 2.5$.

Specificity for the high affinity binding sites has been assessed by observing the competition of metal ions for gadolinium, which binds to carp parvalbumin with a p$K_d(Gd^{3+}) \sim 11.3$ (36). Previously, it was determined that lanthanides such as terbium occupy both parvalbumin-binding sites isomorphously with calcium (38). Other cations compete with gadolinium for the binding site(s) with the following rank order and ionic radii:

$$Gd^{3+}\ (0.94\ \text{Å}) \geq La^{3+}\ (1.06\ \text{Å}) > Cd^{2+}\ (0.92\ \text{Å}) > Ca^{2+}\ (1.06\ \text{Å}) \\ > Sr^{2+}\ (1.27\ \text{Å}) > Zn^{2+}\ (0.69\ \text{Å}) > Mg^{2+}\ (0.78\ \text{Å})$$

The affinity of the lanthanides is about 100 times greater than that of calcium. The enhanced binding of the lanthanides, as compared to calcium, is probably a result of their increased charge, which would result in a larger ion-dipole interaction with negatively charged carboxylate ligands. Rhee et al. (39) reported that the Eu^{3+} ion binds to both parvalbumin binding sites III and IV, retaining one water molecule at each site. Since the lanthanides bind 7 to 9 water molecules, this indicates that the metal ion is losing between 6 and 8 waters of hydration upon binding to the protein. This would certainly increase the free energy of binding relative to calcium, which is reported to be hexahydrate (40, 41) and would lose only 6 waters upon binding.

Strontium has an affinity about 10 times less than calcium, whereas zinc and magnesium have affinities from 10^2 to 10^3 less. Therefore, metal ions with ionic radii either much larger or much smaller than that of calcium bind with much lower affinity. The monovalent cations, Na^+ and K^+, do not bind competitively to the calcium-binding sites in parvalbumin, as assessed by direct binding studies (35). Grandjean et al. (42) using ^{23}Na NMR, however, detected sodium binding to pike parvalbumin with a $pK_d(Na^+) \sim 1.5$; this binding is inhibited by potassium, magnesium, and calcium. They also determined that the bound Na^+ ion has about four water molecules in its coordination shell.

Most investigators have not been able to detect differences in the calcium affinities at sites III and IV, and have concluded that calcium is bound in a noncooperative manner at two independent sites with equal affinities. Donato and Martin (43) did report that the loss of one Ca^{2+} ion can be achieved by treatment of native carp parvalbumin with an excess of EGTA at pH 6. Higher pH values are required to remove a second Ca^{2+} ion. It is difficult, however, to determine whether a specific calcium ion is being removed, or whether an average population of parvalbumin with one calcium is being produced.

Several experiments indicate that cations other than calcium may have affinities for site III different from those for site IV, and hence have a differential affinity relative to calcium. Thus, ^{113}Cd bound to parvalbumin displays two NMR resonances. Drakenberg et al. (44) assigned the resonance containing a water ligand in its calcium form to site IV, on the basis of its pH sensitivity. The $^{113}Cd^{2+}$ ion at site IV is displaced by Gd^{3+} before the $^{113}Cd^{2+}$ at site III is displaced. Thus, site IV apparently has a higher affinity for Gd^{3+} relative to Cd^{2+} than does site III when site III is occupied by Cd^{2+}. Lee et al. (45) obtained similar results by studying the ^{1}H-NMR resonances of carp parvalbumin as ytterbium is added to the protein. These experiments clearly identify sets of resonances affected by the addition of one Yb^{3+} ion, and a set of resonances affected by the addition of a second Yb^{3+}. These results are consistent with the sequential addition of the lanthanide at two binding sites. Recent studies have indicated that the replacement of ^{113}Cd at the two binding sites by lanthanides occurs with a different selectivity, which depends on the ionic radii of the lanthanides (46).

2.2 Troponin-C

2.2.1 Skeletal Troponin-C. Potter and Gergely (47) studied the binding of calcium by S-TNC using an EGTA buffering system to establish low concentrations of free Ca^{2+} ion. This method has subsequently been used to study the calcium-binding characteristics of a number of calcium-binding proteins. Skeletal troponin-C contains four calcium-binding sites (I–IV). At 0.1 *M* KCl these can be separated into two classes of sites: two with a high affinity for calcium $pK_d(Ca^{2+}) = 7.3$, and two with low affinity $pK_d(Ca^{2+}) = 5.5$. The two high affinity sites also bind magnesium with a $pK_d(Mg^{2+}) = 3.6$. Thus, these high affinity sites are similar to those of carp parvalbumin. They bind calcium with high affinity and also bind magnesium at physiological concentrations, albeit with a much lower affinity than calcium. It was originally reported that the two low affinity sites do not bind magnesium; hence, these sites have been referred to as calcium-specific sites. Subsequent studies have determined that the two low affinity sites bind magnesium about 10^4 times more weakly than calcium (48). This binding is, however, too weak to be physiologically important. In the presence of 3.0 m*M* magnesium, the binding constants for the high affinity sites are reduced to $pK_d(Ca^{2+}) = 6.6$; those of the low affinity sites are less affected. Classifying sites into those of high affinity, which bind magnesium, and those with low affinity, which are calcium specific, is useful in discussions of both protein structure and physiological function.

Terbium binds to both the high and the low affinity sites with $pK_d(Tb^{3+})$ of 8.7 and 6.9, respectively. Wang et al. (49) further determined that the coordination sphere of the Eu^{3+} ion at the high affinity sites contains 2 moles of water, and at the low affinity sites contains 3 moles of water. Hence, europium loses about 6 waters of hydration upon binding to S-TNC, as does calcium. The entropic contribution for dehydration of the $Ca(H_2O)_{6-7}^{2+}$ and the $Eu(H_2O)_{8-9}^{3+}$ ions are similar. The greater binding affinities for the lanthanides as compared to calcium are probably due to the greater charge of the lanthanides.

The binding of calcium to isolated TNC does not represent the physiologically relevant binding phenomena, as TNC is a subunit in the whole troponin complex. Troponin-C interacts with troponin-I (TNI) and binds calcium with similar affinities in the 1:1 TNC-I complex as in the heterotrimer containing troponin-T. Skeletal troponin-C in the troponin complex binds calcium with higher affinity than does isolated S-TNC at both classes of sites; $pK_d(Ca^{2+}) = 8.3$ for the high affinity sites and $pK_d(Ca^{2+}) = 6.5$ for the low affinity sites.

2.2.2 Cardiac Troponin-C. Cardiac TNC contains four domains as predicted from its sequence homology with parvalbumin. Domain I, however, has lost two calcium-binding amino acids and is predicted not to bind calcium (Fig. 2). Cardiac troponin-C binds three Ca^{2+} ions at two classes of sites: $n = 2$, $pK_d(Ca^{2+}) = 7.1$, and $n = 1$, $pK_d(Ca^{2+}) = 5.5$ (50). The two high affinity sites display apparent positive cooperativity and also competitively bind magnesium, $pK_d(Mg^{2+}) = 3.1$. Thus, C-TNC contains two high affinity sites, which are similar to those of S-TNC and parvalbumin, and only one low affinity-binding site, in contrast to the two in S-TNC.

The affinities of C-TNC for calcium are also greater in the whole cardiac troponin complex than those of isolated C-TNC; $n = 2$, $pK_d(Ca^{2+}) = 8.6$, and $n = 1$, $pK_d(Ca^{2+}) = 6.4$. This is a similar increase in calcium affinity as observed for S-TNC as a result of being in the skeletal troponin complex. The apparent positive cooperativity for the two high affinity sites in isolated C-TNC is not observed in the whole cardiac troponin complex. The available evidence indicates that calcium and magnesium binding to C-TNC is qualitatively and quantitatively very similar to that observed for S-TNC. The main difference is the loss of metal binding capacity in domain I.

2.2.3 Assignment of Binding Sites. The high and low affinity binding sites in S-TNC have been assigned to domains III and IV and to I and II, respectively, by a number of techniques. Sin et al. (51) monitored the chemical reactivity of carboxylate groups at the binding sites, at different calcium to protein ratios, and found that those at sites III and IV are relatively unreactive when 2 moles of calcium are bound to S-TNC. Thus, these sites were identified as the two high affinity sites, with I and II being the low affinity sites. Studies on tryptic fragments of S-TNC (Table 1) have also shown that fragments containing sites III and IV bind calcium with higher affinities than fragments containing sites I and II (52).

The binding of terbium to S-TNC, as monitored by the sensitized fluorescence enhancement of bound terbium by tyrosine, shows a stoichiometry of two (53). By homology and inferred isology with the structure of parvalbumin, the ring of tyr-109 should overlay site IV, even though its peptide carbonyl oxygen coordinates calcium at site III. Therefore, the first two sites that bind the Tb^{3+} ion are those closest to tyr-109, sites III and IV.

The distance between the two high affinity sites has also been determined by measuring the energy transfer between a lanthanide ion at one of the sites and a lanthanide ion at the other site. The distance obtained between Tb^{3+} and Nd^{3+}, 9.2 Å (54), is in agreement with the distance determined for parvalbumin using the same experimental protocol (9.4 Å) and from its crystal structure (11.8 Å).

Table 1 Properties of Calcium Binding Peptides and Proteolytic Fragments

Protein	Fragment: Name	Fragment: Residues	Fragment: Domain	Residues (total)	$[\Theta]_{222}$ deg cm^2 dmol^{-1}: $-Ca^{2+}/H_2O$	$+Ca^{2+}/H_2O$	CF_3CH_2OH	pK_d (Ca^{2+}): H_2O	CF_3CH_2OH	Reference
MCBP	complete	1-108	II(−8)-IV28	108				7.5 7.5		126
MCBP		1-75	II(−8)-III35	75				2.5		126
MCBP		76-108	IV(−4)-IV28	33				2.5		127
MCBP		38-108	III(−2)-IV29	71				4.9		127
sTNC	complete	1-159	I(−16)-IV30	159	−9600	−14800		7.3 7.3 5.3 5.3		47
sTNC	TR1	9-84	I(−8)-II31	74	−14800	−14800		5.0 5.0		79
sTNC	TR2	89-159	III(−4)-IV30	71	−5900	−12900		7.7 7.7		79
sTNC	TH1	1-120	I(−1)-III27	120	−9000	−12000		5.3 5.0 5.0		79
sTNC	TH2	121-159	IV(−11)-IV30	39	−5800	−9700		4.4		79
sTNC	CB8	46-77	II(−7)-II24	32	−2170	−2670		3.9		79
sTNC	CB9	84-135	III(−9)-IV06	52	−5730	−10800		5.3		76, 79, 128
sTNC		103-123	III10-III30	21	−1900	−6600	−12000	weak		74
AcsTNC		103-123	III10-III30	21	−1900	−6100	−11400	2.5	5.5	74
AcsTNC		98-123	III05-III30	26	−3900	−7700	−12100	4.6	5.7	74
AcsTNC		90-123	III(−3)-III30	34	−4500	−9200	−13400	5.4	6.0	74
cTNC		84-161	II30-IV30	77				8.3 8.3		129
CAM	complete	1-148	I(−9)-IV29	148	−15400	−19750		6.6 6.5 6.4 6.3		10
CAM		129-140	IV10-IV21	12						130
CAM		74-148	II26-IV29	77				4.8, weak		131
CAM	E_1	1-106	I(−9)-III23	106				weak		82
	E_3	107-148	III24-IV29	42	−17300	−19150				82
	E_2	1-90	I(−9)-III07	90	−3550	−13200				82
CAM	TR1-C	1-77	I(−9)-II31	77	−19050	−22150		6.2 6.2		82
	TR2-C	78-148	II32-IV29	71	−15450	−17750		6.2 6.2		82

The two high affinity sites in C-TNC have been assigned to sites III and IV by homology with S-TNC. The ^{1}H-NMR results with gadolinium also identify sites III and IV of C-TNC as the first two sites to bind the lanthanide (55).

2.3 Calmodulin

Numerous studies indicate that calmodulin binds 4 moles of calcium; however, the reported affinities and ranges thereof vary with experimental conditions and techniques (for review, see ref. 10). Crouch and Klee (56) reported that the first two calcium ions bind cooperatively to calmodulin in the presence of potassium with a Hill coefficient of 1.33. Haiech et al. (57) found that in the absence of potassium and magnesium, calcium binds at four sites with $pK_d(Ca^{2+})$'s of 7.2, 6.8, 6.2, and 6.0. Although the differences in calcium affinity between the sites is not large, there are clear distinctions among the sites. The difference in affinity (pK_d 7.2 vs. 6.8), of the two higher affinity sites is reduced in the presence of high potassium. All four sites also bind magnesium with $pK_d(Mg^{2+})$ about 3.0. The concentrations of potassium and magnesium used in various studies are, therefore, very important in determining calcium affinity and in inferring the physiological significance of such affinities.

The small differences in the binding constants for calcium among the four sites in calmodulin, as compared to the difference between the two classes of binding sites in TNC, make it difficult to distinguish populations of calmodulin with specific sites occupied by calcium. Haiech et al. (57) proposed that calcium binds in the order: site II, (I, III), IV. They assumed that the binding affinity at a given site is proportional to the number of carboxylate groups at the binding site; however, there appears to be no simple correlation between the number of carboxylate ligands and calcium affinity at the binding sites.

Experimental evidence for sequential binding of calcium to calmodulin has come from ^{1}H NMR, UV difference spectroscopy, and lanthanide luminescence studies. The ^{1}H NMR spectrum of calmodulin changes during the addition of four equivalents of calcium in two distinct steps (58). One conformation is observed with two Ca^{2+} ions bound; a second conformation is detected with four calciums bound. Unique spectral changes are observed for the resonances of tyr-99, tyr-138, and trimethyllysine-115, as a result of the binding of the first two equivalents of calcium. These residues are contained within sites III and IV. Seamon (58, 59) concluded that sites III and IV were the first to bind calcium. He assumed that these resonance shifts were due to local changes in structure at the binding sites. Calmodulin also exhibits a calcium-induced UV difference spectrum, which detects the change in environment of tyr-138 (60, 61). This change also occurs as the first two equivalents of calcium are added to calmodulin.

Tyr-99 (III16) of calmodulin occupies a position homologous to phe-57 of parvalbumin and tyr-109 of S-TNC. All of these rings should overlay the metal at site IV and should be capable of transferring energy to terbium bound at that site. This has been demonstrated for carp parvalbumin and also for S-TNC. Kilhoffer et al. (62) determined that there is no terbium luminescence when terbium is added to calmodulin up to a molar ratio of 2.0. They concluded, therefore, that the first two Tb^{3+} ions do not bind at either site III or IV. Similar conclusions were also reached in studies on octopus calmodulin, which lacks tyrosine at position 99 (63). In this case, the observed luminescence enhancement results from energy transfer from tyr-138, (IV19). They interpreted their results with octopus calmodulin in terms of a binding order (I, II), III, IV. Similar results were obtained by Wang et al. (64), who also concluded that calcium and terbium bind at sites with the same sequential order.

Sensitized Tb^{3+} luminescence is dependent on the orientation of the donor as well as its proximity to the bound Tb^{3+} ion and, therefore, Tb^{3+} at site III may not be sensitized by tyr-99. The lack of Tb^{3+} luminescence with two Tb^{3+} ions bound does not necessarily imply that site III is not one of the first sites to be occupied. Since, as has been previously discussed, the sites may display different selectivities for various metal ions, one must also be cautious in inferring a sequence for calcium binding from the terbium sequence.

2.4 Spectroscopic Investigations of the Calcium-Binding Sites

Although there are several examples of nitrogen serving as a calcium ligand, oxygen is the only observed or inferred ligand in the various calcium-modulated proteins. It is usually supplied by the side chains of Asp, Asn, Glu, Gln, Ser, or Thr. However, in most of the calcium-modulated proteins, the peptide carbonyl oxygen coordinates calcium at the $-Y$ vertex, position 16. In the intestinal calcium-binding protein, four carbonyl oxygen atoms coordinate calcium in domain I. In at least two instances, IV18($-X$) of parvalbumin and I14(Z) of skeletal TNC, water oxygen coordinates calcium. It is difficult to determine the exact structure of the calcium-binding site by noncrystallographic means. However, several spectroscopic techniques applied to proteins in solution are yielding promising results in studies relating to the coordination of the metal ion at the binding site.

2.4.1 ^{43}Ca NMR. It is possible to observe the spin $I = 7/2$, NMR resonance of ^{43}Ca bound at high affinity calcium-binding sites. Andersson et al. (65, 66) have observed the ^{43}Ca signal corresponding to calcium bound at sites III and IV of parvalbumin, S-TNC, and two sites in calmodulin. The resonances in all three cases consist of broad signals ($\Delta\nu_{1/2}$ of 600 to 800 Hz), which are shifted 10 ppm downfield of the resonance of free ^{43}Ca. The ^{43}Ca bound to the three proteins have resonances at the same chemical shift position and have similar quadrupolar coupling constants. Thus, the environments of the bound ^{43}Ca ions are similar for all three proteins, indicating that the binding sites have similar coordination geometries. The relaxation times of the bound ^{43}Ca nuclei are determined by correlation times similar to those calculated for the rotational correlational times of the proteins. Therefore, the ^{43}Ca is bound relatively tight at the binding sites and is not exchanging rapidly with free ^{43}Ca. It is estimated that k_{off} for ^{43}Ca at the parvalbumin-binding sites is 20 s^{-1} at 23°C.

The ^{43}Ca resonance corresponding to ^{43}Ca bound at low affinity sites is not directly observed because of the more rapid exchange of the bound calcium with free calcium. This rapid exchange results in a broadening of the free ^{43}Ca resonance. Parvalbumin causes a slight broadening of the free ^{43}Ca resonance at 23°C and neutral pH (67, 68). Broadening of the free ^{43}Ca resonance by parvalbumin can be increased when the protein structure is disrupted by incubation at 85°C or at extremes of pH, thereby increasing k_{off} and reducing affinity. This is consistent with the tight binding of ^{43}Ca at the parvalbumin-binding sites, which under normal conditions does not readily exchange with free ^{43}Ca. In contrast, S-TNC significantly broadens the free ^{43}Ca signal at room temperature and neutral pH (65). This broadening is due almost exclusively to the exchange of ^{43}Ca at the low affinity sites with free ^{43}Ca. It is possible to calculate that $k_{off} = 10^{3.0}$ sec^{-1} for ^{43}Ca at the low affinity sites from the temperature dependence of the observed line broadening. This is 50 times faster than k_{off} for the high affinity sites.

Calmodulin also broadens the resonance of free ^{43}Ca (69); however, it is not possible to attribute the broadening of the free ^{43}Ca resonance to only the exchange at two "low affinity sites." An analysis of the temperature dependence of the broadening results in the following parameters: $k^1_{off} = 20 - 50$ sec^{-1} and $k^2_{off} = 1.1 \times 10^3$ sec^{-1} for two pairs of calcium ions. As with

S-TNC, ^{43}Ca is bound at two kinetically distinguishable classes of sites on calmodulin. However, in contrast to S-TNC, ^{43}Ca at the "high affinity" sites of calmodulin does exchange rapidly enough with free ^{43}Ca to broaden the resonance. The rates determined for the exchange of ^{43}Ca at the low affinity sites of calmodulin and the low affinity sites of S-TNC are nearly identical.

2.4.2 ^{113}Cd NMR. ^{113}Cd has a spin $I = 1/2$ nucleus and its resonance is very sensitive to its chemical environment, making it a good probe of the calcium-binding sites. The $(^{113}Cd)_2$-parvalbumin complex exhibits two resonances, at -93.8 and -97.5 ppm, which have been assigned to cadmium bound at sites III and IV, respectively (44). The position of these resonances corresponds to ^{113}Cd bound exclusively to oxygen atoms, consistent with the crystal structure of Ca_2-parvalbumin. The resonances are also relatively insensitive to pH changes from 6 to 9 and are insensitive to Cl^- ion up to 10 mM. The ^{113}Cd resonance assigned to the site IV cadmium is only slightly shifted by high pH. The binding sites are, therefore, relatively rigid and protected from solvent interactions. It has been shown that the ^{113}Cd resonance assigned to site IV in parvalbumins of the β-subfamily are affected by the binding of calcium or magnesium at a secondary site on the protein (37). This is not observed in parvalbumins of the α-subfamily.

Skeletal troponin-C and calmodulin each exhibit two ^{113}Cd resonances at -111 and -107 ppm, and at -118 and -88 ppm, respectively, at a molar ratio of cadmium to protein of two (69–71). These resonances are in the same spectral region as those observed for $(^{113}Cd)_2$-parvalbumin and are consistent with the coordinating ligands being exclusively oxygen atoms. The ^{113}Cd resonances do not change either position or intensity, nor do any other resonances appear as a third and fourth ^{113}Cd ion is added to either S-TNC or calmodulin. The third and fourth $^{113}Cd^{2+}$ ions, therefore, bind at sites that are in fast exchange with free ^{113}Cd, thereby broadening the resonance as a result of the rapid exchange between the different environments. These results are consistent with S-TNC's containing two high affinity sites and two low affinity sites. A similar situation is also observed for ^{113}Cd binding to calmodulin, and is consistent with results observed for calmodulin using ^{43}Ca NMR: two sites that exchange slowly with free calcium and two sites that exchange rapidly with free calcium. The two ^{113}Cd resonances observed for parvalbumin, S-TNC, and calmodulin appear with equal intensity as ^{113}Cd is added to the apoproteins. These sites, therefore, bind ^{113}Cd with equal affinities (or bind ^{113}Cd in a highly cooperative manner).

2.4.3 Lanthanide Luminescence. The lanthanides have been used as spectroscopic probes for many calcium-binding sites. Sowadski et al. (38) showed that terbium could replace calcium at site IV of parvalbumin with very little change in protein structure. Europium bound at sites III and IV of parvalbumin exhibits two peaks in the direct laser-induced excitation spectrum of the $^7F_0 - ^5D_0$ transition at 579.2 and 579.6 nm (39). The different maxima for the two excitation peaks indicate different environments for the bound europium at sites III and IV. Studies on the lifetime of the europium emission in H_2O and D_2O indicate that about 1.2 $-OH$ coordinate the bound europium. Ser-55 contributes an $-OH$ to the binding of calcium at site III (14, *Z*) and would be expected to contribute to the calculated water coordination number. This observation is also consistent with the occurrence of one water molecule at the $-Y$ liganding position at site IV.

The excitation spectrum of S-TNC with europium bound is very similar to that observed for the $(Eu^{3+})_2$-parvalbumin complex. Two peaks are detected in the excitation spectrum at 578.9 and 579.2 nm; these are attributed to Eu^{3+} ions bound at the high and low affinity sites of the protein, respectively (49). The excitation peaks do not exhibit the same maximal intensities; this also reflects differences in the sites. There are no serines, or threonines, and no glycines at the coordinating positions of sites III or IV of S-TNC. The coordination sphere of the Eu^{3+} ion, as deduced from direct laser-induced luminescence, contains two water molecules at the high affinity sites. This implies that the Eu^{3+} ion might be coordinated at the high affinity sites with more than six ligands. Alternatively, the coordination site could be distorted, so that ligands that normally coordinate the Ca^{2+} ion are not able to coordinate the bound Eu^{3+} and water molecules replace the ligands. However, this does not occur for the isomorphous replacement of calcium by terbium in the crystal structure of parvalbumin. The lower affinity sites of S-TNC are also reported to contain three water molecules in the coordination sphere of the bound Eu^{3+}. Both these sites contain a serine at a liganding position; they would contribute to the calculated number of water molecules. Therefore, the low affinity sites may contain two water molecules in the coordination site in addition to the six protein ligands, similar to the high affinity sites. This is consistent with the known preference of the lanthanides for coordination numbers of 8 to 9, in contrast to 6 to 8 for calcium.

Europium binding to calmodulin exhibits only one peak at 579.3 nm in its excitation spectrum (54). Thus, the four binding sites in calmodulin coordinate the Eu^{3+} ion with similar geometries. It is also calculated that there are about 1.5 water molecules coordinating each Eu^{3+} ion in both Eu_2-calmodulin and in Eu_4-calmodulin. Sites I and III of calmodulin have serine and threonine respectively in their coordination sites; however, it is difficult to determine whether the calculated number of water molecules corresponds to one water molecule in each of two sites or two water molecules in one site.

2.4.4 Edge, XANES, and EXAFS. The distance between metals and coordinating ligands has been traditionally determined by x-ray crystallographic techniques. Recent progress in x-ray spectroscopy has allowed these determinations for proteins and complexes in solution (72). Those x-rays having energies near the *K* edge of the absorbing atom can be considered in three regions. The absorption edge is determined by the first allowed transitions from the *K* shell to empty bound states. It is therefore sensitive to the electronegativity, number, geometry, and distance of the first shell ligands. About 40 eV above the *K* edge, the x-ray absorption near edge structure (XANES) spectrum gives information about the number and relationship between ligands. At high energies, the extended x-ray absorption fine structure (EXAFS) results from photoelectron back scattering by ligands within about 4.0 Å of the central atom. From the EXAFS spectrum, one can extract additional information about the radial distribution of ligands. The information from the three spectra is complementary. Ideally, one can calculate ligand number and mean distance as well as infer coordination geometry. Since calcium compounds often have irregular geometries, the spectrum of a protein is regarded as a fingerprint and correlation with spectra of compounds of known coordination geometry is sought. Spectra from lyophilized or from only partially purified material can be valuable.

As reviewed by Powers (72) in initial calibration studies, the edge shift and EXAFS spectra for a series of inorganic calcium salts of known crystal structure were recorded. A correlation was found between edge shift energies, coordination number, and decreasing bond length. Several six coordinate calcium complexes were examined, including $CaCO_3$, Ca(EDTA), and $CaCl_2$, and were found to have bond lengths of 2.36 ± 0.02 Å. These were also determined to be six coordinate in solution. Preliminary data on parvalbumin indicated a coordination number of 6.5 with an average bond length of 2.33 ± 0.05 Å. The crystal structure of parvalbumin has coordination numbers of 6 and 8 in sites III and IV, respectively, with average bond distances of 2.37 ± 0.15 Å. Results with S-TNC indicate a coordination number of 6 with an average bond length of 2.43 ± 0.05 Å, and there is little if any detectable difference in geometry between the high and low affinity sites. Bianconi et al. (73) observed small differences in the XANES spectra between Ca_2-TNC and Ca_4-TNC. They had previously noted the similarity in spectra between TNC and calmodulin and reported an average Ca-O bond length of 2.40 Å for Ca_4-calmodulin.

3 DYNAMICS OF METAL BINDING

To bind calcium selectively, the coordination cavity must have three to five carboxylate groups, be flexible enough to accommodate the dehydration of the $Ca(H_2O)_6^{2+}$ ion, and fit the ionic radius of 1.06 Å. As seen in the preceding section, the homolog domains of the calcium-modulated proteins share these characteristics, even though the details of coordination may vary. The K_d, and of greater interest the k_{off}, of metal dissociation depends on the change in free energy of the entire system; the changes in conformation associated with metal dissociation extend throughout the protein. The ability of calmodulin and of TNC to activate a target enzyme depends on these same changes in conformation. In this section, we turn to the dynamics of metal binding by the homolog domain and the entire protein.

3.1 Metal Ion-Dependent Conformations

3.1.1 Calcium-Binding Peptides. Before considering the effects of metal binding on the entire protein, we discuss the results with small calcium-binding peptides that contain one or two domains (Table 1). The peptide that contains only one *N*-terminal helical segment, and the binding loop AcS-TNC [103–123], binds calcium with low affinity, $pK_d(Ca^{2+}) = 2.5$ (74). This binding of calcium is determined by monitoring calcium-induced formation of α-helix. The peptides that contain both flanking helical regions and the binding loop, AcS-TNC [98–123] and AcS-TNC [90–123], bind calcium with higher affinity, $pK_d(Ca^{2+}) = 4.6$ and 5.4, respectively. As a result of calcium binding a region of α-helix, probably in the *N*-terminal third of the peptide, is formed. These results indicate that the presence of potential binding ligands at the site does not, in itself, guarantee high affinity binding. Rather, the regions flanking both sides of the binding site are necessary to maintain a reasonably high affinity for calcium. The flanking α-helical regions could place steric restraints on the binding sites that retard the dissociation of the metal. This could be due to a more compact or better fitted binding cavity whose formation is coupled to helix formation. This conclusion is not unreasonable since one of the metal ligands, glu-21, is also at the beginning of the *C*-terminal helical segment. The formation of helix and the binding of calcium could then be considered to be coupled to each other.

The effect of the *N*-terminus on the calcium binding by peptides is also emphasized by the high affinity binding of calcium by a cyanogen bromide fragment of S-TNC, CB-9, which contains 52 residues (75). One helical segment, suggested to be at the *C*-terminus, is helical in the absence of calcium, whereas the *N*-terminal segment is helical only in the presence of calcium. These results are similar to those with the AcS-TNC [90–123] peptide; however, the *C*-terminal helical region of the 90–123 peptide is not formed in the absence of calcium.

It is difficult to determine whether there are changes in the tertiary structure of the peptides as a consequence of calcium binding. UV differences and ^{1}H NMR spectroscopy indicate that upon calcium binding, phenylalanines and other nonpolar residues are affected (75–78). However, phe-99, phe-102 in the CB9 peptide, and other hydrophobic residues in helical regions are spaced such that they could be in close contact on the inside surface of the helix, in contrast to being exposed to solvent in a random coil configuration. Therefore, calcium-induced structural changes observed by UV difference spectroscopy and ^{1}H-NMR do not necessarily indicate interaction between the two helical segments but may indicate only increased secondary structure.

It has been observed for parvalbumin and intestinal calcium-binding protein that calcium-binding domains exist as pairs of binding regions related by an approximate twofold symmetry axis. Limited proteolytic cleavage of S-TNC and calmodulin can produce pairs of binding sites. These peptides retain much of the structural characteristics of the native proteins. Limited proteolysis of S-TNC with trypsin in the presence of calcium produces the two peptides TR1 and TR2, which contain sites I and II, and III and IV, respectively (52, 79). Each peptide binds 2 moles of calcium with $pK_d(Ca^{2+}) = 5.0$ and 7.7 for TR1 and TR2, respectively. Thus the TR1 peptide, which contains the two low affinity sites, binds calcium with less affinity than does the TR2 peptide, which contains the two high affinity sites. These two domain fragments bind calcium with higher affinity than do the isolated single domain fragments (Table 1). This emphasizes the importance of interactions between the two domains of each pair in maintaining high affinity calcium binding. The TR1 and TR2 fragments of S-TNC also retain the selectivity that is observed in the intact protein; sites III and IV of TR2 bind magnesium with much higher affinity than do sites I and II of TR1.

The amounts of α-helix of the TR1 and TR2 fragments, in the absence or presence of calcium, are very similar to that observed for the intact S-TNC when summed together (79). This suggests strongly that the secondary structures of the two halves of the protein are not dramatically influenced by each other. The TR2 peptide undergoes a twofold increase in helical structure due to calcium binding. In contrast, there is little if any helix formation induced in the TR1 peptide as a result of calcium binding. This is not due to a lack of helical structure in the peptide; rather, it appears that the TR1 helices are formed even in the absence of calcium, and metal binding does not induce extra helicity. Calcium binding to the TR1 peptide does not require helix formation and, therefore, the decrease in free energy accompanying the binding of calcium to TR1 is not associated with a coil to helix transition. This does not imply that the TR1 does not undergo other changes in structure upon binding calcium.

In studies by Evans et al., ^{1}H NMR revealed that a number of changes in the tertiary structure of the TR1 peptide accompany calcium binding (80). Specific changes in the environment of phenylalanines and aliphatic side chains in the TR1 peptide are detected by ^{1}H NMR as a result of calcium addition. It has been postulated that a change in the orientation between the helical regions flanking each site would affect the environment of residues on the inner surface of the α-helical segments, such as phe-72 and phe-75 of site II and phe-19 and phe-23 of site I (80). This occurs without the formation of extra helical structure. This conformational change in the TR1 peptide is observed by monitoring the chemical shift of anomalously shifted phenylalanine resonances (80). It is interesting that similar resonances are observed in the spectra of S-TNC, C-TNC, and calmodulin (55, 58, 81).

The TR1-C and TR2-C fragments that contain sites I and II and sites III and IV, respectively, of calmodulin have also been produced by limited proteolysis with trypsin (132, 82). The results with these fragments are consistent with the observed calcium- and magnesium-binding parameters of the native protein. Both peptides undergo structural changes as a result of calcium and magnesium binding. There is only a slight difference between the transition midpoints for calcium-induced changes in ellipticity for the two peptides TR1 and TR2—0.6 μM and 0.3 μM, respectively. The calcium-induced changes in ellipticity for the TR1-C and TR2-C peptides are not large and are very similar to that observed for the TR1 peptide of S-TNC (Table 1). Therefore, in many respects both the TR1-C and TR2-C peptides of calmodulin are very similar to the TR1 peptide of S-TNC, which contains the low affinity sites.

3.1.2 Parvalbumin. Parvalbumin contains two functional binding domains, III and IV, and a third domain, II, which is incapable of binding calcium. The helices are formed in the presence of calcium and there is a well defined hydrophobic core that consists of residues from the internal surfaces of the helices. The two binding sites, III and IV, are related by a twofold axis with the metal binding loops, forming a four residue β-sheet of hydrogen bonds. Domain II, which does not bind calcium, contributes to the hydrophobic core of the protein and also to the overall helicity of the structure. The loop of domain II is held to the other two domains by a series of hydrogen bonds to the invariant arg-75 and glu-81 at the beginning of domain IV. When arg-75 is modified by cyclohexanedione, the protein's affinity for calcium is considerably reduced (83, 84). This alteration in a region of parvalbumin more than 20 Å from the binding sites, III and IV, dramatically reduces their ability to bind calcium. Fragments of parvalbumin lacking domain II still bind calcium, however, with lower affinity than the native protein (Table 1).

Calcium removal from parvalbumin results in a 25% loss of α-helix content (43). Magnesium and terbium are able to induce essentially the same amount of helix as does calcium (85, 86). Other effects induced by calcium, such as increased intrinsic fluorescence, protection against proteolysis by trypsin, and decreased reactivity of cys-18, are similarly produced by magnesium (87). Distinct differences between the calcium and magnesium conformers of parvalbumin can be observed by ^{1}H NMR for phenylalanines and other nonpolar amino acid residues. Birdsall et al. attributed these differences to alterations in the hydrophobic core of the protein (88). They have also found that the rates for slowly exchangeable amide N-H protons are different for magnesium and for calcium parvalbumin. Reid and Hodges (89) suggested that because magnesium is smaller than calcium, it will be only four coordinate at the binding sites. This implies that coordination of only four of the ligands is necessary to induce helix formation and most of the tertiary structure. The calcium-specific changes in the hydrophobic core of the protein would then be due to the disposition of the fifth and sixth ligands. These would represent the calcium dependent changes that distinguish it from the magnesium-parvalbumin.

The structure of apoparvalbumin as detected by ^{1}H NMR exhibits a marked loss of tertiary structure as compared to either the magnesium or calcium form of the protein (88). There is very little evidence of close contacts between nonpolar amino acid residues, nor is there any evidence for slowly exchangeable amide N-H protons. Although the loss of metals at the binding sites results in only a 25% reduction in the α-helix, there is still a large alteration in the arrangement of the various helical segments of the protein. This has been reasonably interpreted as reflecting a less tightly packed structure, which results in a loss of close contacts between the groups making up the hydrophobic core of the protein.

3.1.3 Skeletal Troponin-C. Skeletal troponin-C provides a relatively simple system for studying conformational changes. The difference in the calcium-binding constants at the high and low affinity sites allows calcium to occupy the high affinity sites with little occupation of the low affinity sites. It has been demonstrated that S-TNC undergoes a significant conformational change, which is detected by alterations in the UV and CD spectrum and in tyrosine fluorescence (for a review, see ref. 48). Most of these changes were initially attributed to the binding of calcium at the high affinity sites. Many were also similarly induced by magnesium. These include the decreased mobility of a spin label at cys-98, the decreased chemical reactivity of cys-98 (90), the environment of a trifluoroacetonyl derivative of cys-98 (81), and the increase in intrinsic fluorescence of tyrosine residues (91). The ^{1}H NMR is also able to detect the large change in conformation resulting from metal occupation of the high affinity sites (59, 81, 92). The conformations, as reflected by ^{1}H NMR, of magnesium S-TNC and S-TNC with two calciums bound at the high affinity sites are overall similar. The ^{1}H NMR results clearly indicate that the occupation of the high affinity sites, III and IV, of S-TNC by calcium or magnesium produces a greater dispersion in the environment of phenylalanine residues. It is difficult to determine whether this occurs as a result of helix formation or alterations in the hydrophobic core of the protein. There is a unique juxtaposition of phenylalanines in the apoprotein, as indicated by anomalously shifted phenylalanine resonances, which are not affected by calcium or magnesium binding at sites III and IV. Therefore, the conformational change associated with the binding of metals at the high affinity sites does not affect the entire protein structure.

As with parvalbumin, the binding of either calcium or magnesium at sites III and IV produces large changes in structure, including helix formation. This increased content of α-helix could contribute to the free energy of the binding reaction, thereby explaining why metal ions other than calcium, such as magnesium and strontium, can bind at the high affinity sites.

The binding of calcium at the low affinity sites of TNC is ultimately responsible for the initiation of muscle contraction (48). Studies with ^{1}H NMR (59, 81, 92), extrinsic fluorescence probes (93), and more refined CD analyses (94, 95) detect structural alterations resulting from calcium binding at sites I and II. Skeletal troponin-C labeled with dansylaziridine at Met-25 (18) has increased fluorescence owing to the binding of calcium at sites I or II (93). Since Met-25 is at the end of a helical segment near site I, its fluorescence is a sensitive monitor of the environment near that site. Although 25% of the change in α-helix is due to calcium binding at sites I and II, it is difficult to determine whether this increased helicity is responsible for the observed increase in fluorescence of the dansylazirdine, Met-25.

The smaller increases in helical structure produced by calcium binding at sites I and II, as compared to sites III and IV, are consistent with the measured enthalpies and entropies for the binding of calcium at the two classes of sites (96). The enthalpy for binding to each site is identical, $\Delta H = -7.7$ kcal/mol, and therefore the changes in free energy (derived from the binding constants) are due to changes in entropy, $\Delta S^0 = 6.7$ and 8.0 cal/deg/mol for the high and low affinity sites, respectively. If the calcium ions are bound at relatively similar sites with similar geometries and coordination number, the difference in the protein conformation could contribute to the difference in entropy. A large increase in α-helix content, such as that produced by calcium binding at the calcium-magnesium sites, produces an increase in entropy owing to the exclusion of water. A smaller helical change, such as that produced by calcium binding at the low affinity sites, leads to a correspondingly smaller increase in entropy. The increase in entropy may be due, however, to other changes in tertiary structure, which do not involve the formation of helix, but do induce closer associations between hydrophobic groups leading to a greater exclusion of water or, alternatively, to an exclusion of water from the binding sites. Thus, metal binding to organic sequestering agents is also accompanied by a large increase in entropy.

3.1.4 Cardiac Troponin-C. Cardiac troponin-C exhibits qualitatively similar behavior to its skeletal counterpart. It also undergoes an increase in content of α-helix on binding calcium, as well as an increase in tyrosine fluorescence (94, 97). Both of these induced changes occur as a result of the binding of calcium at the two high affinity sites, III and IV. Magnesium (2.0 mM) is able to induce approximately 80% of the CD change and 55% of the change in tyrosine fluorescence. It appears that, as with S-TNC, the amount of α-helix induced by metal binding at sites III and IV is not specific for calcium. The calcium-specific change in structure is the result of changes induced by calcium binding at the low affinity site. The two sulfhydryl groups of C-TNC, cys-35 and 84, can be labeled with a specific fluorescent probe, IADANS; its fluorescence is very sensitive to the binding of calcium at the single low affinity site (98). The fluorescence of this modified C-TNC is only slightly affected by calcium or magnesium binding at the two high affinity sites.

As detected by ^{1}H NMR, the calcium-specific structural alteration due to the binding of calcium at the low affinity site of C-TNC is similar to that of S-TNC (55). Both proteins exhibit changes in tertiary structure surrounding two unique phenylalanines as a result of calcium binding at the low affinity sites (81, 99). Changes in the environment of other phenylalanines are also observed; however, it is difficult to determine whether this is because of increased helical structure or increased interactions in the interior of the protein. The low affinity site change, which is similar to that observed in S-TNC, is the calcium-specific alteration in structure responsible for the initiation of cardiac contraction.

3.1.5 Calmodulin. Calmodulin undergoes conformational changes that are a cross between those observed for metal binding at calcium-magnesium high affinity sites and at calcium-specific low affinity sites of TNC. The conformational changes, as monitored by increased tyrosine fluorescence and increased helical structure, occur as two equivalents of calcium bind to calmodulin (for a review, see ref. 10). Therefore, two binding sites in calmodulin that produce almost all of the calcium-dependent increase in α-helix act in a manner similar to the high affinity sites of S-TNC and C-TNC. The calcium-dependent changes in CD and fluorescence in calmodulin can be mimicked to only a small extent by magnesium (60, 100). About 25% of the fluorescence change and 20% of the CD change are evoked by magnesium in the absence of high salt (61). The sites in calmodulin are, therefore, clearly different from the so-called high affinity sites of parvalbumin, S-TNC, and C-TNC. If magnesium occupies these sites, it does not induce the large rearrangement in structure induced by calcium. Since the calmodulin-binding sites have the same potential ligands as sites III and IV in S-TNC and C-TNC, the observed differences must be related to the overall protein structure.

There is also a specific structural rearrangement that occurs as a third and fourth equivalent of calcium bind to calmodulin. This has been detected by ^{1}H NMR (58) and by absorbance changes of nitrotyrosine-138 (101). The structural rearrangement affects a number of phenylalanines, and aliphatic residues that are expected to interact in the interior of the protein. The changes in conformation are similar to those observed for S-TNC and C-TNC for the binding of calcium at the low affinity sites.

3.2 Tyrosines as Internal Monitors of Conformation

The tyrosines tyr-109 (S-TNC), tyr-111 (C-TNC), and tyr-99 (calmodulin) are homologous with phe-57 of parvalbumin at III16. The carbonyl oxygen coordinates calcium at the $-Y$ vertex. In parvalbumin, the center of the phenyl ring is 8.2 Å from CaIII, and the ring faces CaIV at only 5.2 Å.

All three III16 tyrosines have similar properties in the apoproteins. They are accessible to solvent as assessed by chemical reactivity and p*K*'s (55, 102). Furthermore, they all show relatively unperturbed ^{1}H NMR resonance positions for the ring protons, characteristic of solvent-exposed tyrosine rings (55, 58, 81). All three proteins also exhibit quenched tyrosine fluorescence in the absence of calcium (97, 100). This has been attributed to quenching by the $-COOH$ or $-CONH_2$ groups of site IV in the vicinity of the rings. Following the binding of calcium, the p*K*'s, chemical reactivity, and ^{1}H NMR resonances are only slightly altered; however, none of these changes indicate any gross alteration in the environment of the rings. We can assume, therefore, that they do not enter into the hydrophobic interior of the protein or interact with other residues to restrict their accessibility to solvent. The changes that are observed result from the binding of the first two calciums per mole of protein. These results are consistent with the first binding of two calcium ions at sites III and IV of the proteins. This is certainly well-documented for S-TNC and strongly inferred for C-TNC. The assignment of high affinity sites in calmodulin is not, however, as straightforward as the assignment for the TNC's.

We know that tyr-145 (C-TNC) and tyr-138 (calmodulin) are homologous with val-99 of parvalbumin at IV19. It is at the surface; its α-carbon is 9.8 Å from CaIV. As a function of rotation about the bond between C_α and C_β, a ring at IV19 could be from 7.1 to 12.5 Å from CaIV. The $C_\beta - C_\gamma$ rotation also is not known. At about 10 Å, ring to CaIV distance, the tyrosine ring could turn either its face or edge toward CaIV; at 7.1 or 12.5 Å it would address CaIV mostly with an edge. The tyrosine rings tyr-145 (C-TNC) and tyr-138 (calmodulin) have unique ^{1}H NMR spectral properties (58, 103). In the absence of calcium, the ortho and meta protons of the rings exhibit almost identical chemical shifts. These occur at similar chemical shift positions, 6.86 ppm for tyr-145 and 6.71 ppm for tyr-138. Upon binding calcium, the aromatic proton resonances shift upfield with the meta proton resonances appearing upfield (6.37 ppm) of the ortho proton resonances (6.55 to 6.60 ppm). A normal unperturbed or slightly shifted tyrosine, such as that at III16, exhibits resonances with the ortho protons upfield of the meta protons. This indicates that the environments around tyr-145 in C-TNC and tyr-138 in calmodulin are unique and very similar in both the absence and presence of calcium. The calcium-induced changes in their environments are specific for calcium, and occur as the first two Ca^{2+} ions bind to the protein. The tyrosines are not sensitive to the binding of calcium in excess of 2 moles per mole of protein. The environments of the two tyrosines also exhibit calcium-induced UV difference spectra; tyr-138 of calmodulin and tyr-145 of C-TNC undergo a red shift (94, 104) in the absence of salt. In the presence of 0.1 *M* NaCl, calmodulin exhibits a blue shift owing to calcium binding (60).

3.3 Kinetics of Metal-Ion Binding

The binding of metals by the calcium-binding proteins has been discussed both in terms of the affinity of metals for the sites and the specific conformations induced by the metals. It is possible to generalize certain aspects of these binding characteristics with regard to the affinity, selectivity, and extent of conformational rearrangement. Thus, sites III and IV of parvalbumin, S-TNC, and C-TNC are high affinity sites that bind magnesium competitively with calcium. Sites I and II of S-TNC and II of C-TNC are so-called calcium-specific sites. Sites I to IV of calmodulin exhibit characteristics that are a mixture of the high affinity and low affinity types of sites. These classes of sites also display rates of binding, k_{on}, and dissociation, k_{off}, consistent with their other physical parameters.

3.3.1 High Affinity Calcium–Magnesium-Binding Sites. The high affinity binding sites of S-TNC, C-TNC, and parvalbumin have $pK_d(Ca^{2+})$ 7.0 to 8.5. From our discussion in Section 2 we assume that calcium binding is essentially diffusion controlled, such that k_{on} is $10^{8.3}$ M^{-1} sec^{-1}. Consistent with this, the structural changes that occur upon the binding of calcium are usually too fast to measure and happen within the dead time of the experimental apparatus, < 2 msec. The rate of dissociation, k_{off} can be calculated from the relationship $k_{off}(\sec^{-1}) = K_d(M) \cdot k_{on}(M^{-1} \sec^{-1})$ yielding k_{off} 1.0 to 10 $\sec^{-1}$. This indicates that the half-life for the dissociation is 0.1 to 1.0 sec. Since the dissociation of the metal ion is linked to the conformational change, the rate for the conformational transition should also be slow. In all cases where the conformations of the proteins have been monitored kinetically, the conformational changes all exhibit very slow rates associated with calcium dissociation (Table 2). The slow conformational change occurring upon calcium dissociation from the high affinity sites of S-TNC is observed

Table 2 Rates of Calcium and Magnesium Dissociation and Conformational Relaxation of Calcium Modulated Proteins

Protein	Domain(s)	k_{off} $\sec^{-1}$	Comment	Reference
S-TNC	III, IV	1.0	Dansylaziridine	93
	I, II	230–347	Labeled Met-25	
	III, IV (Mg)	8		
S-TNC	III, IV	10	^{1}H NMR	92
	I, II	100–1000		
S-TNC	III, IV	1000	^{43}Ca NMR	65
	III, IV (Mg)	30	^{1}H NMR	134
S-troponin	III, IV	0.3–0.6	Dansylaziridine	135
	I, II	20–25	Labeled	
	III, IV (Mg)	4.7		
C-TNC	II	300	IADANS[a] labeled	98
C-TNC	III, IV	440	^{1}H NMR	55
	II	700		
C-troponin	II	15–30	IADANS labeled	48
Calmodulin	I, IV	10	Tyr fluorescence	136
Calmodulin	I, II	40	^{1}H NMR	58
	III, IV	200	^{1}H NMR	
Calmodulin	I, II	20–50	^{43}Ca NMR	69
	II, IV	1000		
Parvalbumin	III, IV	1.0, 3.0	IADANS labeled	107
	III, IV (Mg)	1.1, 7.2		
Parvalbumin	III	15	^{1}H NMR	88
	III, IV (Mg)	30	^{1}H NMR	
Parvalbumin	III, IV	20	^{43}Ca NMR	66
Ac-S-TNC (103-123)	III	100	^{1}H NMR	77
Ac-S-TNC (96-123)	III	135	^{1}H NMR	
Ac-S-TNC (90-123)	III	100	^{1}H NMR	

[a]IADANS: 2-(4 iodoacetamido anilino)naphthalene-6-sulfonic acid.

either by monitoring the fluorescence of tyr-99 in domain III (105, 106) or by measuring the fluorescence of dansylaziridine attached to Met-25(I8) in domain I (93, 133). The structural rearrangements that affect these different regions of the protein exhibit essentially the same kinetics, $k_{off} \sim 1.0\ sec^{-1}$. This is especially interesting since the fluorescence of tyr-99 is probably responding to a local effect of the binding of calcium at site III and not a more global conformational change. In this case at least, the off rate of the metal ion and the rate of the conformational change are very similar and are probably determined by the same structural transition. The same slow off rates are also predicted for the high affinity sites of C-TNC and are observed for the sites in parvalbumin, k_{off} about $1.0\ sec^{-1}$ (107). In the case of parvalbumin, however, a $k_{off} \sim 20\ sec^{-1}$ is calculated based on the temperature dependence of the ^{43}Ca-NMR signal (66). For parvalbumin, the dissociation of calcium is significantly faster than the associated conformational transition.

As previously discussed, because of its higher charge to radius ratio magnesium has a much slower rate of substitution, $k_{on} \sim 10^5\ M^{-1}\ sec^{-1}$, which is over a thousand times slower than that of calcium. The onset of the conformational change occurring as a result of the binding of magnesium is too fast to be measured using most techniques, even though it is still three orders of magnitude less than calcium. Although magnesium binds with weaker affinity at the high affinity sites, the calculated k_{off} is still similar to that of calcium, 3, 25, and 80 sec^{-1} for parvalbumin, S-TNC, and C-TNC, respectively. The conformational changes accompanying magnesium dissociation have k_{off} consistent with those calculated above (Table 2).

3.3.2 Low Affinity Calcium-Binding Sites. The low affinity calcium-specific-binding sites have calculated k_{off} rates of 300 and 400 sec^{-1} for S-TNC and C-TNC, respectively. The low affinity sites of TNC clearly have off rates much faster than the high affinity binding sites. This calculated fast dissociation is also consistent with what is observed for the conformational transitions (Table 2). Thus, dansylaziridine labeled met-25 of S-TNC, and fluorescent sulfhydryl probes on cys-35 and cys-84 of C-TNC, both respond to calcium dissociation with k_{off} of about 300 sec^{-1} (108). The k_{off} for calcium determined using ^{43}Ca NMR, 1000 sec^{-1}, is faster than the rates determined for the conformational transitions.

3.3.3 Calmodulin Sites. Most of the published binding affinities for calmodulin range from $pK_d(Ca^{2+})$ 6.0 to 7.2. Although it has not been possible to consider distinct high and low affinity sites for calmodulin, several measures of dissociation rates have indicated distinct groupings. A slow conformational transition occurring as two Ca^{2+} ions are bound can be detected using ^{1}H NMR by the behavior of the trimethyllysine-115 resonance (58). This transition has $k_{off} \leq 40\ sec^{-1}$, which corresponds to a $pK_d(Ca^{2+})$ 6.7. This rate is also consistent with that determined by the sensitivity of tyrosine fluorescence to calcium dissociation and that calculated from ^{43}Ca-NMR (Table 2). A second transition for two more Ca^{2+} ions can also be detected by ^{1}H and ^{43}Ca NMR, with a $k_{off} \sim 200\ sec^{-1}$ and $k_{off} = 1000\ sec^{-1}$, respectively (58, 69). These rates would correspond to $pK_d(Ca^{2+})$ of 6.0 and 5.3, respectively, which are lower than those previously reported from binding experiments. Such differences may be due to the different experimental conditions, or the results with ^{43}Ca may indicate that calcium dissociation is faster than the ensuing conformational change. The results do indicate that, as with S-TNC and C-TNC, calmodulin displays two kinetically distinct conformational transitions owing to calcium dissociation.

For any given site (Table 2), the conformational transition accompanying magnesium dissociation occurs with the same or at a slightly faster rate than does that accompanying calcium dissociation. This close correspondence between the calculated rates and the experimentally determined rates is consistent with the discussion in Section 2. We see that $K_d(M) = k_{off}(sec^{-1})/[k_{on}(M^{-1}\ sec^{-1})]^{-1}$ where $k_{on}(Mg^{2+})$ is determined by the dehydration rate of magnesium ($10^5\ sec^{-1}$), and $k_{on}(Ca^{2+})$ reflects contributions from both dehydration and diffusion ($10^{8.3}\ sec^{-1}$). In some cases, the rates of metal dissociation and conformational change are identical and indicate the very close coupling between metal binding and the conformational transition. One does not preclude the other, rather metal dissociation (and even association) is closely coupled to the conformational change.

3.4 Modulation of Metal Binding

The metal-binding properties and associated changes in conformation of TNC and of calmodulin are affected by their target proteins and by small molecules and ions in the cytosol (48, 109, 110). It has been shown that TNC interacts with TNI in the presence of calcium. Troponin-C binds calcium with similar affinities in the 1:1 TNC:TNI complex as in the heterotrimer containing TNT. A consequence of this calcium-dependent interaction is that TNC binds calcium with higher affinity in the complex than alone in solution; $pK_d(Ca^{2+}) =$

8.3 for the high affinity sites, $pK_d(Ca^{2+}) = 6.5$ for the low affinity sites (47). In the resting muscle cell (pCa ~ 7.5, pMg ~ 2.8), the occupancies of sites I and II versus sites III and IV in whole troponin are calculated to be:

Site	Apo	Magnesium	Calcium
I and II	60%	37%	3%
III and IV	2%	93%	5%

The majority of the S-TNC, therefore, have magnesium at the high affinity sites and no cation at the low affinity sites. This is consistent with the low affinity sites' being those responsible for triggering muscle contraction. The k_{off} determined for the low affinity site conformational change is reduced from 300 sec^{-1} in the isolated S-TNC to 20 sec^{-1} in whole troponin. The k_{off} calculated for the high affinity site conformational changes induced by calcium or magnesium are also reduced in the complex; however, this decrease is less than that for the lower affinity sites (Table 2).

Calmodulin also binds to TNI in the presence of calcium (111). Keller et al. (112) determined the mean dissociation constant for calcium binding to calmodulin; $10^{-4.85}$ M in the absence of TNI and $10^{-5.77}$ M in its presence. They analyzed their results by the formalism mentioned in Section 1.3:

$$\begin{array}{ccc} \text{CMP} + 4\,\text{Ca} & \underset{K_d}{\rightleftharpoons} & \text{CMP*Ca}_4 \\ + & & + \\ \text{ENZ} & & \text{ENZ} \\ \updownarrow {\scriptstyle K_d(\text{ENZ})} & & \updownarrow {\scriptstyle K_d^*(\text{ENZ})} \\ \text{ENZCMP} + 4\,\text{Ca} & \underset{K_d^0}{\rightleftharpoons} & \text{ENZ}^0\text{CMP*Ca}_4 \end{array}$$

1. $K_d K_d^*(\text{ENZ}) = K_d^0 \cdot K_d(\text{ENZ})$
2. $K_d = K_1 \cdot K_2 \cdot K_3 \cdot K_4$ and $K_d^0 = K_1^0 \cdot K_2^0 \cdot K_3^0 \cdot K_4^0$
3. $K_d = (K_{\text{mean}})^4$ and $K_d^0 = (K_{\text{mean}}^0)^4$

Therefore, if the ratio $K_{\text{mean}}/K_{\text{mean}}^0$ is $10^{-4.85}/10^{-5.77} = 8.2$, then K_d/K_d^0 is $8.2^4 = 4500$. This means that an increase in average affinity of each domain for calcium by a factor of only 8.2 increases the affinity of $(Ca)_4$-calmodulin for TNI by $10^{3.65}$.

Huang et al. (113) got similar results for the interaction of cyclic nucleotide phosphodiesterase with calmodulin. They also concluded that the affinity of calmodulin for cyclic nucleotide phosphodiesterase is increased by a factor of 10^4 as a result of only a tenfold increase in the affinity for calcium at each of the four sites of calmodulin. This results in the highly cooperative activation of phosphodiesterase by calcium and calmodulin (56).

A modest increase in $[Ca^{2+}]$, with subsequent increase in [CAM Ca_4], causes a large increase in the fraction of target in the activated state ($ENZ^0CMP^*Ca_4$). This increase in affinity for the enzyme is achieved at only a tenfold reduction in $k_{off}(Ca^{2+})$, from approximately 10^2 to 10 sec^{-1}. This rate is still fast enough for the termination of physiological responses by calcium dissociation.

Phenothiazine and related compounds (114), fluorescent dyes (115), and other hydrophobic molecules (116) bind to calmodulin, as well as to S-100, with calcium dependence. They also inhibit activation of, and supposedly binding to, various target proteins. Forsén et al. (71) found that trifluoperazine lowers the $k_{off}(^{113}Cd)$ and significantly alters the ^{113}Cd NMR of calmodulin. This decrease in k_{off} would result in a higher affinity for the metal ion and can be analyzed in terms of the scheme discussed above.

4 SUMMARY

We can now pose a series of questions about the dynamics of calcium-modulated proteins. The affinities and selectivities of the cation binding cannot be deduced from the structure of the ligands or even the entire loop. There is not yet strong evidence for cooperativity between domains in binding calcium; yet studies on fragments indicate that a pair of domains is required for full affinity. How does the structure of the pair of domains affect the free energy of metal binding?

The calcium and magnesium affinities of all the sites, "high" or "low" affinity, appear to be determined by the dehydration rate of the hydrated metal ion, k_{on}, and by the dissociation rate of the ion from the protein, k_{off} ($K_d = k_{off}/k_{on}$). We infer that the rate of change of protein conformation associated with cation binding is fast, although of course, not nearly so fast as the $10^{8.3}$ sec^{-1} M^{-1} of calcium binding. The rate of the conformational change accompanying metal ion dissociation is slow, and in many cases is close to that determined for the actual off rate of the metal. The differences between the rate of metal dissociation and the ensuing conformational change may have significance in metal ion exchange, such as that between calcium and magnesium.

We have emphasized the homology among the calcium-modulated proteins as well as their close structural similarities. We have discussed numerous experiments on parvalbumin as guides to understanding this information transfer. Yet, after extensive searches, there is little evidence that parvalbumin has any functional interaction with other proteins. Intestinal calcium-binding protein also appears not to interact with other proteins. Robertson et al. (117) and Gilles et al. (118) have presented calculations showing that parvalbumin might well function as a calcium buffer, allowing more rapid relaxation of fast white muscle than could be achieved by the sarcoplasmic reticulum alone. Kretsinger et al. (119) have published calculations and Feher (120) has completed experiments showing the feasibility of the intestinal protein's functioning to facilitate diffusion of calcium across the cytosol of the epithelial cell. Again in this model, ICBP is proposed not to interact with other proteins. If, in fact, parvalbumin and ICBP do not interact with other proteins, why then do they have the same large conformational changes as observed for calmodulin? One explanation, which is hardly proven, is that the apo to calcium change in conformation of the proteins was absolutely essential to their function over a billion years ago when eukaryotic cells evolved. It has been conserved and is common to all eukaryotic cells. It is more common for Nature to modify an existing gene than to create a new one; hence the essential light chains, which bind no calcium. Parvalbumin and ICBP, by this reasoning, have deviated from the "original" function. To what extent is the large conformational change necessary for these suggested buffering and/or transport functions; or is it merely a consequence of removing a Ca^{2+} ion from a cluster of carboxylate groups?

We are concerned with the changes in conformation associated with calcium release and binding. Yet all of this occurs in the presence of relatively constant magnesium, pMg $\sim$ 2.8. The various domains bind magnesium with $pK_d(Mg^{2+})$ from 2.0 to 4.0. As determined by several types of spectroscopy, much of the change in conformation associated with calcium binding also accompanies the binding of magnesium and several other di- and trivalent cations. Whereas magnesium is usually six coordinate, calcium is usually seven or eight, and less frequently six. In the homolog domain, six amino acids contribute ligands. Do all of these ligands coordinate the Mg^{2+} ion? If less than six ligands coordinate magnesium, are these enough to elicit a major conformational change in the protein? Again one can pose the evolutionary question: Is the coordination of magnesium functional—for instance, in permitting binding to TNC to TNI or in reducing proteolysis—or is it a coincidence resulting from the similarity of the two cations?

In the cases of TNC (which binds to TNI and to TNT) and calmodulin (as bound to the α and γ subunit of phosphorylase kinase) the magnesium or apo form is bound to its immediate target in the quiescent cell. For most other calmodulin target proteins, binding to the target occurs only after $(Ca)_n$-calmodulin is formed. Obviously, cells utilize both sequences. In the case of contraction of fast skeletal muscle, it might seem reasonable that S-TNC reside on its target to save diffusion times; no such ad hoc justification seems appropriate for phosphorylase kinase. One can anticipate that apocalmodulin, as well as the apo forms of yet to be discovered homologs, will often exist *in vivo* associated with targets. What is the functional significance of such preassociation?

Although most of our attention has been focused on the dynamics of calmodulin and TNC themselves, ultimately we need to understand the dynamics of the calmodulin target complex. To this end, it appears that calmodulin does not "activate" the target so much as it "relieves an inhibition." In at least four instances—cyclic nucleotide phosphodiesterase (121), phosphorylase kinase (122), calcium ATPase (123), and NAD kinase (124)—one or few site proteolysis activates the enzyme to the same extent as does binding to calmodulin. How do the structural changes in the targets associated with calmodulin binding compare with those caused by limited proteolysis?

About 20 different calmodulin-"activated" proteins have been identified or inferred. Although not all of them occur in significant quantity in a given cell, certainly many cells contain several different calmodulin-binding proteins. It is reasonable to anticipate that, following a calcium pulse, not all targets are turned on coordinately. This may partly reflect compartmentalization or difference in spatial distribution of either target, calmodulin or calcium. We have seen that a tenfold increase in affinity for calcium in each domain can cause a 10^4 gain in affinity of Ca_4-calmodulin, relative to apocalmodulin, for a target. Do any of the targets have a significant affinity for any of (multiple) forms of Ca_3-calmodulin or Ca_2-calmodulin?

Several proteins, such as the phospholipid-dependent *c*-kinase (34), aequorin (33), and the endotoxin receptor from *Limulus* (125), are activated by 10^{-6} *M* calcium. They do not interact with calmodulin. Might a calmodulin domain gene be spliced together with the catalytic region? If so, would it display conformational changes similar to those already observed?

These questions have been generated by a pulse of information consisting of over a thousand papers on calmodulin alone. We anticipate that their answers will be of fundamental importance to understanding the regulation of cell function.

REFERENCES

1. S. Ringer, *J. Physiol.*, **4,** 29 (1882).
2. A. Sandow, *Yale J. Biol. Med.*, **25,** 176 (1952).
3. W. W. Douglas and R. P. Rubin, *J. Physiol.*, **159,** 40 (1961).
4. E. W. Sutherland, I. Öye, and R. W. Butcher, *Rec. Prog. Hormone Res.*, **21,** 623 (1965).
5. Rall, T. W. and Sutherland, E. W., *J. Biol. Chem.*, **232,** 1065 (1958).
6. H. Rasmussen, *Science*, **170,** 404 (1970).
7. S. Ebashi, F. Ebashi, and A. Kodama, *J. Biochem.*, **62,** 137 (1967).
8. P. C. Moews and R. H. Kretsinger, *J. Mol. Biol.*, **91,** 201 (1975).
9. D. M. E. Szebenyi, S. K. Obendorf, and K. Moffat, *Nature*, **294,** 327 (1981).
10. C. B. Klee and T. C. Vanaman, *Adv. in Protein Chemistry*, **35,** 213 (1982).
11. R. H. Kretsinger, *Nature New Biol.*, **240,** 85 (1972).
12. J. H. Collins, J. D. Potter, M. J. Horn, G. Wilshire, and N. Jackman, *FEBS Lett.*, **36,** 268 (1973).
13. M. Eigen and G. G. Hammes, *Adv. Enzymol.*, **25,** 1 (1963).
14. M. Eigen, "The Kinetics of Fast Solvent Substitution in Metal Complex Formation," in *Advances in the Chemistry of the Co-ordination Compounds*, S. Kirschner, Ed., MacMillan, New York, 1961, pp. 371–378.
15. H. Diebler, M. Eigen, G. Ilgenfritz, G. Maass, and R. Winkler, *Pure Appl. Chem*, **20,** 93 (1969).
16. R. H. Kretsinger, *CRC Crit. Rev. Biochem.*, **8,** 119 (1980).
17. M. Goodman, *Prog. Biophys. Mol. Biol.*, **38,** 105 (1981).
18. R. H. Kretsinger and C. D. Barry, *Biochem. Biophys. Acta*, **405,** 40 (1975).
19. M. A. Conti and R. S. Adelstein, *J. Biol. Chem.*, **256,** 3178 (1981).
20. C. R. Bagshaw, *J. Muscle Res. Cell Motil.*, **1,** 255 (1980).
21. C. R. Bagshaw and J. Kendrick-Jones, *J. Mol. Biol.*, **140,** 411 (1980).
22. C. W. Bishop, N. C. Kendrick, and H. F. DeLuca, *J. Biol. Chem.*, **258,** 1305 (1983).
23. T. Isobe and T. Okuyama, *Eur. J. Biochem.*, **116,** 79 (1981).
24. D. R. Marshak, D. M. Watterson, and L. J. van Eldik, *Proc. Natl. Acad. Sci.*, **78,** 6793 (1981).
25. P. Calissano, S. Alema, and P. Fasella, *Biochemistry*, **13,** 4553 (1974).

26. R. S. Mani, B. E. Boyes, and C. M. Kay, *Biochemistry*, **21,** 2607 (1982).
27. W. Wnuk, J. A. Cox, and E. A. Stein, *J. Biol. Chem.*, **256,** 11538 (1981).
28. J. P. MacManus, J. F. Whitfield, A. L. Boynton, J. P. Durkin, and S. H. H. Swierenga, *Oncodev. Biol. Med.*, **3,** (1982) in press.
29. A. L. Boynton, J. P. MacManus, and J. F. Whitfield, *Exp. Cell Res.*, **138,** 454 (1982).
30. R. H. Kretsinger, "Hypothesis: Calcium Modulated Proteins Contain EF Hands," in *Calcium Transport in Contraction and Secretion*, E. Carafoli, Ed., North-Holland Publishing Co., 1975, pp. 469–478.
31. T. D. Pollard and R. R. Weihing, *CRC Crit. Rev. Biochem.*, **2,** 1 (1974).
32. W. R. Dayton, *Biochim. Biophys. Acta*, **709,** 166 (1982).
33. F. G. Prendergast and K. G. Mann, *Biochemistry*, **17,** 3448 (1978).
34. Y. Takai, A. Kishimoto, Y. Iwasa, Y. Kawahara, T. Mori, and Y. Nishizuka, *J. Biol. Chem.*, **254,** 3692 (1979).
35. J. Haiech, J. Derancourt, J.-F. Pechére, and J. G. Demaille, *Biochemistry*, **13,** 2752 (1979).
36. A. Cavé, M.-F. Daures, J. Parello, A. Saint-Yves, and R. Sempere, *Biochimie*, **61,** 755 (1979).
37. A. Cavé, A. Saint-Yves, J. Parello, M. Swärd, E. Thulin, and B. Lindman, *Biochemistry*, **44,** 161 (1982).
38. J. Sowadski, G. Cornick, and R. H. Kretsinger, *J. Mol. Biol.*, **124,** 123 (1978).
39. M.-J. Rhee, D. R. Sudneck, V. K. Arkle, and W. D. Horrocks, *Biochemistry*, **20,** 3328 (1981).
40. G. Licheri, G. Piccaloga, and G. Pinna, *J. Chem. Phys.*, **64,** 2437 (1976).
41. S. Cummings, J. E. Enderby, and R. A. Kowe, *J. Phys. Chem. C.*, **13,** 1 (1980).
42. J. Grandjean, P. Laszlo, and C. Gerday, *FEBS Lett.*, **81,** 376 (1977).
43. H. Donato and B. R. Martin, *Biochemistry*, **13,** 4575 (1974).
44. T. Drakenberg, B. Lindman, A. Cavé, and J. Parello, *FEBS Lett.*, **92,** 346 (1978).
45. L. Lee, B. D. Sykes, and E. R. Birnbaum, *FEBS Lett.*, **98,** 169 (1979).
46. H. J. Vogel, T. Drakenberg, and S. Forsén, in "NMR of Newly Accessible Nuclei (Laszlo, P., Ed.) Academic Press, New York, in press. (1983).
47. J. D. Potter and J. Gergely, *J. Biol. Chem.*, **250,** 4628 (1975).
48. J. D. Potter and D. J. Johnson, *Calcium Cell Funct.*, **11,** 145 (1982).
49. C.-L.A. Wang, P. C. Leavis, W. D. Horrocks, and J. Gergely, *Biochemistry*, **20,** 2439 (1981).
50. M. J. Holroyde, S. P. Robertson, D. J. Johnson, J. R. Solaro, and J. D. Potter, *J. Biol. Chem.*, **255,** 1168 (1980).
51. I. L. Sin, R. Fernandes, and D. Mercola, *Biochem. Biophys. Res. Comm.*, **82,** 1132 (1978).
52. P. C. Leavis, S. Rosenfeld, and R. C. Lu, *Biochem. Biophys. Acta.*, **535,** 281 (1978).
53. P. C. Leavis, B. Nagy, S. S. Lehrer, H. Bialkowska, and J. Gergely, *Arch. Biochem. Biophys.*, **200,** 17 (1980).
54. C.-L.A. Wang, T. Tao, and J. Gergely, *J. Biol. Chem.*, **257,** 8372 (1982).
55. M. T. Hincke, B. D. Sykes, and C. M. Kay, *Biochemistry*, **20,** 3286 (1981).

56. T. H. Crouch and C. B. Klee, *Biochemistry*, **19,** 3692 (1980).
57. J. Haiech, C. B. Klee, and J. G. Demaille, *Biochemistry*, **20,** 3890 (1981).
58. K. B. Seamon, *Biochemistry*, **19,** 207 (1980).
59. K. B. Seamon, "Structural Studies on High Affinity Calcium Binding Proteins: Skeletal Troponin-C and Brain Calmodulin," in *Biochemical Structure Determination by NMR*, Bothner-By et al., Eds., Marcel Dekker, New York, 1982, pp. 199–224.
60. C. B. Klee, *Biochemistry*, **16,** 1017 (1977).
61. P. G. Richman and C. B. Klee, *J. Biol. Chem.*, **254,** 5372 (1979).
62. M.-C. Kilhoffer, J. G. Demaille, and D. Gerard, *FEBS Lett.*, **116,** 269 (1980).
63. M.-C. Kilhoffer, D. Gerard, and J. G. Demaille, *FEBS Lett.*, **120,** 99 (1980).
64. C.-L. A. Wang, R. R. Aquaron, P. C. Leavis, and J. Gergely, *Eur. J. Biochem.*, **124,** 7 (1982).
65. T. Andersson, T. Drakenberg, S. Forsén, and E. Thulin, *FEBS Lett.*, **125,** 39 (1981).
66. T. Andersson, T. Drakenberg, S. Forsén, E. Thulin, and M. Swärd, *J. Am. Chem. Soc.*, **104,** 576 (1982).
67. J. Parello, H. Lilja, A. Cavé, and B. Lindman, *FEBS Lett.*, **87,** 191 (1978).
68. S. Forsén, T. Andersson, T. Drakenberg, E. Thulin, and M. Swärd, *Fed. Proc.*, **41,** 2981 (1982).
69. T. Andersson, T. Drakenberg, S. Forsén, and E. Thulin, *Eur. J. Biochem.*, **126,** 501 (1982).
70. S. Forsén, E. Thulin, and H. Lilja, *FEBS Lett.*, **104,** 123 (1979).
71. S. Forsén, *FEBS Lett.*, **117,** 189 (1980).
72. L. Powers, *Biochim. Biophys. Acta*, **683,** 1 (1982).
73. A. Bianconi, A. Giovanelli, and L. Castellani, *J. Mol. Biol.*, **165,** 125 (1983).
74. R. E. Reid, J. Gariepy, A. K. Saund, and R. S. Hodges, *J. Biol. Chem.*, **256,** 2742 (1981).
75. B. Nagy, J. D. Potter, and J. Gergely, *J. Biol. Chem.*, **253,** 5971 (1978).
76. E. R. Birnbaum and B. D. Sykes, *Biochemistry*, **17,** 4965 (1978).
77. J. Gariepy, B. D. Sykes, R. E. Reid, and R. S. Hodges, *Biochemistry*, **21,** 1506 (1982).
78. P. C. Leavis, J. S. Evans, and B. A. Levine, *J. Inorg. Biochem.*, **16,** 257 (1982).
79. P. C. Leavis, S. S. Rosenfeld, J. Gergely, Z. Grabarek, and W. Drabikowski, *J. Biol. Chem.*, **253,** 5452 (1978).
80. J. S. Evans, B. A. Levine, P. C. Leavis, J. Gergely, Z. Corabarek, and W. Drabikowski, *Biochim. Biophys. Acta*, **623,** 10 (1980).
81. K. B. Seamon, D. J. Hartshorne, and A. A. Bothner-By, *Biochemistry*, **16,** 4039 (1977).
82. W. Drabikowski, H. Brzeska, and S. Y. Venyaminov, *J. Biol. Chem.*, **257,** 11584 (1982).
83. C. Gosselin-Rey, N. Bernard, and C. Gerday, *Biochim. Biophys. Acta*, **303,** 90 (1973).
84. C. J. Coffee and C. Solano, *Biochim. Biophys. Acta*, **453,** 67 (1976).
85. J. Closset and C. Gerday, *Biochim. Biophys. Acta*, **405,** 228 (1975).
86. A. Cavé, J. Parello, T. Drakenberg, E. Thulin, and B. Lindman, *FEBS Lett.*, **100,** 148 (1979).
87. J. A. Cox, D. R. Winge, and E. A. Stein, *Biochimie*, **61,** 601 (1979).
88. W. J. Birdsall, B. A. Levine, R. J. P. Williams, J. G. Demaille, J. Haiech, and J.-F. Pechère, *Biochimie*, **61,** 741 (1979).

89. R. E. Reid and R. S. Hodges, *J. Theor. Biol.*, **84,** 401 (1980).
90. J. D. Potter, J. C. Seidel, P. Leavis, S. S. Lehrer, and J. Gergely, *J. Biol. Chem.*, **251,** 7551 (1976).
91. J. P. Van Eerd and Y. Kawasaki, *Biochem. Biophys. Res. Commun.*, **47,** 859 (1972).
92. B. A. Levine, D. Mercola, D. Coffman, and J. M. Thornton, *J. Mol. Biol.*, **115,** 743 (1977).
93. D. J. Johnson, J. H. Collins, and J. D. Potter, *J. Biol. Chem.*, **253,** 6451 (1978).
94. M. T. Hincke, W. D. McCubbin, and C. M. Kay, *Can. J. Biochem.*, **56,** 384 (1978).
95. D. J. Johnson and J. D. Potter, *J. Biol. Chem.*, **253,** 3775 (1978).
96. J. D. Potter, F.-J. Hsu, and H. J. Pownall, *J. Biol. Chem.*, **252,** 2452 (1977).
97. P. C. Leavis and E. L. Kraft, *Arch. Biochem. Biophys.*, **186,** 411 (1978).
98. D. J. Johnson, J. H. Collins, S. P. Robertson, and J. D. Potter, *J. Biol. Chem.*, **255,** 9635 (1980).
99. M. T. Hincke, B. D. Sykes, and C. M. Kay, *Biochemistry*, **20,** 4185 (1981).
100. J. R. Dedman, J. D. Potter, R. L. Jackson, J. D. Johnson, and A. R. Means, *J. Biol. Chem.*, **252,** 8415 (1977).
101. W. D. McCubbin, M. T. Hincke, and C. M. Kay, *Can. J. Biochem.*, **57,** 15 (1979).
102. P. G. Richman and C. B. Klee, *Biochemistry*, **17,** 923 (1978).
103. M. T. Hincke, B. D. Sykes, and C. B. Kay, *Biochemistry*, **20,** 3286 (1981).
104. M. Yazawa, M. Sakuma, and K. Yagi, *J. Biochem.*, **87,** 1313 (1980).
105. T. Iio, K. Mihashi, and H. Kondo, *J. Biochem.*, **83,** 961 (1978).
106. T. Iio, K. Mihashi, and H. Kondo, *J. Biochem.*, **85,** 97 (1979).
107. J. D. Potter, S. P. Robertson, F. Mandel, and D. J. Johnson, *J. Biol. Chem.*, (1983) in press.
108. D. J. Johnson, S. C. Charlton, and J. D. Potter, *J. Biol. Chem.*, **254,** 3497 (1979).
109. J. A. Cox, A. Malnoë, and E. A. Stein, *J. Biol. Chem.*, **256,** 3218 (1981).
110. J. A. Cox, M. Comte, and E. A. Stein, *Proc. Natl. Acad. Sci.*, **79,** 4265 (1982).
111. G. W. Amphlett, T. C. Vanaman, and S. V. Perry, *FEBS Lett.*, **72,** 163 (1976).
112. C. H. Keller, B. B. Olwin, D. C. LaPorte, and D. R. Storm, *Biochemistry*, **21,** 156 (1982).
113. C. Y. Huang, V. Chau, P. B. Chock, J. H. Wang, and R. K. Sharma, *Proc. Natl. Acad. Sci.*, **78,** 871 (1981).
114. R. M. Levin and B. Weiss, *Biochim. Biophys. Acta*, **540,** 197 (1978).
115. D. C. LaPorte, B. M. Wierman, and D. R. Storm, *Biochemistry*, **19,** 3814 (1980).
116. H. Hidaka, T. Yamaki, T. Totsuka, and M. Asano, *Mol. Pharmacol.*, **15,** 49 (1979).
117. S. P. Robertson, D. J. Johnson, and J. D. Potter, *Biophys. J.*, **34,** 559 (1981).
118. J.-M. Gilles, D. B. Thomason, J. LeFevre, and R. H. Kretsinger, *J. Muscle Res. Cell Motil.*, **3,** 377 (1982).
119. R. H. Kretsinger, J. E. Mann, and J. G. Simmonds, "Evaluation of the Role of Intestinal Calcium Binding Protein in the Transcellular Diffusion of Calcium," in *Proceedings of the Fifth Workshop on Vitamin D*, A. W. Norman, Ed., 1982, in press.
120. J. J. Feher, *Am. J. Physiol.*, **244,** C303 (1983).
121. M. M. Tucker, J. B. Robinson, and E. Stellwagen, *J. Biol. Chem.*, **256,** 9051 (1981).

122. M. W. Killiman and L. M. G. Heilmeyer, *Biochemistry*, **21,** 1727 (1982).
123. V. Niglli, E. S. Adunyah, and E. Carafoli, *J. Biol. Chem.*, **256,** 8588 (1981).
124. L. Meijer and P. Guerrier, *Biochem. Biophys. Acta*, **702,** 143 (1982).
125. S. M. Liang, C. M. Liang, and T. Y. Liu, *J. Biol. Chem.*, **256,** 4968 (1981).
126. J. Derancourt, J. Haiech, and J.-F. Pechère, *Biochim. Biophys. Acta*, **532,** 373 (1978).
127. E. E. Maximov and Y. V. Mitin, *Biochimie*, **61,** 751 (1979).
128. R. A. Weeks and S. V. Perry, *Biochem. J.*, **173,** 449 (1978).
129. N. V. Barskaya and N. B. Gusev, *Biochem. J.*, **207,** 185 (1982).
130. L. J. Van Eldik and D. M. Watterson, *J. Biol. Chem.*, **256,** 4205 (1981).
131. W. E. Schreiber, T. Sasagawa, K. Titani, R. D. Wade, D. Malencick, and E. H. Fischer, *Biochemistry*, **20,** 5239 (1981).
132. M. Walsh, F. C. Stevens, J. Kuznicki, and W. Drabikowski, *J. Biol. Chem.*, **252,** 7440 (1977).
133. D. J. Johnson and A. Schwartz, *J. Biol. Chem.*, **253,** 5243 (1978).
134. B. A. Levine, J. Thornton, R. Fernandes, C. M. Kelly, and D. Mercola, *Biochem. Biophys. Acta*, **535,** 11 (198).
135. D. L. Johnson, M. J. Holroyde, T. H. Crouch, J. R. Solaro, and J. D. Potter, *J. Biol. Chem.*, **256,** 12194 (1981).
136. D. A. Malencik, S. R. Anderson, Y. Shalitin, and M. I. Schimerilik, *Biochem. Biophys. Res. Commun.*, **101,** 390 (1981).

CHAPTER 2

Calmodulin: The Regulation of Cellular Function

MILTON J. CORMIER

Department of Biochemistry
University of Georgia
Athens, Georgia

CONTENTS

1 INTRODUCTION

The regulatory functions of recently discovered Ca^{2+}-binding proteins have attracted much notice during the past few years. Calmodulin, because of its apparent ubiquity in eukaryotes and its numerous regulatory functions in both animals and plants, has received a great deal of this attention. Many recent reviews (1–6), books (7, 8), and special journal issues (9, 10) have been devoted to calmodulin- and Ca^{2+}-binding proteins. This chapter is not a comprehensive review of the calmodulin literature; the references given above provide more detailed information. Here I attempt to bring to the attention of a wide audience a subject about which there is much current interest. I also summarize the data that have led investigators to accept free Ca^{2+} as a second messenger in both animal and plant cells.

1.1 Ca^{2+} and Cyclic Nucleotides as Second Messengers

The original criteria described by Sutherland and colleagues (11) for establishing the second messenger function of cAMP cannot be used for free Ca^{2+} owing to differences in the ways that the intracellular concentrations of the two second messengers are regulated. Nevertheless, there are certain basic similarities in the ways second messengers function. A cellular component must exhibit at least three characteristic features to qualify as a second messenger:

1. In the unstimulated cell, the concentration of the second messenger, X, must be maintained at a level that does not result in a significant interaction of X with its target(s).
2. In the stimulated cell, the concentration of X must rise to a physiologically significant level in response to a stimulus, and eventually decline in the absence of the stimulus at a rate dependent upon the factors regulating the response time.
3. There must be a cellular target(s) for X, that is, an effector(s) subsequently leading to a cellular response. The targets are proteins and the cellular response is generally the result of a complex cascade of events.

There are two known kinds of second messengers. These are cyclic nucleotides (cAMP and cGMP) and free Ca^{2+} (12, 13). The second messenger function of cyclic nucleotides is expressed by the binding of these nucleotides to their target proteins, a ubiquitous class of protein kinases. These protein kinases are activated by cyclic nucleotides and regulate a variety of cellular processes via the phosphorylation of specific proteins. The second messenger function of Ca^{2+} is expressed by the binding of Ca^{2+} to a number of functionally distinct, but structurally related, Ca^{2+}-binding proteins (15). In some cases, these Ca^{2+}-protein complexes may function directly to initiate a cellular response. An example is the binding of Ca^{2+} to troponin-C in the initiation of muscle contraction. Alternatively, a Ca^{2+}-protein complex, such as Ca^{2+}–calmodulin, may in turn bind to its target proteins (enzymes), resulting in an increase in their activities that leads to a cellular response.

The second messenger functions of cyclic nucleotides and Ca^{2+} are interrelated by multiple feedback relationships. For example, Ca^{2+} may regulate the intracellular levels of cyclic nucleotides, and cyclic nucleotides may in turn regulate Ca^{2+} homeostasis. Specific examples are outlined below. It is beyond the scope of this chapter to focus on the regulatory roles of both kinds of second messengers. Rather, I focus on the calmodulin-dependent roles of Ca^{2+}. For more details on the roles of cyclic nucleotides, refer to Berridge (13) and Rasmussen and Goodman (14).

1.2 Metabolic Control of Intracellular Ca^{2+}

The regulation of intracellular Ca^{2+} homeostasis is complex (12, 13). Understanding this process is vital to the overall understanding of the second messenger function of Ca^{2+}. The following salient features may help orient the reader.

The regulation of free Ca^{2+} levels in the cytosol is most important in regulating the duration of the response to Ca^{2+} as a second messenger. The protein targets for Ca^{2+} are localized primarily in the cytosol of animal cells, or on membranes exposed to the cytosol.

We know that Ca^{2+} is actively removed from the cytosol. This follows from Donnan equilibrium considerations. For example, the extracellular Ca^{2+} concentrations range from 1 to 5 mM, whereas the membrane potential of the unstimulated animal cell generally ranges from 50 to 70 mV, inside negative. This results in a large electrochemical gradient, and if the intracellular concentration of Ca^{2+} were determined passively, the Donnan equilibrium predicts that its concentration would be about 1 M (16). Although plasma membranes are reasonably impermeable to ions such as

Ca^{2+}, some passive influx does occur as a result of this large concentration gradient. The consensus is that the intracellular total Ca^{2+} levels are approximately millimolar, and the free Ca^{2+} in cytoplasm is less than 0.1 μM. These values depend upon the cell type used and the method of measurement. It follows that mechanisms must exist for the active removal of free Ca^{2+} from the cytosol to the outside and to intracellular compartments. Two energy dependent mechanisms account for the active removal of Ca^{2+} through the plasma membrane, the Ca^{2+} pump (Ca^{2+}-ATPase) and a Na^+-Ca^{2+} exchange process that depends upon a Na^+ gradient (12). The concentration of free Ca^{2+} required for half-maximal activation of both of these processes is in the range of 1 μM.

Ca^{2+} is equilibrated between the cytosol and intracellular compartments, such as the mitochondria and the endoplasmic or sarcoplasmic reticulum (12, 13, 16). In addition to the active extrusion of Ca^{2+} through the plasma membrane, cells use intracellular buffering systems to regulate the magnitude and the duration of a stimulus-induced Ca^{2+} transient. These subcellular compartments are well-suited for the rapid regulation of cytosolic Ca^{2+} levels, because their total Ca^{2+}-transporting membrane area is generally much greater than that of the plasma membrane. They also contain significant pools of exchangeable Ca^{2+}. Although estimates of free Ca^{2+} concentrations in these compartments range from 10 to 100 μM, the value for total calcium falls in the millimolar range. Therefore, most of the calcium is bound. For example, in mitochondria Ca^{2+} combines with phosphate to form an osmotically inactive calcium phosphate complex. This allows millimolar amounts of Ca^{2+} to be taken up. It is important to note that a considerable fraction (20 to 60%) of the mitochondrial calcium is rapidly exchangeable with extramitochondrial pools. The K_m for the mitochondrial Ca^{2+} uptake system has been estimated to be between 2 and 10 μM, and the driving force is an electrochemical gradient formed by substrate oxidation or ATP hydrolysis. The K_m for the sarcoplasmic reticulum Ca^{2+} transport system has been estimated to be between 0.1 and 1.0 μM, and the driving force for uptake is the hydrolysis of ATP by a Ca^{2+}-ATPase. In the sarcoplasmic reticulum, Ca^{2+} appears to be bound to a low affinity Ca^{2+} binding protein ($K_d = 0.8$ mM) with multiple Ca^{2+} sites. In cardiac muscle, Ca^{2+} can be released from its sarcoplasmic reticulum stores by a mechanism involving Ca^{2+} itself as the trigger (17). The rise of Ca^{2+} in the cytosol, in response to a stimulus, triggers the release of the additional Ca^{2+} needed to initiate contraction. This Ca^{2+}-dependent Ca^{2+} release appears to operate in certain nonmuscle cells, for example, the release of Ca^{2+} from subcellular pools into the cytosol during the fertilization wave in the medaka egg (18).

The steady state levels of cytosolic free Ca^{2+} in the unstimulated cell are determined by the balance between the passive influx of Ca^{2+} into the cytosol from extracellular and subcellular pools and the active removal of Ca^{2+} from the cytosol by energy-dependent removal through the plasma membrane and energy-dependent uptake into subcellular compartments. The relative importance of these processes varies with the cell type. The subcellular compartmentalization of Ca^{2+} may serve the following functions: (a) to aid in maintaining low levels of free Ca^{2+} in the cytosol of the unstimulated cell; (b) to act as intracellular buffers during the rapid removal of free Ca^{2+} in the cytosol following a stimulus; or (c) to serve as sources of cytosolic free Ca^{2+} following a stimulus, as is the case in muscle and in the medaka egg.

Following a stimulus, localized increases in the concentration of free Ca^{2+} might occur in the cytosol (12, 13, 15, 16, 19, 20). How a receptor-mediated change in cytosolic Ca^{2+} occurs is not known. During the last 15 years, numerous cell types exhibiting Ca^{2+}-dependent responses have been examined by various techniques that measure the concentration of cytosolic free Ca^{2+} (19–25). For a comprehensive review of the techniques used to measure free Ca^{2+} in cells, see reference 30.

According to Rose and Lowenstein (26), the stimulus-induced rise in cytosolic Ca^{2+} probably does not result in a uniform distribution of cytosolic Ca^{2+}. They monitored the salivary gland's ability to sequester localized high levels of microinjected Ca^{2+}. They concluded that cytosolic Ca^{2+} remains highly restricted because of the efficient energy-dependent buffering capacity of intracellular compartments in response to acute changes in cytosolic Ca^{2+} concentration. Since Ca^{2+} entering the cytosol in response to a stimulus can be restricted to a small area of the total cytosolic space, they suggested that two or more message domains can coexist without interfering with one another.

1.3 Ca^{2+}-Binding Proteins as Targets for Free Ca^{2+}

One century ago Ringer (27, 28) demonstrated that Ca^{2+} was required to sustain the contractile properties of frog heart preparations, and in 1894 Locke (29) demonstrated that Ca^{2+} was necessary for neuromuscular transduction. Thus, before the turn of the century Ca^{2+} had been shown to play at least two distinct roles in excitation-contraction coupling. Since these pioneering studies, there have been numerous observations suggesting the importance of Ca^{2+} in a variety of physiological processes. A summary of many of these can be found in a recent book by Rasmussen (49).

Until a few years ago it was difficult to understand Ca^{2+}-mediated events. The work of Heilbrunn (31–33) four to five decades ago led to the suggestion that Ca^{2+} is important in controlling a number of key cellular functions. Heilbrunn's proposals were not generally accepted by his colleagues. During the past two decades, several discoveries have been made and new concepts proposed that have resulted in significant advances in our understanding of the molecular roles of Ca^{2+} in cellular function.

The discovery of Ca^{2+}-binding proteins. In 1959 Krebs et al. (34) demonstrated that the activity of phosphorylase *b* kinase was enhanced when Ca^{2+} was added to extracts of rabbit skeletal muscle. Since they were dealing with a complex mixture of proteins, it was not possible to pinpoint the site of Ca^{2+} action. It was not until the mid-1960s, using pure phosphorylase *b* kinase, that the enzyme was shown to be a high affinity Ca^{2+}-binding protein (35–37). About this time Ebashi and his colleagues were elucidating the molecular basis for the regulatory role of Ca^{2+} in smooth muscle contraction (38). They demonstrated that the regulation of myosin-actin interactions by Ca^{2+} is mediated by two proteins, troponin and tropomyosin. They subsequently showed, in 1967, that troponin was a high affinity Ca^{2+}-binding protein complex capable of binding 5 moles Ca^{2+} per 100,000 grams of protein with a binding constant of $6 \times 10^5\ M^{-1}$ in the presence of 4 mM Mg^{2+} (39). They further showed that most of this bound Ca^{2+} was rapidly exchangeable, an important requirement for a Ca^{2+}-dependent regulatory process.

The Ca^{2+}-binding protein aequorin, although well-known for its usefulness today as a Ca^{2+} indicator, was overlooked in terms of its historical importance as a Ca^{2+}-binding protein. In 1962 Shimomura et al. (40) isolated aequorin, the bioluminescent protein, from the luminous hydromedusa, *Aequorea*. They showed that this low molecular weight protein produced visible light in the presence of Ca^{2+}. In 1963 they further showed that micromolar levels of Ca^{2+} were sufficient for light production (41). Although it was not known until much later that aequorin is an oxygen-containing protein substrate complex that is stable in the absence of Ca^{2+} (42–44), it was clear by 1963 that aequorin was a high affinity Ca^{2+}-binding protein. Historically, aequorin was thus the first high affinity Ca^{2+}-binding protein to be isolated.

Additional Ca^{2+}-binding proteins have since been isolated and studied in great detail. One example is calmodulin, a low molecular weight protein which appears to be ubiquitous among eukaryotes.

The development of the concept by Kretsinger (15, 24) that Ca^{2+}-modulated proteins are responsible for transducing the intracellular Ca^{2+} message. As Kretsinger points out, proteins are the only macromolecules of the cell that have the appropriate selectivity and affinity for Ca^{2+}. Although cellular components, such as phospholipids and carbohydrates, bind Ca^{2+}, their affinities are too low for them to bind the micromolar levels of free Ca^{2+} produced in the cytosol following a stimulus. These Ca^{2+}-modulated proteins have evidently evolved recognition sites that bind Ca^{2+} with high affinity.

Kretsinger has also suggested that the only function of Ca^{2+} in the cytosol is the transmission of information and that the target of Ca^{2+}, as a second messenger, is a Ca^{2+}-modulated protein. He (15) defined Ca^{2+}-modulated proteins as those existing in the cytosol, or attached to membranes facing the cytosol, and which contain high affinity Ca^{2+}-binding sites ($K_d = \mu M$). As illustrated in Figure 1, the transient binding of Ca^{2+} to a Ca^{2+}-modulated protein confers Ca^{2+}-modulation on cytosolic enzymes and cellular processes.

The development by Rasmussen (48) of the concept that Ca^{2+} acts as a second messenger. He proposed that in many tissues Ca^{2+} and cAMP serve related and interdependent messenger functions. This thesis is now generally accepted as a result of the advances made in our understanding of Ca^{2+} metabolism and because of the discovery of Ca^{2+} receptor proteins.

2 STRUCTURE AND PROPERTIES OF ANIMAL AND PLANT CALMODULIN

Calmodulin was discovered as a protein activator of mammalian brain cyclic nucleotide phosphodiesterase simultaneously by Cheung (50) and Kakiuchi et al. (51). Kakiuchi et al. (51) observed that the phosphodiesterase activity was Ca^{2+}-dependent and Teo and Wang (52) later demonstrated that calmodulin was a Ca^{2+}-binding protein, thus explaining the earlier observation.

Calmodulin has been isolated from a variety of mammalian tissues and other eukaryotic sources, including protozoans, coelenterates, annelids, flowering plants, fungi, green algae, and amoebae (1–6, 54). The molecular weight of bovine brain calmodulin, based upon its amino acid sequence, is

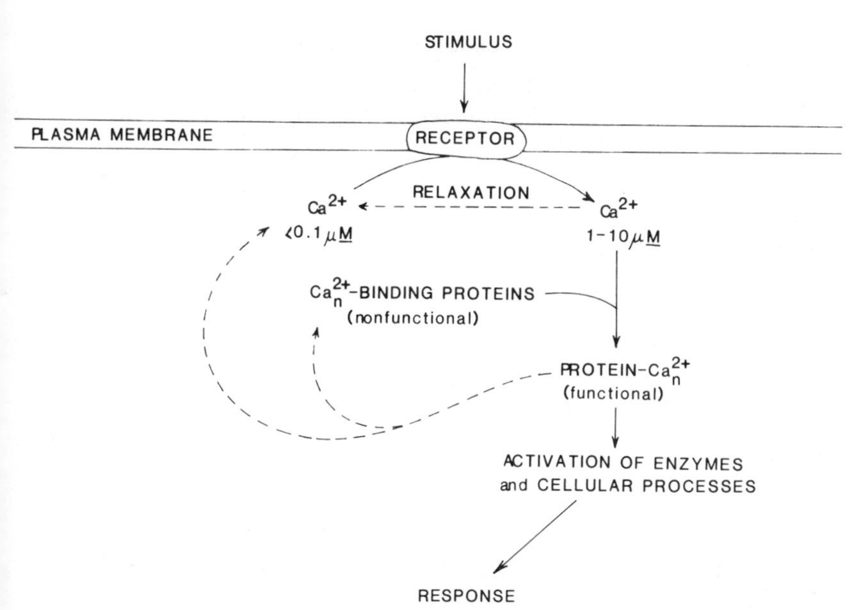

Figure 1 Model for Ca^{2+}-dependent regulation in eukaryotic cells. In the unstimulated cell, the cytosolic levels of Ca^{2+} remain low and thus Ca^{2+} cannot bind to its target proteins. Upon stimulation, the Ca^{2+} levels rise and Ca^{2+} binds to a group of structurally related Ca^{2+}-binding proteins. The protein-Ca^{2+} complex is the functional species that modulates the activities of numerous enzymes and cellular processes. During relaxation (broken lines) the Ca^{2+} levels fall, resulting in the dissociation of the protein-Ca^{2+} complexes with subsequent deactivation of enzyme activities and cellular processes.

16,790 (53), and molecular weights of other calmodulins range from 15,000 to 19,000, based upon methods such as sedimentation equilibrium and gel electrophoresis in the presence of sodium dodecylsulfate. The protein in solution must, therefore, exist primarily as a monomer. Animal and plant calmodulins share other similarities, including stability to acid and heat treatment, amino acid composition, highly acidic nature (isoelectric point 3.9 to 4.3), and ability to bind Ca^{2+} at micromolar concentrations. The tissue content of calmodulin is relatively high from diverse tissue sources (Table 1).

2.1 Structural Features of Calmodulin

A unique feature of calmodulin is its high degree of structural conservation, as noted in those isolated from diverse species (Fig. 2). There are 148 amino

Table 1 Tissue Content of Calmodulins from Diverse Sources

Source	Percent Total Soluble Protein	Content (mg/kg)	Yield (mg/kg)	References
Vertebrates				
Bovine brain	1.0	310	41 to 170	53, 114
Rat testis	0.15	137	96	55
Rabbit uterus	0.4	—	110	114
Ox aorta	—	—	60	114
Chicken gizzard	—	—	180	114
Rabbit liver	—	—	140	114
Hamster kidney (BHK-21)	—	—	83	120
Bovine pancreas	0.05	90	21	115
Electroplax (Electric eel)	3 to 4	1000	700	121, 122
Spermatozoa	0.2	—	—	56
Invertebrates				
Coelenterate (*Renilla*)	0.04	14	6	117
Echinoderms				
Sea urchin eggs	0.1 to 0.2	—	—	56, 57
Sea urchin spermatozoa	0.3	—	70	56, 116
Protozoa				
Tetrahymena pyriformis	1.0	—	17	118
Fungi				
Aqaricus Sp.	—	24[a]	5[a]	
Dictyostelium discoideum	0.04	26	5	119
Plants				
Peanut seeds	0.12	42	7	58
Spinach leaves	0.6	26	8	59

[a]Data from reference 374.

acid residues in the amino acid sequence of bovine brain calmodulin (60). When the amino acid sequence of this protein is compared to calmodulins isolated from two marine invertebrates, only three replacements are found (61, 62). These occur at residues 99 (tyr→phe), 143 (gln→lys), and 147 (ala→ser). The only other differences appear to be in several of the amide assignments. Another invertebrate (Scallop) calmodulin also showed identical replacements at positions 99 and 147, but glutamine is replaced by threonine at position 143 (368). Calmodulin isolated from the protozoan *Tetrahymena* has eleven substitutions and one deletion when compared to bovine brain calmodulin (63). These differences are concentrated on the *C*-terminal end of the molecule. The complete amino acid sequence of a

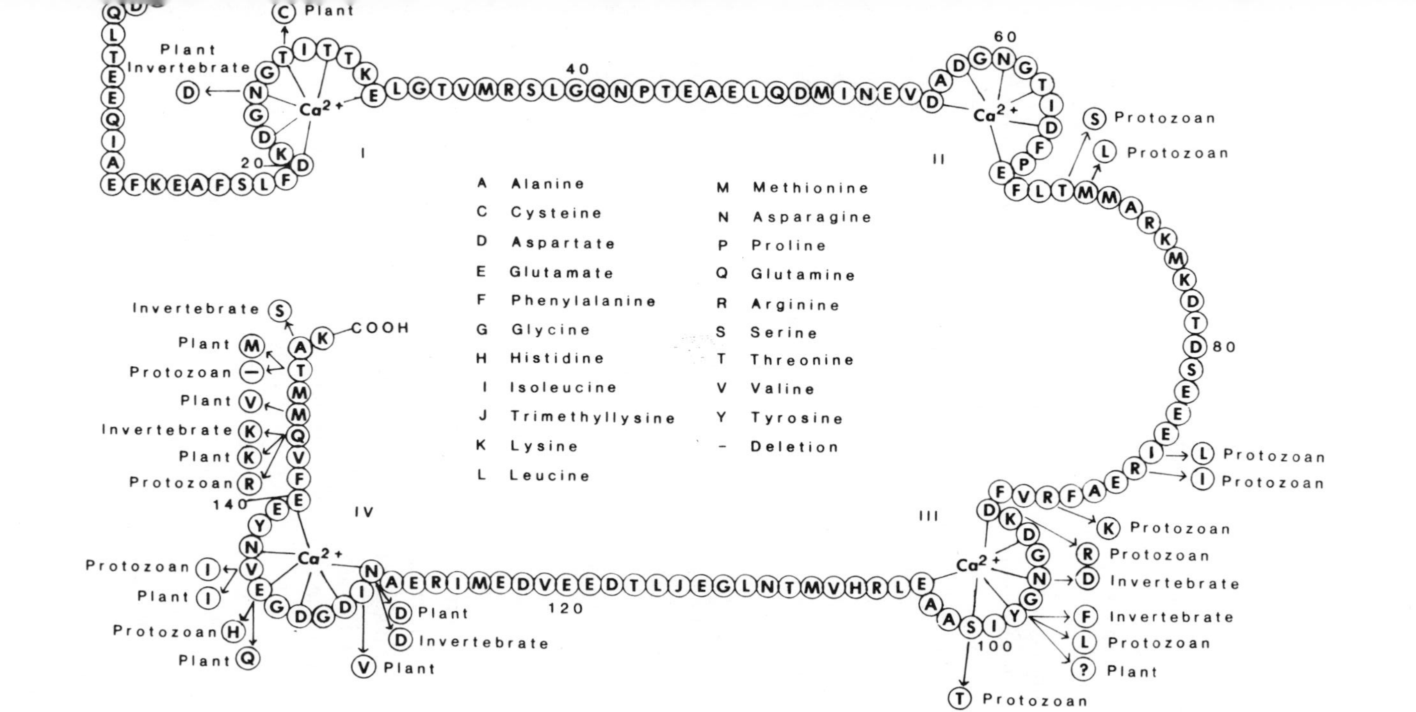

Figure 2 Comparison of the primary structures of calmodulins isolated from diverse species. The amino acid sequence of bovine brain calmodulin is shown (60). The amino acid sequences of calmodulins isolated from a protozoan, *Tetrahymena* (63), a sea anemone (61), and a coelenterate, *Renilla reniformis* (62) are identical to bovine brain calmodulin, except for the indicated substitutions. The possible differences in amide assignments were not determined for sea anemone calmodulin (61). Some substitutions are also indicated for plant calmodulin (64, 65), although the complete amino acid sequence is not yet available. The question mark at position 99 in plant calmodulin refers to the fact that plant calmodulin contains a single tyrosine residue (66) located at position 138 (64). The four putative Ca^{2+}-binding domains are also indicated (46, 47).

plant calmodulin is not yet available, but recent data (64, 65) suggest that plant calmodulin is also highly conserved. As shown in Figure 2, the known substitutions in plant calmodulin are also concentrated on the *C*-terminal end of the molecule.

Both animal and plant calmodulins thus far examined are characterized by the absence of tryptophan, a high content of glutamate and aspartate residues, and a single residue of trimethyllysine located between domains III and IV in the molecule (Fig. 2; refs. 59–61, 63, 64, 66, 67). Certain calmodulins do not contain trimethyllysine, however. These include calmodulin isolated from *Dictyostelium* (163), *Agaricus* (mushroom) (164), and *Neurospora* (165). Octopus calmodulin was found to contain approximately 0.1 mole (166) or 1.0 mole (167) of trimethyllysine per molar equivalent, depending upon whether calmodulin was obtained from extracts of the whole organism or of the optic lobe. All vertebrate calmodulins thus far examined contain two residues of tyrosine, located at positions 99 and 138, and have a phenylalanine/tyrosine ratio of 4. Invertebrate and higher plant calmodulins contain a single tyrosine residue, located at position 138, and have a phenylalanine/tyrosine ratio of 8 to 9. The $\epsilon^{1\%}_{276nm}$ values for these calmodulins vary accordingly. They are 1.8 to 2.0 for vertebrate calmodulin and 0.9 to 1.0 for plant calmodulin. Since calmodulin is devoid of tryptophan, the difference in the phenylalanine/tyrosine ratios found in the protein isolated from diverse species affects the shape of the absorption spectra (Fig. 3).

2.2 Ca^{2+} Binding to Calmodulin

From crystal structure analysis of parvalbumin, Moews and Kretsinger (45) have shown that the conformation of the Ca^{2+} binding domain is distinctive. They named this conformation the EF-hand. The EF-hand domain contains 31 consecutive amino acid residues. Amino acid sequences similar to those found in EF-hand domains have been observed in a number of other Ca^{2+}-binding proteins, including troponin-C, intestinal Ca^{2+}-binding protein, S-100, regulatory myosin light chains, essential myosin light chains, and calmodulin (46, 47). These homologous amino acid sequences represent putative EF-hand domains. The proteins listed above contain at least as many putative EF-hand domains as there are high affinity Ca^{2+}-binding sites on the protein. For example, calmodulin has four putative EF-hand domains and four high affinity Ca^{2+}-binding sites (4). Kretsinger has postulated that all intracellular proteins that are modulated by Ca^{2+} contain EF-hands (15, 46).

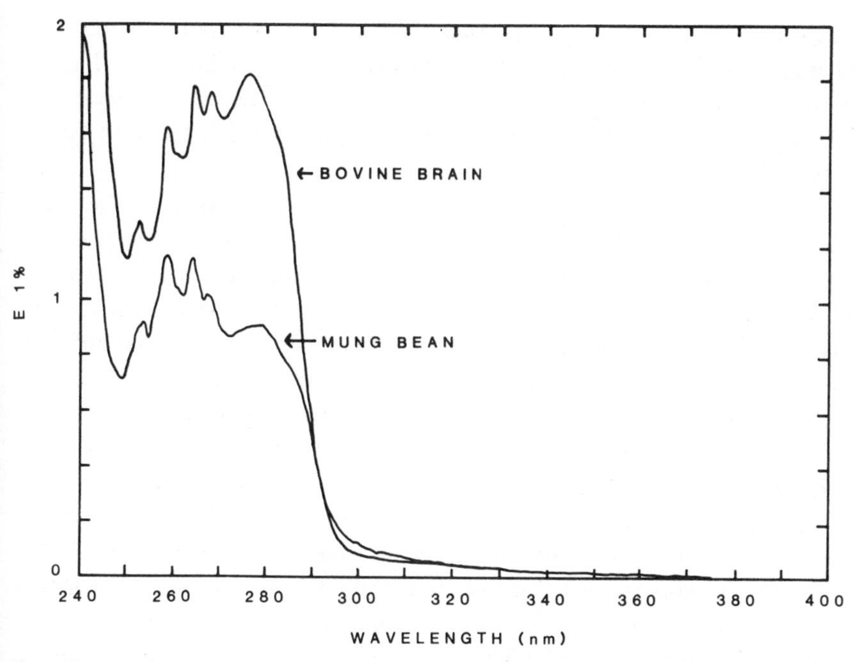

Figure 3 Comparison of the absorption spectra of bovine brain and mung bean calmodulins.

Calmodulin will bind 4 moles Ca^{2+} per mole with high affinity (4, 6). These Ca^{2+} sites have been designated Ca^{2+}-binding domains I through IV (Fig. 2). Note that the amino acid sequences around each of the domains have been highly conserved, and that the homology is greatest between domains I and III, and II and IV. This suggests that two successive doublings of a primordial gene coding for a Ca^{2+}-binding domain could have been important in the evolution of the calmodulin molecule (60).

The binding of Ca^{2+} to calmodulin results in significant conformational changes in the molecule. There is an increase in α-helical content, as determined by circular dichroism and optical rotatory dispersion (55, 68–70). There are also changes in reactivity toward group-specific chemical modification agents (71–74), altered susceptibility to cleavage by trypsin (75), as well as changes in the spectral properties of the protein (55, 68–70, 74, 76–78). A number of investigators have found that there is about a three-fold enhancement of tyrosine fluorescence that occurs upon Ca^{2+}-binding to mammalian, octopus, and plant calmodulin (55, 66, 79). Mammalian calmodulin contains two tyrosyl residues, whereas octopus and plant calmodulins contain a single tyrosyl residue, suggesting that the bulk of the tyrosine fluorescence change is associated with tyrosyl residue 138 located in domain IV (Fig. 2). Studies on the effects of Ca^{2+} binding on the UV-difference spectrum (72, 74, 80) also indicate that these spectral changes are associated with tyrosyl residue 138.

The intrinsic dissociation constants for the four Ca^{2+} sites vary over the range 0.1 to 1.0 μM (82). The value of these constants for each of the four sites were 67, 170, 600, and 900 nM. Hence, there appear to be two sites of relatively high and two sites of relatively low affinity. Recent evidence suggests that there is a positive cooperativity of Ca^{2+}-binding to calmodulin, at low Ca^{2+} concentrations, in the presence or absence of 3 mM Mg^{2+} (81).

Studies on the binding of Tb^{3+} to mammalian calmodulin have suggested that Ca^{2+}-binding domains I and II are sites of highest affinity. Titration of the protein with Tb^{3+} showed that Tb^{3+} fluorescence increased (whereas tyrosine fluorescence decreased) only after the binding of two Tb^{3+} equivalents, presumably as a result of energy transfer from tyrosine to Tb^{3+} (83, 369). The authors suggest that the first two Tb^{3+} bind to sites that lack a tyrosine, namely domains I and II (Fig. 2). Large increases in Tb^{3+} fluorescence are produced by Tb^{3+} binding to domains III and IV. Similar studies have been done with octopus calmodulin, which contains a single tyrosine in domain IV (84). Titration of octopus calmodulin with Tb^{3+} showed that Tb^{3+} fluorescence increased only after the binding of about three Tb^{3+} equivalents. A substantial increase in Tb^{3+} fluorescence occurred when the fourth site, which contains the single tyrosyl residue, was occupied.

These studies suggest that the order of Ca^{2+} binding to calmodulin is as follows: domains I and II (the highest affinity sites) are filled first, followed by binding to domain III, and finally domain IV (the lowest affinity site) is filled (81–84, 369). Similar conclusions have been reached by studying the competitive binding of K^+, Mg^{2+}, and Ca^{2+} to the Ca^{2+}-binding sites of calmodulin (82).

Studies of the physical changes that occur in calmodulin as a function of the number of Ca^{2+} bound per mole of protein clearly show that multiple conformational states exist in the protein. A number of the Ca^{2+}-dependent spectral changes associated with calmodulin occur upon filling the first two Ca^{2+} sites (4). The effects of Ca^{2+} binding on the solution conformation of calmodulin has been studied using 360 MHz proton nuclear magnetic resonance (77). This study showed that at least three distinct conformations of calmodulin may exist: metal free, calmodulin–$(Ca^{2+})_2$, and calmodulin–

$(Ca^{2+})_{3-4}$. Other investigators, who have studied the Ca^{2+}-dependence of the near-UV circular dichroic spectra of modified calmodulin, also came to similar conclusions (85, 86). Studies on the positive cooperative binding of Ca^{2+} to calmodulin (81) also suggest the existence of at least three conformations, and possibly a fourth, calmodulin–$(Ca^{2+})_1$.

2.3 Drug Binding to Calmodulin

A number of antipsychotic drugs bind to calmodulin in a Ca^{2+}-dependent manner (87). For example, 2 moles of a phenothiazine, trifluoperazine, will bind per mole of calmodulin, in the presence of Ca^{2+}, with an apparent dissociation constant of 1 μM (88). When Ca^{2+} is removed, by the addition of a Ca^{2+} chelator such as EGTA, the drug-calmodulin complex dissociates. In contrast to the two high affinity sites for trifluoperazine observed in the presence of Ca^{2+}, only low affinity sites are observed in the absence of Ca^{2+}. In the presence of Ca^{2+}, the affinity of the phenothiazine sulfoxides for calmodulin is several orders of magnitude lower than the corresponding phenothiazine (89).

In addition to the phenothiazines, another group of compounds also binds to calmodulin in a Ca^{2+}-dependent manner. For example, a number of the naphthalene sulfonic acid derivatives, such as *N*-(6-aminohexyl)-5-chloro-1-naphthalene sulfonamide (W-7), binds to high affinity sites on calmodulin in the presence of Ca^{2+} (90, 91). Since 8-anilino-1-naphthalenesulfonate, 9-anthroylcholine, and *N*-phenyl-1-naphthylamine also bind to calmodulin in a Ca^{2+}-dependent manner (92), it seems that the drug binding domain is hydrophobic, and becomes available as a result of conformational changes in calmodulin which take place upon Ca^{2+}-binding (see discussion following). Similar conclusions were reached from studies on the binding of several chlorpromazine analogs to calmodulin (93).

All of the drugs mentioned also inhibit the activation of calmodulin-dependent enzymes purified from both animal and plant tissue sources (1–6, 54, 87). A new inhibitor of calmodulin function, R 24571, appears to be the most potent yet reported (95). Certain derivatives of these compounds such as the phenothiazine sulfoxides and W-5, a chloro-deficient analog of W-7, show little or no inhibition of calmodulin-dependent enzyme activation when compared to the parent compound. For these reasons, such drug pairs have been used as probes for calmodulin-linked processes both *in vitro* and *in vivo* (1–6, 54, 90, 94).

The use of these compounds, especially as *in vivo* probes for calmodulin function, should be viewed with caution. This is especially true for the phenothiazines, which are known to behave as amphiphathic detergents and, as such, bind to numerous membrane sites with resultant effects on membrane structure and function (96–100). The analogs W-7 and W-5 may be more useful than the phenothiazines as *in vivo* probes for calmodulin function, since their hydrophobicities are similar and they appear to have less drastic effects on membrane structure.

The Ca^{2+}-dependent binding property of these drugs to calmodulin has been exploited for use in the development of Ca^{2+}-dependent affinity chromatography procedures (58, 94, 101–103). The purification of calmodulin from crude extracts can be greatly simplified by using these affinity resins. In certain cases, other Ca^{2+}-binding proteins may also bind to these affinity resins and may copurify with calmodulin (94, 102, 103, 366).

3 MODE OF CALMODULIN-DEPENDENT ENZYME ACTIVATION

The binding of Ca^{2+} to calmodulin is highly cooperative and the evidence suggests that this process results in an equilibrium mixture of several conformational states of calmodulin (CaM) as indicated in eqs. (1) and (2).

$$\text{CaM} + 2\,\text{Ca}^{2+} \rightleftharpoons \text{CaM-Ca}_2^{2+} \tag{1}$$

$$\text{CaM-Ca}_2^{2+} + 1\text{-}2\,\text{Ca}^{2+} \rightleftharpoons \text{CaM-Ca}_{3\text{-}4}^{2+} \tag{2}$$

This scheme may be oversimplified since the binding of each Ca^{2+} may produce a distinct conformational state of the protein, some of which may be difficult to detect by present methods, for example, CaM-Ca_1^{2+} and CaM-Ca_3^{2+} versus CaM-Ca_4^{2+} (4, 77, 81). Studies of Ca^{2+} binding to calmodulin show that the binding of the first two Ca^{2+}, to domains I and II (Fig. 2), produces a distinct conformation of the protein. Binding of the third and fourth Ca^{2+}, to domains III and IV, produces yet another conformer(s) of calmodulin.

The appropriate Ca^{2+}-dependent conformer of calmodulin (CaM*-Ca_n^{2+}) then interacts with its target enzyme(s) converting the inactive (or partially active) enzyme to an active (or more active) one, as indicated in eq. (3).

$$\underset{\text{(INACTIVE)}}{\text{Enzyme}} + \text{CaM*-Ca}_n^{2+} \rightleftharpoons \text{Enzyme-CaM*-Ca}_n^{2+} \rightleftharpoons \underset{\text{(ACTIVE)}}{\text{Enzyme*-CaM*-Ca}_n^{2+}} \quad (3)$$

The inactive enzyme interacts with CaM*-Ca_n^{2+} to form a complex, which presumably induces a conformational change in the enzyme (indicated by the asterisk) resulting in activation. In eq. (3), CaM*-Ca_n^{2+} represents the active conformer of calmodulin, where n could range in value from 1 to 4.

As suggested by Klee et al. (4), it will be of interest to determine whether a specific conformer of calmodulin is capable of recognizing one calmodulin-dependent enzyme over another. This may be important in the selective recognition of enzymes in the same cell as a function of localized changes in Ca^{2+} concentrations. Cyclic nucleotide phosphodiesterase is an example of an enzyme that is not activated by CaM*-Ca_2^{2+}, but is activated by CaM*-Ca_{3-4}^{2+} (69, 81, 104, 105). It is technically difficult to determine whether three (or all four) Ca^{2+} sites must be occupied on calmodulin to activate cyclic nucleotide phosphodiesterase. Recent kinetic evidence suggests that all four sites must be occupied (105).

On the other hand, some reports suggest that myosin light chain kinase is activated by CaM*-Ca_{1-2}^{2+} (106–107). Another report proposes that CaM*-Ca_4^{2+} is the functioning species in the activation of myosin light chain kinase (108). The difference in these two reports may simply reflect differences in experimental conditions. In particular, physiological concentrations of Mg^{2+} and K^+ influence both the affinity and apparent positive cooperativity of Ca^{2+} for calmodulin by competing with Ca^{2+} for the same sites (4, 6, 81, 82). Under physiological conditions, calmodulin could selectively recognize its various target enzymes by making use of its various Ca^{2+}-dependent conformers. Thus, calmodulin could recognize quantitative differences in intracellular Ca^{2+} concentrations, and subsequently translate these differences into qualitatively different cellular responses (4).

Some salient features of calmodulin-dependent enzyme activation are beginning to emerge from studies on the activation of cyclic nucleotide phosphodiesterase, myosin light chain kinase, and phosphorylase kinase.

A complex is formed between the inactive enzyme and CaM*-Ca_n^{2+}, the active conformer of calmodulin [see eq. (3)]. Such a complex has been demonstrated during the activation of both animal and plant enzymes (4, 6,

54). The techniques used include the demonstration of: (a) a reversible Ca^{2+}-dependent calmodulin-enzyme complex during gel filtration; (b) a Ca^{2+}-dependent calmodulin-enzyme complex by the Hummel-Dryer chromatographic technique; (c) a Ca^{2+} and calmodulin-dependent increase in enzyme tryptophan fluorescence; and (d) a reversible Ca^{2+}-dependent binding of an enzyme to calmodulin during chromatography on calmodulin-Sepharose. The presence of Ca^{2+} is an absolute requirement for the formation of calmodulin-enzyme complexes. In the absence of Ca^{2+}, calmodulin-enzyme complexes are not detected. Attempts to crosslink calmodulin to calmodulin-dependent enzymes, at enzyme concentrations of 10 μM, fail to detect complex formation in the absence of Ca^{2+} (109, 123). The concentration of free Ca^{2+} required to activate a calmodulin-dependent enzyme would be lowered as the calmodulin concentration increases (52, 105, 107, 108, 111–113). Thus, the cytosolic concentration of calmodulin is an important factor in determining the concentration of free Ca^{2+} levels required to achieve activation of a cytosolic enzyme.

Most calmodulin-dependent enzymes interact specifically with calmodulin. Other structurally analogous Ca^{2+}-binding proteins, troponin-C and parvalbumin, for example, do not activate enzymes such as cyclic nucleotide phosphodiesterase, myosin light chain kinase, or erythrocyte Ca^{2+}–ATPase even at concentrations that are several orders of magnitude higher than calmodulin (128–131), even though troponin-C has approximately 70% sequence homology with calmodulin (53, 136). An exception is the activation of phosphorylase kinase by troponin-C (Section 4.2.1).

The enzyme may recognize only one of several possible $CaM^*\text{–}Ca_n^{2+}$ conformers (see discussion given above).

As outlined in Section 2.3, the phenothiazines and a number of hydrophobic compounds bind to calmodulin in a Ca^{2+}-dependent manner (87–92; 134, 135). These drugs inhibit the calmodulin-dependent activation of enzymes. It has been proposed that Ca^{2+}-binding to calmodulin opens up a hydrophobic domain(s) on the protein, which is recognized by both the inhibitory drug and the target enzymes (92).

Attempts have been made to identify the region(s) of the calmodulin molecule responsible for binding to and activating its target protein by chemical modification of functional groups. For example, modification of lysine residues decreased the ability of calmodulin to activate phosphodiesterase, but had no effect on its activation of adenylate cyclase. The reverse was found after modification of arginine residues (71, 137). Modification of histidine had no effect on the ability of calmodulin to activate either phosphodiesterase or adenylate cyclase (71, 137). Modification of tyrosine residues also had no effect on the activation of both enzymes (71–73, 137–140). Alteration of methionine residues has the most pronounced effect on the biological properties of calmodulin. For example, carboxymethylation of one or two residues decreased the ability of calmodulin to activate both phosphodiesterase and adenylate cyclase; oxidation of methionine residues eliminated the ability of the protein to activate either enzyme (71, 137).

Peptides derived from calmodulin have also been examined for their ability to activate phosphodiesterase. A peptide has recently been isolated and shown by sequence data to correspond to peptide 72-148 of calmodulin (141). This peptide was unable to activate phosphodiesterase even at a 1000-fold higher concentration than that required for activation by calmodulin. It was able to bind one Ca^{2+}, rather than the 2 moles expected, possibly owing to an altered conformation in one of the domains following cleavage (141). These results are consistent with another report in which calmodulin fragments, tentatively identified by amino acid composition as peptides 1-77 and 78-148, were also found to be inactive in the activation of phosphodiesterase (142). A mixture of peptides of molecular weights ranging from 6000 to 8000, derived from the *C*-terminal end of plant calmodulin, has also been isolated (143). These fragments were shown to contain at least one Ca^{2+} and one phenothiazine binding site, and to activate the erythrocyte Ca^{2+}-ATPase with reduced affinity (3% that of calmodulin).

The results on chemical modification of calmodulin, and the properties of calmodulin fragments, demonstrate that the calmodulin molecule must be intact to retain significant biological activity. Furthermore, it appears that different enzymes may recognize different domains on the calmodulin molecule.

4 CALMODULIN-DEPENDENT REGULATORY PROCESSES IN ANIMAL CELLS

As shown in Figure 4, there are numerous calmodulin regulated enzymes. The number and kinds of calmodulin-dependent enzymes found in eukaryotes varies from tissue to tissue. One example is the Ca^{2+} and calmodulin-dependent form of cyclic nucleotide phosphodiesterase first reported in extracts of brain tissue (50, 51, 144). There are, however, calmodulin-independent

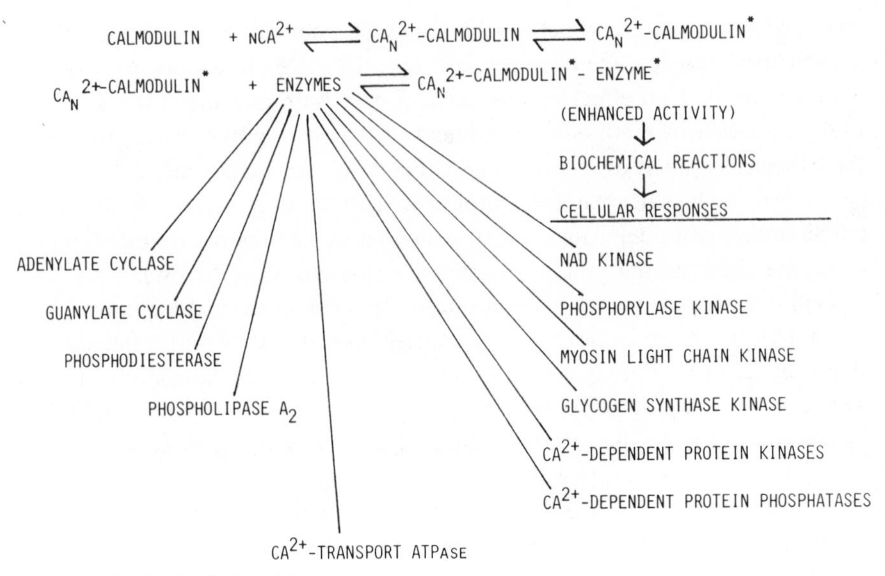

PROCESSES REGULATED BY CALMODULIN:

1. CELLULAR MOTILITY (ACTIN-MYOSIN INTERACTION; ACTIVATION OF DYNEIN ATPASES)
2. CYTOSKELETAL FUNCTION
3. GLYCOGEN METABOLISM
4. CYCLIC NUCLEOTIDE METABOLISM
5. CALCIUM METABOLISM
6. PHOTOSYNTHESIS

Figure 4 Schematic representation of calmodulin function in eukaryotic organisms. The various enzymes activated by calmodulin–Ca_n^{2+} are listed together with some cellular processes that are apparently under calmodulin control. (See Section 4 for details.)

forms of the enzyme (145, 146). This is especially true among invertebrates, which oftentimes are devoid of the calmodulin-dependent form of the enzyme, but do contain calmodulin (57, 117, 118, 147, 148).

The list of proteins that depend upon the binding of calmodulin–Ca_n^{2+} for expression of their function, shown in Figure 4, is incomplete. For example, several proteins from various tissues that bind to calmodulin in a Ca^{2+}-dependent manner have been detected (149–159). One such protein is calcineurin (252). In *in vitro* assays, such proteins may compete with the target enzyme for calmodulin and behave as inhibitors of enzyme activation. Some may represent calmodulin-dependent enzymes of unknown function.

A variety of mammalian tissues have been examined to determine the distribution of calmodulin between the cytosol and particulate fractions (113, 146, 160–162). In the particulate fraction, calmodulin is distributed among the microsomal, mitochondrial, and nuclear fractions (160). This distribution is affected by the presence or absence of Ca^{2+} in the extraction buffers. For example, in brain tissue approximately 70% of the total calmodulin was found in the particulate fraction when Ca^{2+} was included in the extraction buffer. When Ca^{2+} was replaced by EGTA, only 20% of the total calmodulin was found. This Ca^{2+}-dependent distribution of calmodulin was reversible, although the distribution of total protein was not affected by Ca^{2+}. It is possible that an intracellular redistribution of calmodulin in response to a Ca^{2+} transient may be of physiological importance.

4.1 Regulation of Cyclic Nucleotide Metabolism

Adenylate cyclase is a protein complex associated with the plasma membrane of most animal cells. This enzyme is responsible for the synthesis of cAMP, and is under hormonal and guanine nucleotide regulatory control (168). The adenylate cyclase complex consists of a hormone receptor, a catalytic subunit, and a guanine nucleotide regulatory unit. In brain tissue this enzyme is also regulated by calmodulin-Ca_n^{2+} (169, 170). Several other tissues and cell types contain an adenylate cyclase whose basal activity is stimulated by calmodulin-Ca_n^{2+} (1–6). This stimulation of adenylate cyclase activity by calmodulin-Ca_n^{2+} is not due to a change in its K_m for MgATP, but rather to an increase in the maximum velocity of the enzyme (171). As shown by eq. (3) (Section 3), adenylate cyclase forms a reversible Ca^{2+}-dependent complex with calmodulin (172). Recent evidence suggests that calmodulin-Ca_n^{2+} interacts directly with the catalytic subunit of the enzyme (174). In these studies, the catalytic and guanine nucleotide regulatory units of solubilized brain adenylate cyclase were separated. Using the isolated components, it was shown that both calmodulin-Ca_n^{2+} and the guanine nucleotide regulatory units activated the catalytic unit in an independent fashion, and these activities were additive, not synergistic (174).

Multiple forms of cyclic nucleotide phosphodiesterase occur in most animal tissues (173), but only the Ca^{2+}-dependent activity is stimulated by calmodulin (50, 51). The soluble enzymes are responsible for the degradation of cAMP and cGMP and are widely distributed among vertebrates and invertebrates (1–6).

The calmodulin-dependent phosphodiesterase isolated from brain and heart tissue consists of a dimer of two identical subunits, each having a M_r of approximately 60,000 and 57,000, respectively. The native enzyme apparently consists of a dimer of two identical subunits (124–127). This enzyme also forms a complex with calmodulin-Ca_n^{2+}, [eq. (3)], (175, 176). Each subunit binds 1 mole of calmodulin-Ca_n^{2+}, generating the active species A_2C_2, where A represents the phosphodiesterase subunit and C represents calmodulin-Ca_n^{2+} (126, 127). The formation of an active A_2C_2 complex requires that calmodulin-Ca_n^{2+} be in the form calmodulin-Ca_3^{2+} or calmodulin-Ca_4^{2+} (81, 104, 105). The apparent affinity constant of the enzyme for calmodulin-Ca_{3-4}^{2+} is between 0.1 and 1.0 nm (105, 123, 127).

Studies on the catalytic properties of the calmodulin-dependent form of phosphodiesterase have shown that the active complex has a K_m value of about 100 μM for cAMP and a K_m value of 5 to 10 μM for cGMP. Because its K_m for cAMP is much higher than the other forms of phosphodiesterase, the enzyme is often referred to as the "high K_m cyclic nucleotide phosphodiesterase" (173). Depending upon the preparation, the addition of calmodulin-Ca_{3-4}^{2+} to the enzyme results in a 10- to 50-fold increase in maximum velocity and a 2- to 5-fold decrease in the K_m values for cAMP and cGMP (123, 177–179).

The variable levels of basal activities noted in different preparations of phosphodiesterase may be caused by the irreversible proteolytic conversion of the enzyme to the calmodulin-independent form. This form is fully active, has a subunit M_r of about 36,000, but will not form a complex with calmodulin-Ca_{3-4}^{2+} (6). Thus, native calmodulin-dependent phosphodiesterase contains a domain that recognizes cyclic nucleotides and a second regulatory domain that recognizes calmodulin-Ca_{3-4}^{2+}. In contrast, limited proteolysis of Ca^{2+}-dependent adenylate cyclase has no effect on basal activity, but results in a loss of sensitivity to calmodulin-Ca_n^{2+} (180).

If we assume that the activations of both adenylate cyclase and cyclic nucleotide phosphodiesterase by calmodulin-Ca_n^{2+} occur in the same cell type, then such simultaneous activations present a paradox. Nevertheless, the evidence suggests that the activations of both enzymes could occur in a sequential fashion. For example, in the presence of calmodulin, adenylate cyclase is fully activated at about 0.1 μM free Ca^{2+} and is strongly inhibited at 1 μM free Ca^{2+}. However, under the same conditions, phosphodiesterase is activated at 1 μM free Ca^{2+} but is unaffected at 0.1 μM free Ca^{2+} (181, 182). Thus, as the intracellular free Ca^{2+} levels rise in response to a stimulus, adenylate cyclase could be activated first and then inhibited, concomitant with the activation of phosphodiesterase. Furthermore, the high K_m feature of the Ca^{2+}-dependent phosphodiesterase suggests that the levels of cyclic nucleotides would have to rise substantially above basal levels prior to observing significant catalytic activity by this enzyme.

4.2 Regulation of Protein Phosphorylation-Dephosphorylation

Many diverse cellular functions are regulated by protein phosphorylation (183). Both second messenger systems (cyclic nucleotides and Ca^{2+}) are interconnected in these regulatory processes at the level of protein kinases. This is shown in Figure 5, which also summarizes several important features of these interrelationships. For example, in mammalian cells, the cAMP system activates a protein kinase which phosphorylates relatively few proteins at specific sites. Enzymes dependent upon calmodulin-Ca_n^{2+}, such as phosphorylase kinase and myosin light chain kinase, may often be phosphorylated by cAMP-dependent protein kinase, resulting in altered properties. Furthermore, both cAMP and Ca^{2+}-dependent protein kinases frequently phosphorylate the same protein, although at different regulatory sites. Examples are glycogen synthetase and phospholambdan.

4.2.1 Ca^{2+}-Dependent Regulation of Glycogen Metabolism. Protein kinases are involved in the regulation of glycogen metabolism, that is, in the breakdown and synthesis of glycogen, processes that are regulated in part through calmodulin-Ca_n^{2+}. These are phosphorylase kinase and glycogen synthase kinase.

Phosphorylase kinase catalyzes the phosphorylation of glycogen phosphorylase, resulting in the activation of glycogen phosphorylase and a subsequent increase in the rate of glycogenolysis. This enzyme is an oligomeric protein made up of four different polypeptide chains with the subunit composition $(\alpha\beta\gamma\delta)_4$ (6, 183). The activity of phosphorylase kinase may be regulated by cAMP-dependent protein kinases, as well as by Ca^{2+}. For example, serine residues on the α and β subunits are phosphorylated by cAMP-dependent protein kinases (184, 185), resulting in greatly enhanced activity in the presence of Ca^{2+} (186). The δ subunit is calmodulin (187), thus explaining the sensitivity of the enzyme to Ca^{2+} (188, 189). The γ subunit appears to be the catalytic subunit (190, 191).

Unlike most calmodulin-dependent enzymes, the δ subunit (calmodulin) of phosphorylase kinase remains attached to the protein in the absence of

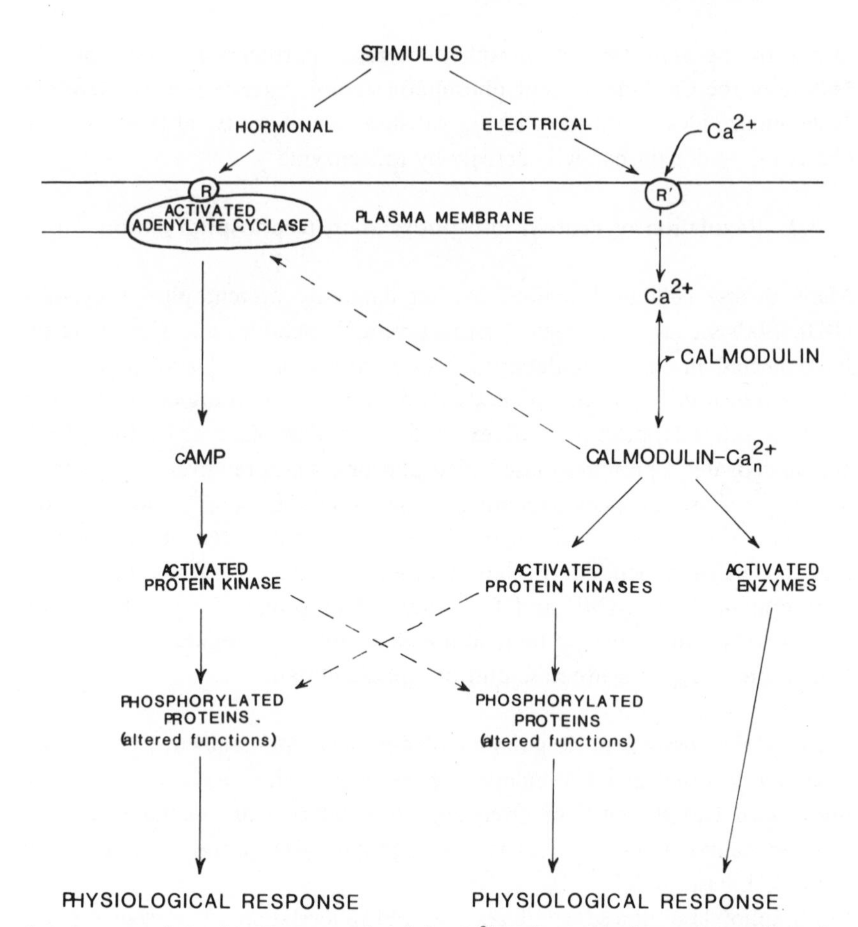

Figure 5 Interrelationships between calmodulin–Ca_n^{2+} and cAMP in the regulation of stimulus-induced cellular responses. Interconnections between the two second messenger systems, cAMP and Ca^{2+}, are indicated by the broken lines. The symbols *R* and *R'* represent membrane receptors for hormonal and electrical stimuli, respectively. (Adapted from ref. 183.)

Ca^{2+} (187), and is apparently complexed with the γ subunit (192, 193). In the presence of Ca^{2+}, a second molecule of calmodulin complexes with the α and β subunits, resulting in additional activation of the dephosphorylated form of the enzyme, yet having no significant effect on the phosphorylated form (186, 192–194). In the presence of μM concentrations of Ca^{2+}, the troponin complex activates the dephosphorylated form of phosphorylase kinase about 20-fold. It is believed that the Ca^{2+}-binding protein, troponin-C, rather than calmodulin, is the physiological activator (186, 195, 196). That is, the δ subunit (calmodulin) confers Ca^{2+} sensitivity to the phosphorylated form of the enzyme, and troponin-C confers Ca^{2+} sensitivity to the dephosphorylated form of phosphorylase kinase.

There are several glycogen synthase kinases which catalyze the phosphorylation of glycogen synthase at specific sites (183). In addition, the enzyme is phosphorylated by phosphorylase kinase (197–201) and cAMP-dependent protein kinase (202–204). These various kinases phosphorylate seven different sites on glycogen synthase (183). This multisite phosphorylation results in a decrease in the activity of glycogen synthase.

Thus, it appears that during muscle contraction the two second messengers, cAMP and Ca^{2+}, may regulate a group of protein kinases that cause the simultaneous activation of glycogenolysis and the inhibition of glycogen synthesis through multisite phosphorylation of glycogen phosphorylase and glycogen synthase.

4.2.2 Ca^{2+}-Dependent Regulation of Myosin Light Chain Kinases. Myosin light chain kinases have been isolated from skeletal, cardiac, and smooth muscle, as well as from a variety of nonmuscle cells (6). Regardless of the tissue source, the enzyme consists of two different subunits, a large catalytic subunit (M_r 80,000 to 124,000 depending on the source) and a small dissociable regulatory subunit identified as calmodulin (205–213). The catalytic subunit possesses a single binding site for calmodulin-Ca_n^{2+} (106–110). In the absence of Ca^{2+} the subunits dissociate, [eq. (3)] and the enzyme becomes completely inactive.

Myosin light chain kinase catalyzes the transfer of the gamma phosphate of ATP to its protein substrate, the 20,000 M_r regulatory light chains of myosin. In the presence of substrate, the affinity of calmodulin-Ca_n^{2+} for the enzyme is high, since the concentration required for 50% activation of the enzyme is about 1 nM (106, 108, 109). In the absence of substrate, the association constant of calmodulin-Ca_n^{2+} for myosin light chain kinase is about $2 \times 10^7\ M^{-1}$ (106, 107). Thus, the affinity of calmodulin-Ca_n^{2+} for the enzyme is enhanced in the presence of substrate.

Myosin light chain phosphorylation, in both striated and smooth muscle, occurs on serine residues near the N-terminal ends of the polypeptide chains (214, 215). The role of phosphorylation of the myosin light chains in regulating muscle contraction in striated muscle by myosin light chain kinase is not known. However, such phosphorylation appears to be a prerequisite to contraction in smooth muscle.

Although phosphorylation of the myosin light chains occurs in striated muscle, the regulation of striated muscle contraction by Ca^{2+} is mediated by the troponin system (216, 217). In skeletal and cardiac muscles, the proteins troponin and tropomyosin bind stoichiometrically to actin. In the absence of Ca^{2+}, the actin-induced activation of myosin ATPase activity is inhibited. The binding of Ca^{2+} to troponin-C removes this inhibition, resulting in the interaction of actin and myosin, a marked increase in myosin ATPase activity, and subsequent contraction. The phosphorylation of myosin light chains in straited muscle myosin has no effect on the actin-activated ATPase activity measured *in vitro* (218), although this observation does not rule out a possible *in vivo* role in the regulation of contraction.

In contrast, the prevailing theory of the regulation of smooth muscle contraction involves the phosphorylation-dephosphorylation of the myosin light chains (208, 219–223). As shown in Figure 6, the phosphorylation of the 20,000 M_r myosin light chains appears to be a requirement for actin-myosin interaction and the observed increase in myosin ATPase activity. Activation of the myosin light chain kinase presumably occurs by its interaction with calmodulin-Ca_n^{2+} as discussed above. During relaxation, the Ca^{2+} levels fall to below 0.1 μM, resulting in the dissociation of the myosin light chain-calmodulin-Ca_n^{2+} complex, thereby inactivating the kinase. Relaxation is presumably achieved by dephosphorylation of the 20,000 M_r myosin light chains by a Ca^{2+}-independent phosphatase.

An alternative theory of the regulation of smooth muscle contraction has been proposed (229–231). This kind of regulation does not involve the reversible phosphorylation of myosin, but rather the participation of another Ca^{2+}-binding protein, leiotonin. Whereas the role of calmodulin-Ca_n^{2+} in the phosphorylation of myosin, outlined above, appears to be one of the necessary regulatory components of smooth muscle contraction, it is also possible that multiple Ca^{2+}-dependent steps, in which leiotonin participates, are involved.

As discussed in Section 4.2.1, phosphorylase kinase and glycogen synthase are substrates for cAMP-dependent protein kinase. This is also true for myosin light chain kinase (224–227). In the absence of calmodulin-Ca_n^{2+}, cAMP protein kinase incorporates phosphate into two sites on myosin light chain kinase, resulting in a decrease in activity of the enzyme. This decreased activity is due to a 10- to 20-fold lowered affinity of the enzyme for calmodu-

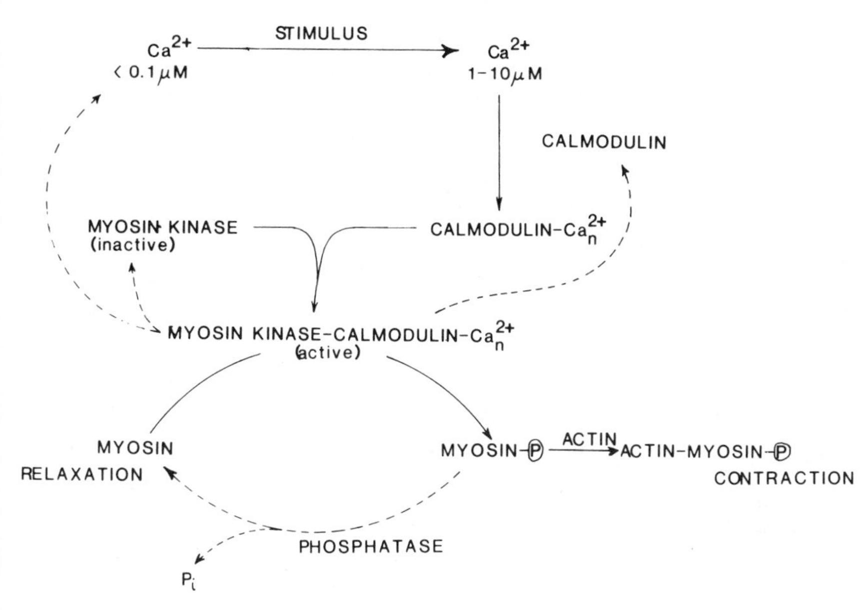

Figure 6 Schematic representation of how Ca^{2+} may be involved in the regulation of smooth muscle contraction. A stimulus changes the intracellular Ca^{2+} levels from $<0.1\ \mu M$ (relaxed state) to 1 to 10 μM. Ca^{2+} binds to calmodulin and calmodulin–Ca_n^{2+}, in turn, binds to myosin light chain kinase (MLCK), converting it from an inactive to an active enzyme. The MLCK–calmodulin–Ca_n^{2+} complex then catalyzes the phosphorylation of the 20,000 M_r light chains of myosin, initiating the interaction between actin and phosphorylated myosin, and leading to contraction. During relaxation (broken lines), the Ca^{2+} levels return to $<0.1\ \mu M$, followed by the dissociation of the MLCK–calmodulin–Ca_n^{2+} complex and the dephosphorylation of myosin by a Ca^{2+}-independent phosphatase.

lin–Ca_n^{2+}. In the presence of calmodulin–Ca_n^{2+}, phosphate is incorporated into a single site on the enzyme with no effect on myosin light chain kinase activity.

4.2.3 Ca^{2+}-Dependent Regulation of Other Protein Kinases. The release of neurotransmitters from the presynaptic nerve terminal involves Ca^{2+} (232), but the role(s) of Ca^{2+} in this process has been elusive. A possible function for calmodulin has recently become apparent with the discovery that synaptic vesicles contain calmodulin-dependent protein kinases (233–235). These kinases phosphorylate a number of presynaptic nerve terminal proteins in both synaptic vesicles and intact synaptosomes. Using synaptic vesicles it has been shown that the phosphorylation of two of these nerve terminal proteins (M_r 51,000 and 62,000) depends upon calmodulin-Ca_n^{2+} (233–235). These same proteins are also phosphorylated in intact synaptosomes during the depolarization-dependent uptake of Ca^{2+} (235). In intact synaptosomes, the calmodulin-dependent phosphorylation of the 51,000 and 62,000 M_r proteins, which occurs in response to depolarization, is concomitant with neurotransmitter release (235).

Protein Ia (M_r 86,000) and protein Ib (M_r 80,000) are also presynaptic nerve terminal proteins that are phosphorylated by cAMP and Ca^{2+}-dependent protein kinases (236–238). Moreover, phosphorylation occurs on two different regions of the polypeptide chains (239). Three distinct protein kinases have been identified, one is cAMP-dependent (237), the others are calmodulin-Ca_n^{2+} dependent (240). One of the Ca^{2+}-dependent protein kinases catalyzes the phosphorylation of proteins Ia and Ib at a site also phosphorylated by the cAMP-dependent protein kinase. The other Ca^{2+}-dependent protein kinase phosphorylates these proteins at a site not phosphorylated by the cAMP-dependent kinase. These multisite phosphorylations suggest that these proteins may play important roles in neurosecretion. These observations also provide another example of a regulatory process affected by both cAMP and Ca^{2+}.

Postsynaptic density fractions isolated from canine cerebral cortex also contain calmodulin, a number of calmodulin-binding proteins, and a protein kinase that phosphorylates two proteins of M_r 51,000 and 62,000 (246–249). The protein kinase is dependent upon calmodulin-Ca_n^{2+} for activity, but the role of this enzyme and other calmodulin-binding proteins in the postsynaptic densities is not known.

In addition to an apparent role for Ca^{2+}-dependent phosphorylation in neurosecretion, other Ca^{2+}-dependent protein kinases may play important parts in the biosynthesis of certain neurotransmitters. For example, tryptophan 5-monooxygenase and tyrosine 3-monooxygenase are the rate-limiting enzymes in the biosynthesis of serotonin and noradrenaline in brain tissue, respectively. Both are regulated through a protein kinase that depends on calmodulin-Ca_n^{2+} for activity (241–244). It has also been shown that enzyme phosphorylation alone is insufficient to generate an active enzyme. Activation of the phosphorylated enzymes requires the addition of a protein activator (M_r 70,000) isolated to homogeneity from rat brain, which occurs in most invertebrate tissues (245).

4.2.4 Ca^{2+}-Dependent Regulation of Protein Dephosphorylation. A protein that binds calmodulin in a Ca^{2+}-dependent manner and, consequently, inhibits the activation of calmodulin-dependent enzymes was originally isolated from brain tissue (150, 151, 250, 251). This protein, now termed calcineurin (123), is composed of two subunits, calcineurin A (M_r 61,000) and calcineurin B (M_r 15,000). Calcineurin A binds calmodulin in a Ca^{2+}-dependent manner; calcineurin B is itself a Ca^{2+}-binding protein and, like calmodulin, binds 4 moles of Ca^{2+} per mol (123), with high affinity (K_d, μM).

Recent evidence suggests that calcineurin and protein phosphatase-2B are the same protein (252). Protein phosphatase-2B was recently isolated from rabbit skeletal muscle and was shown to preferentially dephosphorylate the α-subunit of phosphorylase kinase (252). Calmodulin-Ca_n^{2+} stimulates the activity of protein phosphatase-2B (calcineurin) by increasing the V_{max} about tenfold (252). This protein phosphatase-2B resembles phosphorylase kinase in the way it is regulated through Ca^{2+}. That is, two different Ca^{2+}-binding proteins are involved, one as an integral part of its oligomeric structure, and the other interacting in the presence of Ca^{2+}.

4.3 Regulation of Ca^{2+}-Transport ATPase

As discussed in Section 1.2, the levels of cytosolic free Ca^{2+} are regulated at <0.1 μM in the unstimulated cell. A plasma membrane Ca^{2+}–ATPase appears to be regulated by direct interaction with calmodulin–Ca_n^{2+}. Activation of this Ca^{2+}–ATPase by calmodulin–Ca_n^{2+} also leads to activation of Ca^{2+} transport. It is generally believed that the plasma membrane Ca^{2+}–ATPase provides one mechanism whereby calmodulin may modulate the intracellular levels of free Ca^{2+}. The sarcoplasmic reticulum Ca^{2+}–ATPase is regulated indirectly through protein phosphorylation involving both cAMP and Ca^{2+}-dependent protein kinases. The properties of both of these Ca^{2+}–ATPases are briefly described below.

4.3.1 Plasma Membrane Ca^{2+}-Transport ATPase. The first plasma membrane Ca^{2+}–ATPase was discovered in erythrocyte plasma membranes (259), and it was suggested that this enzyme was responsible for the removal of intracellular free Ca^{2+} in these cells (260). Although most of the data have been derived from studies on the erythrocyte-derived enzyme, similar Ca^{2+}–ATPases have also been found in the plasma membranes of rat luteal cells (254), adipocytes (255), pancreatic islet cells (256), epithelial cells of the small intestine (257), human lymphocytes (258), and cardiac sarcolemma (277).

The erythrocyte membrane Ca^{2+}–ATPase is a monomeric glycoprotein with a molecular weight between 125,000 and 145,000 (261-265). This Ca^{2+}-ATPase is clearly different from the Mg^{2+}-ATPase, since the two proteins have been separated and the purified Ca^{2+}-ATPase uses Ca-ATP, rather than Mg-ATP, as the substrate (265–267). Calmodulin-Ca_n^{2+} was found to be a requirement for maximal activity of this enzyme (268, 269) following reports of the presence of a protein activator in erythrocytes (270, 271). Calmodulin-Ca_{3-4}^{2+} appears to be the functional species in the activation of this Ca^{2+}-ATPase (372).

The interaction of calmodulin with the membrane-bound Ca^{2+}-ATPase is Ca^{2+}-dependent and occurs on the cytoplasmic face of the membrane (272–274). Because of this property, preparation of membrane-bound Ca^{2+}-ATPase deficient in calmodulin have been made by a number of investigators (see ref. 6 for a review). Such preparations contain a basal level of ATPase activity that is enhanced about threefold upon the addition of saturating levels of calmodulin-Ca_n^{2+}. Using sealed membrane vesicles, it has been shown that activation of the Ca^{2+}-ATPase by calmodulin-Ca_n^{2+} also leads to an activation of Ca^{2+} transport, as measured by $^{45}Ca^{2+}$ uptake into the vesicles (see ref. 6).

Calmodulin-Ca_n^{2+} stimulates the activity of the purified Ca^{2+}-ATPase incorporated into phosphatidylcholine-containing liposomes (267, 275). In the absence of calmodulin, the enzyme has a relatively low affinity for Ca^{2+} ($K_{0.5}$ = 10 to 14 μM) and a relatively low V_{max}. In the presence of saturating levels of calmodulin, the affinity for Ca^{2+} is increased ($K_{0.5}$ = 0.8 to 0.9 μM), as is V_{max}. The increase in V_{max} is highly dependent upon the free Ca^{2+} concentration. At Ca^{2+} concentrations above 1 μM, V_{max} was increased 5- to 10-fold, but at Ca^{2+} concentrations below 1 μM, V_{max} was increased by as much as 100-fold. As has been found for erythrocyte ghosts, ATP hydrolysis is linked to Ca^{2+} transport in these liposomes. The ratio of ATP hydrolyzed to Ca^{2+} transported was found to be approximately 1.0.

Studies on the purified Ca^{2+}-ATPase have shown that a variety of acidic phospholipids substitute for calmodulin, resulting in an equivalent stimulation of the V_{max} and decrease in the K_m for Ca^{2+} (261, 263, 267, 275). Furthermore, when the purified enzyme is incorporated into phosphatidylcholine liposomes (and thus becomes calmodulin sensitive), limited proteolysis generates a fully active, calmodulin-insensitive Ca^{2+}-ATPase. It is not known whether acidic phospholipids play a functional role in the *in vivo* regulation of the plasma membrane Ca^{2+}-ATPase. One interpretation of these experiments is that acidic phospholipids, because of their net negative charge and hydrophobicity, somehow mimic the effects of calmodulin, and that limited proteolysis removes a regulatory site on the enzyme resulting in enzyme activation.

The role of calmodulin in the plasma membrane Ca^{2+} pump is still poorly understood. Recent reports show that Ca^{2+}-ATPases activated by protein activators distinct from calmodulin exist in the erythrocyte membranes of certain species (6, 276).

4.3.2 Sarcoplasmic Reticulum Ca^{2+}-Transport ATPase. The transport of Ca^{2+} in cardiac sarcoplasmic reticulum appears to be regulated by both cAMP and Ca^{2+}-dependent processes. The sarcoplasmic reticulum contains a membrane-associated Ca^{2+}-ATPase (M_r = 100,000) that may be involved in the regulation of free Ca^{2+} homeostasis in internal membrane systems (see ref. 278 for review). The simultaneous stimulation of this Ca^{2+}-ATPase and Ca^{2+} transport by cAMP in these membrane systems is due to the activation of a cAMP-dependent protein kinase (279–283). This cAMP-dependent protein kinase catalyzes the ATP-dependent phosphorylation of phospholambdan, a 22,000 M_r membrane protein located in the sarcoplasmic reticulum (280, 281). The phosphorylation of a serine residue on phospholambdan results in a severalfold enhancement of the rates of Ca^{2+} transport and Ca^{2+}-ATPase (279–281). A stoichiometry of 2 moles of Ca^{2+} transported per mole ATP hydrolyzed was also observed (278). These activations are reversed by the dephosphorylation of phospholambdan catalyzed by a phosphoprotein phosphatase (281).

In addition to the effects of cAMP, calmodulin-Ca_n^{2+} also appears to be involved in the regulation of Ca^{2+} transport in cardiac sarcoplasmic reticulum. Calmodulin has recently been found to stimulate Ca^{2+} transport into vesicles prepared from cardiac sarcoplasmic reticulum (284–286). Stimulation of Ca^{2+} transport by calmodulin results from the activation of a protein kinase by calmodulin-Ca_n^{2+}, which phosphorylates phospholambdan at a serine site distinct from the one phosphorylated by cAMP-dependent protein kinase (285). This phosphorylation was found to increase the rates of Ca^{2+} transport and Ca^{2+}-ATPase activity (286, 287), and the stimulation of these rates by calmodulin occurs at physiological concentrations of K^+ (288).

The regulation of Ca^{2+} transport in the sarcoplasmic reticulum by these two protein kinases occurs by two separate pathways, each involving the phosphorylation of a specific site on phospholambdan. Both the cAMP and Ca^{2+}-dependent phosphorylation of phospholambdan independently activate Ca^{2+} transport and Ca^{2+}-ATPase, and the effects of these phosphorylations are additive (285–287). This appears to be another example of regulation by the two second messenger systems in which a common molecular target is involved.

Skeletal muscle sarcoplasmic reticulum also contains a calmodulin-Ca_n^{2+} dependent protein kinase (363). Phosphorylation of a 60,000 M_r protein was observed in purified sarcoplasmic reticulum vesicles; serine and threonine are the sites of phosphorylation (363). Whereas the stoichiometry between the cardiac sarcoplasmic reticulum Ca^{2+}-ATPase and phospholambdan is 1:1 (364), the stoichiometry between the skeletal muscle sarcoplasmic reticulum Ca^{2+}-ATPase and the phosphorylated 60,000 M_r protein is about 60:1 (363). For this reason, the 60,000 M_r protein is probably not involved in Ca^{2+} uptake. The authors have proposed that it may be involved in Ca^{2+} release.

4.4 Ca^{2+}-Dependent Regulation of Other Enzymes

As discussed in Section 5, the activation of NAD kinase by calmodulin-Ca_n^{2+} was discovered in plants (289). More recently, a similar activation of NAD kinase has been observed in extracts of sea urchin eggs (290). The NAD kinase activity is completely dependent upon calmodulin-Ca_n^{2+} (365) and, in this regard, is similar to plant NAD kinase (154). Shortly after sperm-egg fusion, a rapid increase in cytosolic free Ca^{2+} triggers a series of events crucial to the onset of embryonic development (18, 291). Two such events that are triggered following the rise in cytosolic free Ca^{2+} include cortical vesicle exocytosis and the activation of NAD kinase (292, 293). Calmodulin has recently been identified in sea urchin gametes (56, 57, 116, 294, 362). Since cortical vesicle exocytosis is Ca^{2+}-dependent (292), it may also be calmodulin dependent (370).

Calmodulin has also been implicated as an activator of phospholipase A_2 in platelet membrane preparations (295). Phospholipase A_2 catalyzes the hydrolysis of fatty acid ester linkages at position 2 of phosphoglycerides, such as phosphatidylcholine, releasing arachidonic acid. A severalfold stimulation of arachidonic acid production from phosphatidylcholine by phospholipase A_2, in the presence of calmodulin-Ca_n^{2+}, was noted in platelet membrane preparations, suggesting a role for calmodulin in the regulation of prostaglandin biosynthesis (295). But, in the absence of added calmodulin, purified phospholipase A_2 is activated by Ca^{2+} alone (296). Furthermore, a protein inhibitor of phospholipase A_2 activity has been reported (297). This 40,000 M_r protein, termed lipomodulin, has been isolated from rabbit neutrophils and can be phosphorylated *in vitro* by a cAMP-dependent protein kinase (297). When lipomodulin is phosphorylated, it loses its ability to inhibit phospholipase A_2. In contrast to the *in vitro* system, phosphorylation of lipomodulin in intact neutrophils is Ca^{2+}-dependent (297). A similar protein inhibitor of phospholipase A_2, which also undergoes phosphorylation, has been observed in brain synaptic vesicles (298). These observations suggest that calmodulin-Ca_n^{2+} may act indirectly in the regulation of phospholipase A_2 activity, that is, through the regulation of a protein kinase that phosphorylates lipomodulin, resulting in the reversal of phospholipase A_2 inhibition by lipomodulin.

4.5 Regulation of Cytoskeletal Function

The molecular basis of motility in smooth and nonmuscle cells involves the cellular cytoskeleton, which is composed of several filamentous structures. Among the best characterized of these are microtubules and the actomyosin-based microfilaments. The involvement of the actomyosin system in contractility and its regulation by calmodulin-Ca_n^{2+} has been discussed in Section 4.2.2. The microtubule system is responsible for ciliary and flagellar movement and has been implicated in other cellular processes, such as chromosome movement and fast axonal transport.

4.5.1 Ca^{2+}-Dependent Regulation of Axonemal ATPases. Ciliary and flagellar motility is based upon the interaction of microtubules and dynein. Microtubules provide a unique structure which undergoes sliding motions in response to energy derived from the dynein ATPases. Micromolar levels of Ca^{2+} affect motility patterns of cilia and flagella (299, 300). Furthermore, a number of reports show that, depending upon the species, either calmodulin or a closely related protein is present within the axoneme of ciliated or flagellated eukaryotic organisms (see refs. 3 and 6 for reviews). In addition, evidence suggests that calmodulin-Ca_n^{2+} is involved in the regulation of cilia-derived dynein ATPases (301). The 14S dynein ATPase fraction, obtained from *Tetrahymena pyriformis* by KCl extraction, was activated up to tenfold upon the addition of calmodulin-Ca_n^{2+} (301). In the presence of Ca^{2+}, half-maximal activation occurred at 0.1 μM calmodulin. No activation of dynein ATPase by calmodulin occurred in the absence of Ca^{2+}. These observations suggest a role for calmodulin-Ca_n^{2+} in the regulation of motility in cilia. Similar observations have not been reported for flagellar organisms, although it has been shown that calmodulin is a major protein of spermatozoa. Calmodulin is associated with the acrosome and is also localized in a band across the lower third of the head, and at the base, tip, and various other points throughout the axoneme in mammalian spermatozoa (3, 56, 302, 303). For these reasons, a potential role was suggested for calmodulin in the Ca^{2+}-dependent sperm acrosome reaction, in sperm motility, and in gamete fusion (302).

4.5.2 Ca^{2+}-Dependent Regulation of Microtubule Polymerization. Microtubules are composed of tubulin, a mixed dimer ($\alpha\beta$) of two closely related subunits of M_r 60,000. The factors responsible for the *in vivo* regulation of microtubule assembly-disassembly have been the focus of much current interest, since microtubules are involved in important cellular functions, such as chromosome movement, fast axonal transport, and movement of cilia and flagella.

It has been known for some time that Ca^{2+} causes the depolymerization of microtubules *in vitro* and also inhibits the polymerization process (304–309). It has also been implicated in the *in vivo* regulation of microtubule assembly-disassembly (310–312). Purified tubulin contains both high and low affinity Ca^{2+} binding sites. One mole of Ca^{2+} will bind, per mole of $\alpha\beta$ tubulin dimer, with a K_d in the micromolar range, whereas multiple Ca^{2+} sites are available with K_d values in the 300 μM range (313–314). Presumably, the interaction of Ca^{2+} with these sites is responsible for the observed Ca^{2+}-dependent inhibition of tubulin polymerization *in vitro*. However, the effects of Ca^{2+} on microtubule assembly-disassembly are complex and the levels of Ca^{2+} required to inhibit polymerization, as well as to cause depolymerization, reportedly vary from 10 μM to 1 mM. In addition, the degree of Ca^{2+} sensitivity of these processes is now known to depend upon the concentration of tubulin, the presence of microtubule associated proteins (MAPs), and the presence of calmodulin (315–318).

The degree of inhibition of microtubule polymerization at micromolar levels of Ca^{2+} is substantially increased by the addition of calmodulin (318). This effect is presumably due to the interaction of calmodulin-Ca_n^{2+} with one or more of the MPAs since, in the presence of MAPs, calmodulin confers increased Ca^{2+} sensitivity on tubulin polymerization (316, 317). In this regard, one of the MAPs, the tau factor, has been shown to interact with calmodulin-Ca_n^{2+} (319).

Recent evidence suggests that calmodulin might play a role in the regulation of microtubule assembly-disassembly during mitosis; thus calmodulin could be involved in the regulation of chromosome movement. Immunocytochemical evidence has revealed that, during mitosis, calmodulin becomes concentrated in the spindle apparatus, with the highest concentrations observed near the polar regions (320–322). The localization of calmodulin corresponds to regions where microtubule assembly-disassembly apparently occurs and is thus consistent with a role for calmodulin in this process.

5 CALMODULIN-DEPENDENT REGULATORY PROCESSES IN PLANT CELLS

As outlined in Section 1.3, the role of Ca^{2+} in physiological processes in animal systems was first recognized near the turn of the century and since that time a large number of such processes have been identified in animal cells. The molecular basis for these Ca^{2+}-dependent regulatory processes was established with the discovery of Ca^{2+}-binding proteins and the realization that these proteins serve as intracellular receptors for micromolar levels of free Ca^{2+} generated in response to a stimulus.

In contrast, the recognition that Ca^{2+} is important to the physiology of plant cells has been recent. Examples of processes affected by Ca^{2+} in plants are provided in Table 2. As the references in Table 2 indicate, most of these observations were made during the last decade.

Calmodulin from plants was originally isolated and characterized from peanut seeds, pea seedlings, and mushrooms; this was the first demonstra-

Table 2 Examples of Ca^{2+}-Linked Processes in Plants

Type of Response	References
Phytochrome mediated regulation of transmembrane Ca^{2+} flux	331–333
Stimulation of chloroplast rotation in *Mougeotia*	334
Phytochrome dependent depolarization of *Nitella* cells	335
Inhibition of cytoplasmic streaming in *Nitella*	329
Regulation of directional growth	336–339
Stimulation of cyclic nucleotide phosphodiesterase in *Phaseolus*	340
Inhibition of chloroplast fructose bisphosphatase	341
Stimulation of glutamate dehydrogenase	342
Activation of NAD kinase	289
Stimulation of cytokinin-like mitosis in *Funaria*	343

tion that high affinity Ca^{2+}-binding proteins existed in plants (58, 66, 289, 323). Subsequently, calmodulin was isolated from other plants and fungi (59, 119, 163, 324–327). The structural features of the plant protein clearly demonstrate the high degree of structural conservation that exists between animal and plant calmodulins (Fig. 2).

The presence of calmodulin provided the molecular basis for suggesting that Ca^{2+} serves a second messenger role in plants, and that Ca^{2+}-dependent metabolic regulation in plant cells may be mediated by Ca^{2+}-binding proteins, such as calmodulin (66, 289). The finding that calmodulin and Ca^{2+} were required for the activation of NAD kinase in plants (289) strengthened this proposal and led to the suggestion that Ca^{2+} and Ca^{2+}-binding proteins may be important in mediating some of the plant responses to stimuli, such as light, auxins, gibberellins, and cytokinins (54, 66, 289).

As discussed in Section 1.2, in order for Ca^{2+} to function as a second messenger, its cytosolic concentration must remain low ($<0.1\ \mu M$) in the unstimulated cell and transiently rise following a stimulus. Until recently, no information on the levels of free Ca^{2+} in plant cells has been available. It has now been shown, using aequorin injected cells, that the free Ca^{2+} concentration in the green algae *Chara* and *Nitella* is in the $0.1\ \mu M$ range (328). Upon electrical stimulation, these cells propagate action potentials along the plasma and tonoplast membranes, with concomitant inhibition of cytoplasmic streaming (329, 330). Following electrical stimulation of these algal cells, it was found that the cytosolic concentration of free Ca^{2+} increased to approximately $10\ \mu M$ (328), a concentration at which cytoplasmic streaming is inhibited. Thus, the concentration requirements for Ca^{2+} serving as a second messenger in green algae are fulfilled.

It is also clear that a number of enzymes in plants are regulated by calmodulin-Ca_n^{2+}. This list will almost certainly grow, as it has in animal systems. The evidence discussed in this section suggests that regulatory processes in plant cells are similar to those studied in animal cells, particularly with regard to Ca^{2+}.

5.1 Regulation of NAD Kinase

It was found that NAD kinase activity in higher plants is stimulated by a protein activator that could be separated from the enzyme by ion-exchange chromatography (344). The activation of NAD kinase by the protein activa tor requires micromolar levels of Ca^{2+} (289). This protein activator was isolated and characterized and found to be calmodulin (58, 66, 289).

NAD kinase has been purified approximately 4100-fold from extracts of pea seedlings (345). As shown in Table 3, this enzyme is completely dependent upon Ca^{2+} and calmodulin for activity (154). Inhibitors of calmodulin function, such as trifluoperazine, inhibit the calmodulin-dependent activation of NAD kinase. This is expected since plant calmodulin contains Ca^{2+}-dependent phenothiazine binding sites (58, 59; see Section 2.1).

The reversible Ca^{2+}-dependent binding of NAD kinase to calmodulin-sepharose has also been shown (154, 346). These data are consistent with the scheme shown by eq. (3), in which an NAD kinase-calmodulin–Ca_n^{2+} complex is formed.

Throughout the purification of NAD kinase, it was noted that more than 97% of the activity was Ca^{2+}-dependent; a Ca^{2+}-independent NAD kinase activity was not detectable. In algae, such as *Euglena* and *Chlamydamonas*, all of the NAD kinase activity appears to be Ca^{2+}-independent and thus not regulated by calmodulin (373). In plants such as pea seedlings, NAD kinase represents about 0.01% of the total soluble protein in crude extracts (345); calmodulin from several species of higher plants represents about 0.1 to 0.6% (58, 59). Thus, there is at least a tenfold molar excess of calmodulin over NAD kinase in the pea seedlings, suggesting that the bulk of it is used for other regulatory functions.

Is NAD kinase regulated *in vivo* by calmodulin–Ca_n^{2+} as it is *in vitro*? The

Table 3 Requirements for Pea NAD Kinase Activity

Additions or Deletions	NAD Kinase Activity (pmoles min^{-1} mg^{-1})
Complete system[a]	153.6 ± 8.1
−ATP	2.2 ± 1.5
$-NAD^+$	2.9 ± 2.2
$-Ca^{2+}$	1.6 ± 1.8
−Calmodulin	1.2 ± 2.7
+Trifluoperazine (50 μM, final concentrations)	7.2 ± 6.7

[a]The complete assay medium contained 50 mM Tris, 50 mM KCl, 10 mM $MgCl_2$, 2 mM NAD, 3 mM ATP, 0.2 mM $CaCl_2$, 20 μg/ml bovine brain calmodulin, and 0.2 ml of calmodulin-Sepharose purified NAD kinase. Additions and deletions from this standard assay mixture are as noted. When $CaCl_2$ was deleted, it was replaced by 0.2 mM EGTA.

evidence outlined in Table 4 suggests that it is. We know that NAD is converted to NADP by a light-dependent process *in vivo*. The data suggests that NAD kinase is regulated *in vivo* by light-induced changes in the levels of free Ca^{2+} (54, 66, 289). That is, the cell could undergo a receptor-mediated increase in the level of intracellular free Ca^{2+} in response to light, resulting in the formation of calmodulin-Ca_n^{2+}. Such a scheme could explain how light causes an increase in the NADP/NAD ratio *in vivo*. A possible candidate for the light receptor in this process is phytochrome (54, 289), since it has been implicated in the intracellular regulation of free Ca^{2+} (331–333, 335).

5.1.1 Calmodulin-Ca_n^{2+} as a Regulatory Component of Photosynthesis. Since light is the stimulus for the observed increases in the NADP/NAD ratios, it is possible that the *in vivo* regulation of NAD kinase occurs in the chloroplast. Table 5 summarizes the data, which suggest that such regulation occurs exclusively in the chloroplast. As shown in Table 5, most of the NAD kinase is in chloroplasts, and the light induced conversion of NAD to NADP also occurs in the chloroplasts (348, 350). Furthermore, isolated chloroplasts exhibit a rapid uptake of Ca^{2+} in response to light (351). Finally, we have recently shown that a protein activator of NAD kinase, found in the stroma of isolated intact pea chloroplasts, is indistinguishable from plant calmodulin (352). It's concentration in the stroma was estimated to be 0.1 to 0.2 μM. Numerous other physical and chemical similarities to calmodulin were reported (352), and we have suggested that the chloroplastic activator is indeed calmodulin. Wheat chloroplasts have also been shown to contain calmodulin (371).

There is additional evidence that NAD kinase is regulated via calmodulin–

Table 4 Evidence that Regulation of NAD Kinase Occurs *In Vivo*

Evidence	Reference
Illumination of green leaves, or of *Chlorella*, results in an increase in the NADP/NAD ratio.	347, 348
At least 90% of all NAD kinase found in extracts of higher plants depends on calmodulin-Ca_n^{2+} for activity.	345
There is at least a tenfold molar excess of camodulin over NAD kinase in extracts of pea seedlings.	58, 345
The light-dependent conversion of NAD to NADP is too fast to suggest *de novo* synthesis of NAD kinase.	345, 349, 350

Table 5 Evidence that Regulation of NAD Kinase Occurs in the Chloroplasts of Higher Plants

Evidence	References
Approximately 90% of the NAD kinase is localized in the chloroplasts of most higher plants.	350
A light induced conversion of NAD to NADP occurs in isolated chloroplasts.	348, 350
Chloroplasts exhibit a rapid uptake of Ca^{2+} in response to light.	351
At least 90% of all NAD kinase found in higher plants depends upon calmodulin-Ca_n^{2+} for activity.	345
Calmodulin is found in the stroma of pea and wheat chloroplasts.	352, 371

Ca_n^{2+} in the chloroplasts. For example, 50 to 100 μM levels of trifluoperazine completely inhibit the light-induced conversion of NAD to NADP in pea protoplasts (352). Furthermore, Figure 7 shows that chlorpromazine and trifluoperazine are potent inhibitors of photosynthetic O_2 evolution. At approximately 28 μM, these drugs caused 50% inhibition of O_2 evolution, whereas chlorpromazine sulfoxide produced very little inhibition at 100 μM levels. Similarly, W-7, but not W-5, was found to be a potent inhibitor of photosynthetic O_2 evolution (352). The inhibition of photosynthetic O_2 evolution by these drugs might be related to the observation that electron transport in photosystem II is inhibited by phenothiazine drugs in spinach chloroplasts (367). Because photosynthesis is dependent upon membrane-mediated phenomena, caution must be used in the interpretation of these results (see Section 2.1).

The data outlined above suggest that the light-induced conversion of NAD to NADP observed in isolated chloroplasts can be explained by the scheme shown in Figure 8. In this scheme, there is a light receptor-mediated increase in the concentration of free Ca^{2+} in the stroma, resulting in the formation of calmodulin-Ca_n^{2+}. Then, NAD kinase is converted from an inactive to an active form by complex formation with calmodulin-Ca_n^{2+}. This results in an increase in the levels of NADP. The inhibition of photosynthetic O_2 evolution by inhibitors of calmodulin function (Fig. 7) suggests that the photosynthetic process utilizes the NADP pool produced by the calmodulin-dependent pathway shown in Figure 8. When viewed in this way, calmodulin-Ca_n^{2+} is a regulatory component of the photosynthetic process, since NADP is the pri-

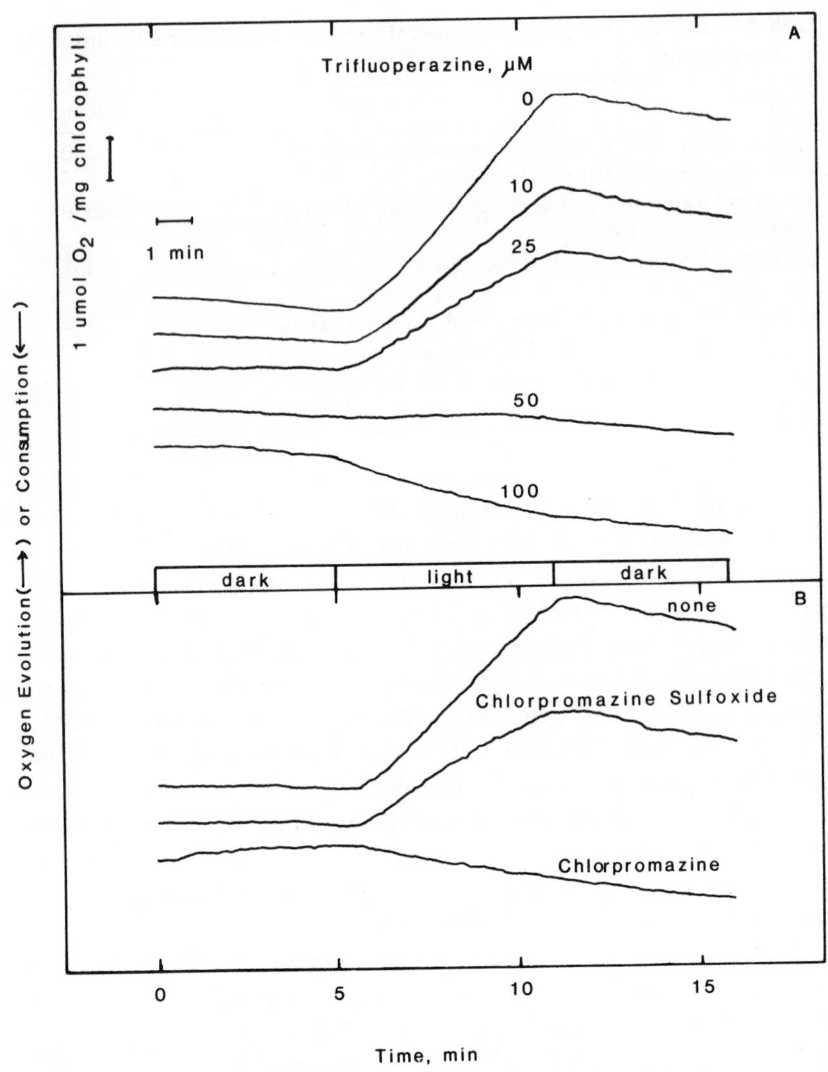

Figure 7 Effect of phenothiazine derivatives on photosynthetic O_2 evolution from pea protoplasts. (See Section 5.1.1 for details.)

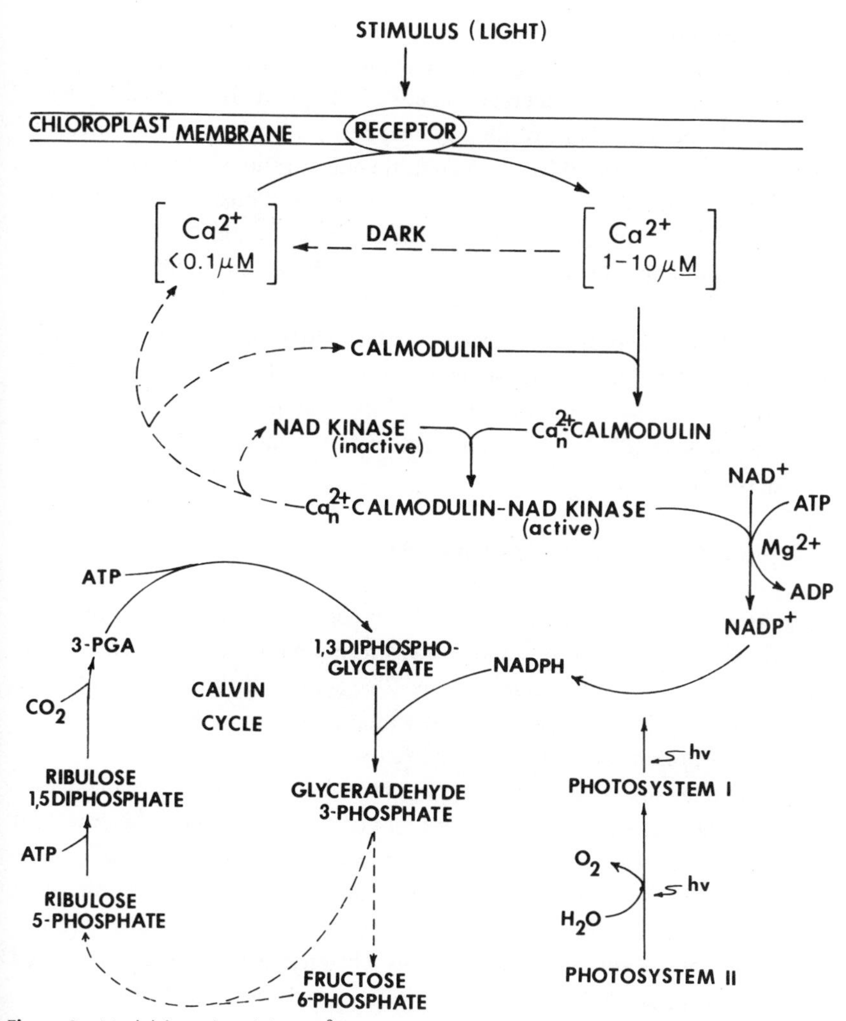

Figure 8 Model for calmodulin–Ca_n^{2+} as a regulatory component of photosynthesis. (See Section 5.1.1 for details.)

mary acceptor of the electrons produced from water by photosystems II and I. As illustrated in Figure 8, NADPH is a required component of the Calvin cycle, and its concentration is determined by the available concentration of NADP produced by the calmodulin-dependent pathway.

The production of NADPH by the pathway shown in Figure 8 may also be important in the regulation of glycolysis in the chloroplast. When starch is being produced during photosynthesis, it is important to inhibit its glycolytic breakdown. During the dark cycle, it is equally necessary to turn glycolysis on, in order to meet the bioenergetic needs of the plant. It has recently been shown that the activity of one of the key enzymes of glycolysis, phosphofructokinase, is strongly inhibited by physiological concentrations of NADPH (353). The enzyme was obtained from spinach chloroplasts. Thus, NADP produced by the calmodulin-dependent pathway in the light would be converted to NADPH, resulting in the activation of the Calvin cycle and simultaneous inhibition of glycolysis. In the dark, the levels of NADPH would decline. As a consequence, glycolysis would be turned on. Of course other regulatory pathways, such as the thioredoxin-mediated inhibition of the oxidative pentose phosphate cycle (354–355), and the inhibition of phosphofructokinase by 2-phosphoglycolate (356), could complement the regulation of starch breakdown by NADPH.

5.2 Regulation of Ca^{2+}-Transport ATPase

As discussed in Section 4.3, one of the cellular components apparently involved in the regulation of free Ca^{2+} homeostasis in animal cells is a plasma membrane-associated Ca^{2+}-transport ATPase. The activity of this Ca^{2+}–ATPase is enhanced by its direct interaction with calmodulin–Ca_n^{2+}. A similar enzyme appears to be present in plant cell membranes.

For example, an ATP-dependent uptake of Ca^{2+} into a plasma membrane enriched microsomal fraction, prepared from etiolated zucchini squash seedlings, was stimulated approximately twofold by the addition of partially purified squash or bovine brain calmodulin (357). In addition, a microsomal ATPase from corn, which was detergent solubilized and partially purified by calmodulin-Sepharose affinity chromatography, was stimulated about twofold upon the addition of saturating levels of calmodulin (358).

5.3 Regulation of Protein Phosphorylation

Although plant tissues contain protein kinases (359, 360), there has never been any evidence that these protein kinases are regulated through cyclic nucleotides (359). For this reason we suggested (54) that protein kinases may be important mediators of Ca^{2+}-dependent regulation in plants.

Protein kinases, which depend upon micromolar levels of free Ca^{2+}, have recently been observed in pea shoot membrane preparations (361). Some stimulation of protein phosphorylation, upon the addition of calmodulin, was also observed. In addition, protein dephosphorylation by a Ca^{2+}-dependent process was observed, suggesting the possibility that the membrane preparations contained a Ca^{2+}-dependent phosphatase (361). By analogy to animal systems, I predict that numerous enzyme activities in plant cells will be found to be regulated via Ca^{2+}-dependent protein phosphorylation-dephosphorylation mechanisms.

In the past, there has been a tendency to think of animal and plant cell regulation in distinct terms or concepts. However, it is now becoming clear that many similarities exist in the ways in which animal and plant cells utilize Ca^{2+} as a second messenger.

REFERENCES

1. J. H. Wang and D. M. Waisman, *Curr. Top. Cell. Reg.*, **15,** 47 (1979).
2. W. Y. Cheung, *Science*, **207,** 19 (1980).
3. A. R. Means, J. R. Tash, and J. G. Chafouleas, *Physiol. Rev.*, **62,** 1 (1982).
4. C. B. Klee, T. H. Crouch, and P. G. Richman, *Ann. Rev. Biochem.*, **49,** 489 (1980).
5. D. J. Wolff and C. O. Brostrom, *Adv. Cyclic Nucleo. Res.*, **11,** 27 (1979).
6. C. B. Klee and T. C. Vanaman, *Adv. Protein Chem.*, **35,** 213 (1982).
7. W. Y. Cheung, Ed., *Calcium and Cell Function*, Vol. 1, Academic Press, New York, 1980.
8. F. L. Siegel, E. Carafoli, R. H. Kretsinger, D. H. MacLennan, and R. H. Wasserman, Eds., *Calcium-Binding Proteins: Structure and Function*, Elsevier/North Holland, New York, 1980.
9. D. M. Watterson and F. F. Vincenzi, Eds., "Calmodulin and Cell Functions," *Ann. NY Acad. Sci.*, **356,** 1980.
10. R. M. Case, Ed., *Cell Calcium*, Vol. 2, Churchill Livingstone, New York, 1980.
11. G. A. Robison, R. W. Butcher, and E. W. Sutherland, *Cyclic AMP*, Academic Press, New York, 1971.
12. H. Rasmussen and D. Waisman, *Bioch. Actions Horm.*, **8** (1981).
13. M. J. Berridge, "The Interaction of Cyclic Nucleotides and Calcium in the Control of Cellular Activity," in *Advances in Cyclic Nucleotide Research*, Vol. 6, P. Greengard and G. A. Robison, Eds., Raven Press, New York, 1975, p. 1.
14. H. Rasmussen and D. B. P. Goodman, *Physiol. Rev.*, **57,** 422 (1977).
15. R. H. Kretsinger, "Evolution of the Informational Role of Calcium in Eukaryotes," in *Calcium-Binding Proteins and Calcium Function*, R. Wasserman, R. A. Corradino, E. Carafoli, R. H. Kretsinger, D. H. MacLennan and F. L. Siegel, Eds., Elsevier/North Holland, New York, 1977, p. 63.
16. P. F. Baker, *Prog. Biophys. Mol. Biol.*, **24,** 177 (1972).
17. A. Fabiato and F. Fabiato, *Ann. NY Acad. Sci.*, **307,** 491 (1978).

18. J. C. Gilkey, L. F. Jaffe, E. B. Ridgway, and G. T. Reynolds, *J. Cell Biol*, **76,** 448 (1978).
19. J. R. Blinks, F. G. Prendergast, and D. G. Allen, *Pharmacol. Rev.*, **28,** 1 (1976).
20. A. Scarpa, B. J. Brinley, T. Tiffert, and G. R. Dubyak, *Ann. NY Acad. Sci.*, **307,** 86 (1978).
21. A. H. Caswell and J. D. Hutchinson, *Biochem. Biophys. Res. Commun.*, **42,** 43 (1971).
22. R. Y. Tsien, *Nature, Lond.*, **290,** 527 (1981).
23. P. F. Baker, A. L. Hodgkin, and E. B. Ridgway, *J. Physiol.*, **218,** 709 (1971).
24. R. H. Krestinger, *Adv. Cyclic Nucleo. Res.*, **11,** 1 (1979).
25. C. C. Ashley and E. B. Ridgway, *Nature*, **219,** 1168 (1968).
26. B. Rose and W. H. Lowenstein, *Science*, **190,** 1204 (1975).
27. S. Ringer, *J. Physiol.*, **3,** 380 (1882).
28. S. Ringer, *J. Physiol.*, **4,** 29 (1883).
29. F. S. Locke, *Zent. Physiol.*, **8,** 166 (1894).
30. J. R. Blinks, W. G. Wier, P. Hess, and F. G. Prendergast, *Prog. Biophys. Mol. Biol.*, **40,** 1 (1982).
31. L. V. Heilbrunn, *Physiol. Zool.*, **13,** 88 (1940).
32. L. V. Heilbrunn and F. J. Wiercinski, *J. Cell. Comp. Physiol.*, **29,** 15 (1947).
33. L. V. Heilbrunn, *The Dynamics of Living Protoplasm*, Academic Press, New York, 1956.
34. E. G. Krebs, D. J. Graves, and E. H. Fischer, *J. Biol. Chem.*, **234,** 2867 (1959).
35. W. L. Meyer, E. H. Fischer, and E. G. Krebs, *Biochemistry*, **3,** 1033 (1964).
36. E. G. Krebs, D. S. Love, G. E. Bratvald, K. A. Trayser, W. L. Mayer, and E. H. Fischer, *Biochemistry*, **3,** 1022 (1964).
37. E. Ozawa, K. Hosoi, and S. Ebashi, *J. Biochem.*, **61,** 531 (1967).
38. S. Ebashi and M. Endo, *Prog. Biophys. Mol. Biol.*, **18,** 123 (1968).
39. S. Ebashi, F. Ebashi, and A. Kodema, *J. Biochem.*, **62,** 137 (1967).
40. O. Shimomura, F. H. Johnson, and Y. Saiga, *J. Cell. Comp. Physiol.*, **59,** 223 (1962).
41. O. Shimomura, F. H. Johnson, and Y. Saiga, *J. Cell. Comp. Physiol.*, **62,** 1 (1963).
42. W. W. Ward and M. J. Cormier, *Proc. Natl. Acad. Sci., USA*, **72,** 2530 (1975).
43. O. Shimomura and F. H. Johnson, *Nature*, **256,** 236 (1975).
44. M. J. Cormier, "Renilla and Aequorea Bioluminescence," in *Bioluminescence and Chemiluminescence*, M. DeLuca and W. D. McElroy, Eds., Academic Press, New York, 1981, p. 225.
45. P. C. Moews and R. H. Kretsinger, *J. Mol. Biol.*, **91,** 201 (1975).
46. R. H. Kretsinger, *Ann. Rev. Biochem.*, **45,** 239 (1976).
47. R. H. Kretsinger, *CRC Crt. Rev. Biochem.*, **8,** 119 (1980).
48. H. Rasmussen, *Science*, **170,** 404 (1970).
49. H. Rasmussen, *Calcium and cAMP as Synarchic Messengers*, John Wiley and Sons, New York, 1981.
50. W. Y. Cheung, *Biochem. Biophys. Res. Commun.*, **38,** 533 (1970).
51. S. Kakiuchi, R. Yamazaki, and H. Nabajima, *Proc. Japan Acad.*, **46,** 587 (1970).
52. T. S. Teo and J. H. Wang, *J. Biol. Chem.*, **248,** 5950 (1973).

53. D. M. Watterson, W. G. Harrelson, P. M. Keller, F. Sharief, and T. C. Vanaman, *J. Biol. Chem.*, **251,** 4501 (1976).
54. M. J. Cormier, H. Charbonneau, and H. W. Jarrett, *Cell Calcium*, **2,** 313 (1981).
55. J. R. Dedman, J. D. Potter, R. L. Jackson, J. D. Johnson and, A. R. Means, *J. Biol. Chem.*, **252,** 8415 (1977).
56. H. P. Jones, M. M. Bradford, R. A. McRorie, and M. J. Cormier, *Biochem. Biophys. Res. Commun.*, **82,** 1264 (1978).
57. J. F. Head, S. Mạder, and B. Kaminer, *J. Cell. Biol.*, **80,** 211 (1979).
58. H. Charbonneau and M. J. Cormier, *Biochem. Biophys. Res. Commun.*, **90,** 1039 (1979).
59. D. M. Watterson, D. B. Iverson, and L. J. Van Eldik, *Biochemistry*, **19,** 5762 (1980).
60. D. M. Watterson, F. Sharief, and T. C. Vanaman, *J. Biol. Chem.*, **255,** 962 (1980).
61. T. Takagi, T. Nemoto, K. Konishi, M. Yazawa, and K. Yagi, *Biochem. Biophys. Res. Commun.*, **96,** 377 (1980).
62. G. A. Jamieson, A. Hayes, J. J. Blum, and T. C. Vanaman, "Structure and Function Relationships Among Calmodulins From Divergent Eukaryotic Organisms" in *Calcium-Binding Proteins: Structure and Function*, F. L. Siegel et al., Eds., Elsevier/North Holland, New York, 1980, p. 165.
63. M. Yazawa, K. Yagi, H. Toda, K. Kondo, K. Narita, R. Yamazaki, K. Sobue, S. Kakiuchi, S. Nagao, and Y. Nozawa, *Biochem. Biophys. Res. Commun.*, **99,** 1051 (1981).
64. D. B. Iverson, M. Schleicher, J. G. Zendegui, and D. M. Watterson, *Fed. Proc. Abst.*, **40,** 1738 (1981).
65. H. Charbonneau, T. C. Vanaman, and M. J. Cormier, *Fed. Proc. Abst.*, **41,** 504 (1982).
66. J. M. Anderson, H. Charbonneau, H. P. Jones, R. O. McCann, and M. J. Cormier, *Biochemistry*, **19,** 3113 (1980).
67. R. L. Jackson, J. R. Dedman, W. E. Schreiber, P. K. Bhatnagar, R. D. Knapp, and A. R. Means, *Biochem. Biophys. Res. Commun.*, **77,** 723 (1977).
68. C. B. Klee, *Biochemistry*, **16,** 1017 (1977).
69. D. J. Wolff, P. G. Poirier, C. O. Brostrom, and M. A. Brostrom, *J. Biol. Chem.*, **252,** 4108 (1977).
70. Y. P. Liu and W. Y. Cheung, *J. Biol. Chem.*, **251,** 4193 (1976).
71. M. Walsh and F. C. Stevens, *Biochemistry*, **16,** 2742 (1977).
72. P. G. Richman and C. B. Klee, *Biochemistry*, **17,** 928 (1978).
73. P. G. Richman, *Biochemistry*, **17,** 3001, (1978).
74. P. G. Richman and C. B. Klee, *J. Biol. Chem.*, **254,** 5372 (1979).
75. M. Walsh, F. C. Stevens, J. Kuznicki, and W. Drabikowski, *J. Biol. Chem.*, **252,** 7440 (1977).
76. M. Walsh, F. C. Stevens, K. Oikawa, and C. M. Kay, *Biochemistry*, **17,** 3924 (1978).
77. K. Seamon, *Biochemistry*, **19,** 207 (1980).
78. J. H. Wang, T. S. Teo, H. C. Ho, and F. C. Stevens, *Adv. Cyclic Nucleotide Res.*, **5,** 179 (1975).
79. M. C. Kilhoffer, D. Gerard, and J. G. Demaille, *FEBS Lett.*, **120,** 99 (1980).
80. M. Yazawa, M. Subuma, and K. Yagi, *J. Biochem.*, **87,** 1313 (1980).

81. T. H. Crouch and C. B. Klee, *Biochemistry*, **19,** 3692 (1980).
82. J. Haiech, C. B. Klee, and J. G. Demaille, *Biochemistry*, **20,** 3890 (1981).
83. M. C. Kilhoffer, J. G. Demaille, and D. Gerard, *FEBS Lett.*, **116,** 269 (1980).
84. M. C. Kilhoffer, D. Gerard, and J. G. Demaille, *FEBS Lett.*, **120,** 99 (1980).
85. W. D. McCubbin, M. T. Hincke, and C. M. Kay, *Can. J. Biochem.*, **57,** 15 (1979).
86. M. Walsh, F. C. Stevens, K. Oikawa, and C. M. Kay, *Can. J. Biochem.*, **57,** 267 (1979).
87. B. Weiss and R. M. Levin, *Adv. Cyclic Nucleotide Res.*, **9,** 285 (1978).
88. R. M. Levin and B. Weiss, *Mol. Pharmacol.*, **13,** 690 (1977).
89. R. M. Levin and B. Weiss, *J. Pharmacol. Exper. Ther.*, **208,** 454 (1979).
90. H. Hidaka, T. Yamaki, T. Totsuka, and M. Asano, *Mol. Pharmacol.*, **15,** 49 (1979).
91. H. Hidaka, T. Yamaki, M. Naka, T. Tanaka, H. Hayashi, and R. Kobayashi, *Mol. Pharmocol.*, **17,** 66 (1980).
92. D. C. LaPorte, B. M. Wierman, and D. R. Storm, *Biochemistry*, **19,** 3814 (1980).
93. B. D. Roufogalis, *Biochem. Biophys. Res. Commun.*, **98,** 607 (1981).
94. T. Endo, T. Tanaka, T. Isobe, H. Kasai, T. Okuyama, and H. Hidaka, *J. Biol. Chem.*, **256,** 12485, (1981).
95. H. Van Belle, *Cell Calcium*, **2,** 483 (1981).
96. B. Deuticke, *Biochim. Biophys. Acta*, **163,** 494 (1968).
97. M. P. Sheetz and S. J. Singer, *Proc. Natl. Acad. Sci., USA*, **71,** 4457 (1974).
98. P. Seeman, *Pharm. Rev.*, **24,** 583 (1972).
99. R. W. Lenz and M. J. Cormier, *Ann. NY Acad. Sci.*, **383,** 85 (1982).
100. Y. Landry, M. Amellal, and M. Ruckstuhl, *Biochem. Pharmarcol.*, **30,** 2031 (1981).
101. G. A. Jamieson and T. C. Vanaman, *Biochem. Biophys. Res. Commun.*, **90,** 1048 (1979).
102. D. R. Marshak, D. M. Watterson, and L. J. Van Eldik, *Proc. Natl. Acad. Sci., USA*, **78,** 6793 (1981).
103. H. Charbonneau, R. E. Hice, R. C. Hart, and M. J. Cormier, *Methods in Enzymology*, **102,** 17 (1983).
104. J. A. Cox, A. Malnoe, and E. A. Stein, *J. Biol. Chem.*, **256,** 3218 (1981).
105. C. Y. Huang, V. Chau, P. B. Chock, J. H. Wang, and R. K. Sharma, *Proc. Natl. Acad. Sci., USA*, **78,** 871 (1981).
106. T. H. Crouch, M. J. Holroyde, J. H. Collins, R. J. Solaro, and J. D. Potter, *Biochemistry*, **20,** 6318 (1981).
107. J. D. Johnson, M. J. Holroyde, T. H. Crouch, R. J. Solaro, and J. D. Potter, *J. Biol. Chem.*, **256,** 12194 (1981).
108. D. K. Blumenthal and J. T. Stull, *Biochemistry*, **19,** 5608 (1980).
109. R. S. Adelstein and C. B. Klee, *J. Biol. Chem.*, **256,** 7501 (1981).
110. A. C. Nairn and S. V. Perry, *Biochem. J.*, **179,** 89 (1979).
111. Y. M. Lin, Y. P. Liu, and W. Y. Cheung, *J. Biol. Chem.*, **249,** 4943 (1974).
112. C. O. Brostrom and D. J. Wolff, *Arch. Biochem. Biophys.*, **165,** 715 (1974).
113. S. Kakiuchi, R. Yamazaki, Y. Teshima, K. Uenishi, S. Yasuda, A. Kashiba, K. Sobue, M. Ohshima, and T. Nabajima, *Adv. Cyclic Nucleotide Res.*, **9,** 253 (1978).
114. R. J. A. Grand, S. V. Perry, and R. A. Weeks, *Biochem. J.*, **177,** 521 (1979).

115. A. Vandermeers, M. C. Vandermeers-Piret, J. Rathe, R. Kutzner, A. Delforge and J. Christophe, *Eur. J. Biochem.*, **81**, 379 (1977).
116. D. L. Garbers, J. R. Hansbrough, E. W. Radany, R. V. Hyne, and G. S. Kopf, *J. Reprod. Fert.*, **59**, 377 (1980).
117. H. P. Jones, J. C. Matthews, and M. J. Cormier, *Biochemistry*, **18**, 55 (1979).
118. G. A. Jamieson, T. C. Vanaman, and J. J. Blum, *Proc. Natl. Acad. Sci., USA*, **76**, 6471 (1979).
119. M. Clarke, W. L. Bazari, and S. C. Kayman, *J. Bacteriol.*, **141**, 397 (1980).
120. M. J. Yerna, D. J. Hartshorne, and R. D. Goldman, *Biochemistry*, **18**, 673 (1979).
121. S. P. Chock, W. R. Albers, and G. Catravas, *Fed. Proc., Fed. Am. Soc. Exp. Biol.*, **41**, 503 (1982).
122. S. R. Childers and F. L. Siegel, *Biochim. Biophys. Acta*, **405**, 99 (1975).
123. C. B. Klee, T. H. Crouch, and M. H. Krinks, *Proc. Natl. Acad. Sci., USA*, **76**, 6270 (1979).
124. C. B. Klee, T. H. Crouch, and M. H. Krinks, *Biochemistry*, **18**, 722 (1979).
125. M. E. Morrill, S. T. Thompson, and E. Stellwagen, *J. Biol. Chem.*, **254**, 4371 (1979).
126. R. K. Sharma, T. H. Wang, E. Wirch, and J. H. Wang, *J. Biol. Chem.*, **255**, 5916 (1980).
127. D. C. LaPorte, W. A. Toscano, and D. R. Storm, *Biochemistry*, **18**, 2820 (1979).
128. D. J. Wolff and C. O. Brostrom, *Arch. Biochem. Biophys.*, **173**, 720 (1976).
129. M. P. Walsh, B. Vallet, J. C. Cavadore, and J. G. Demaille, *J. Biol. Chem.*, **255**, 335 (1980).
130. R. M. Gopinath and F. F. Vincenzi, *Biochem. Biophys. Res. Commun.*, **77**, 1203 (1977).
131. H. W. Jarrett and J. T. Penniston, *Biochem. Biophys. Res. Commun.*, **77**, 1210 (1977).
132. P. Cohen, A. Burchell, J. G. Foulkes, P. T. W. Cohen, T. C. Vanaman, and A. C. Nairn, *FEBS Lett.*, **92**, 287 (1978).
133. S. Shenolikan, P. T. W. Cohen, P. Cohen, A. C. Nairn, and S. V. Perry, *Eur. J. Biochem.*, **100**, 329 (1979).
134. K. Gietzen, A. Wüthrich, and H. Bader, *Biochem. Biophys. Res. Commun.*, **101**, 418 (1981).
135. R. K. Sharma and J. H. Wang, *Biochem. Biophys. Res. Commun.*, **100**, 710 (1981).
136. J. R. Dedman, R. L. Jackson, W. E. Schreiber, and A. R. Means, *J. Biol. Chem.*, **253**, 343 (1978).
137. P. Thiry, A. Vandermeers, M. C. Vandermeers-Piret, J. Rathe, and J. Christophe, *Eur. J. Biochem.*, **103**, 409 (1980).
138. P. G. Richman and C. B. Klee, *J. Biol. Chem.*, **253**, 6323 (1978).
139. E. Graf, A. G. Filoteo, and J. T. Penniston, *Arch. Biochem.*, **203**, 719 (1980).
140. D. C. LaPorte and D. R. Storm, *J. Biol. Chem.*, **253**, 3374 (1978).
141. W. E. Schreiber, T. Sasagawa, K. Titani, R. D. Wade, D. Malencik, and E. H. Fischer, *Biochemistry*, **20**, 5239 (1981).
142. M. Walsh, F. C. Stevens, J. Kuznicki, and W. Drabikowski, *J. Biol. Chem.*, **252**, 7440 (1977).
143. H. Charbonneau, H. W. Jarrett, R. O. McCann, and M. J. Cormier, "Calmodulin in Plants and Fungi," in *Calcium-Binding Proteins: Structure and Function*, F. L. Siegel et al., Eds., Elsevier/North Holland, New York, 1980, p. 155.

144. S. Kakiuchi and R. Yamazaki, *Proc. Japan Acad.*, **46,** 387 (1970).

145. S. Kakiuchi, R. Yamazaki, Y. Teshima, K. Uenishi, and E. Miyamoto, *Biochem. J.*, **146,** 109 (1975).

146. J. C. Egrie, J. A. Campbell, A. L. Flangas, and F. L. Siegel, *J. Neurochem.*, **28,** 1207 (1977).

147. S. R. Childers and F. L. Siegel, *J. Neurochem.*, **28,** 1229 (1977).

148. D. Waisman, F. C. Stevens, and J. H. Wang, *Biochem. Biophys. Res. Commun.*, **65,** 975 (1975).

149. P. Cohen, C. Picton, and C. B. Klee, *FEBS Lett.*, **104,** 25 (1979).

150. J. H. Wang and R. Desai, *J. Biol. Chem.*, **252,** 4175 (1977).

151. C. B. Klee and M. H. Krinks, *Biochemistry*, **17,** 120 (1978).

152. R. W. Wallace, T. J. Lynch, E. A. Tallant, and W. Y. Cheung, *J. Biol. Chem.*, **254,** 377 (1979).

153. K. Sobue and S. Kakiuchi, *Biochem. Biophys. Res. Commun.*, **93,** 850 (1980).

154. H. W. Jarrett, H. Charbonneau, M. J. Anderson, R. O. McCann, and M. J. Cormier, *Ann. NY Acad. Sci.*, **356,** 119 (1980).

155. K. Sobue, Y. Muramoto, M. Fujita, and S. Kakiuchi, *Biochem. Biophys. Res. Commun.*, **100,** 1063 (1981).

156. K. Sobue, Y. Muramoto, M. Fujita, and S. Kakiuchi, *Proc. Natl. Acad. Sci, USA*, **78,** 5652 (1981).

157. S. Kakiuchi, K. Sobue, and M. Fujita, *FEBS Lett.*, **132,** 144 (1981).

158. H. C. Palfrey, W. Schiebler, and P. Greengard, *Proc. Natl. Acad. Sci., USA*, **79,** 3780 (1982).

159. J. R. Glenney, P. Glenney, and K. Weber, *Proc. Natl. Acad. Sci., USA*, **79,** 4002 (1982).

160. J. A. Smooke, S-Y. Song, and W. Y. Cheung, *Biochim. Biophys. Acta*, **341,** 402 (1974).

161. T. C. Vanaman, F. Sharief, J. L. Awaramik, P. A. Mandel, and D. M. Watterson, "Chemical and Biological Properties of the Ubiquitous Troponin-C Like Protein from Non-Muscle Tissues, A Multifunctional Ca^{2+}-Dependent Regulatory Protein," in *Contractile Systems in Non-Muscle Tissues*, S. V. Perry, A. Margreth, and R. S. Adelstein, Eds., Elsevier/North Holland, New York, 1976, p. 165.

162. R. Dabrowska, J. M. Sherry, D. K. Aromatorio, and D. J. Hartshorne, *Biochemistry*, **17,** 253 (1978).

163. W. L. Bazari and M. Clarke, *J. Biol. Chem.*, **256,** 3598 (1981).

164. R. J. A. Grand, A. C. Nairn, and S. V. Perry, *Biochem. J.*, **185,** 755 (1980).

165. J. A. Cox, C. Ferraz, J. G. Demaille, R. O. Perez, D. Van Tuinen, and D. Marmé, *J. Biol. Chem.*, **257,** 10694 (1982).

166. A. Molla, M. C. Kilhoffer, C. Ferraz, E. Audemard, M. P. Walsh, and J. G. Demaille, *J. Biol. Chem.*, **256,** 15 (1981).

167. K. B. Seamon and B. W. Moore, *J. Biol. Chem.*, **255,** 11644 (1980).

168. E. M. Ross and A. C. Gilman, *Ann. Rev. Biochem.*, **49,** 533 (1980).

169. C. O. Brostrom, Y.-C. Huang, B. McL. Breckenridge, and D. J. Wolff, *Proc. Natl. Acad. Sci., USA*, **72,** 64 (1975).

170. W. Y. Cheung, L. S. Bradham, T. J. Lynch, Y. M. Lin, and E. A. Tallant, *Biochem. Biophys. Res. Comm.*, **66,** 1055 (1975).

171. T. J. Lynch, E. A. Tallant, and W. Y. Cheung, *Arch. Biochem. Biophys.*, **182,** 124 (1977).
172. K. R. Westcott, D. C. Laporte, and D. R. Storm, *Proc. Natl. Acad. Sci., USA*, **76,** 204 (1979).
173. J. N. Wells and J. G. Hardman, *Adv. Cyclic Nucleotide Res.*, **8,** 119 (1977).
174. R. S. Salter, M. H. Krinks, C. B. Klee, and E. J. Neer, *J. Biol. Chem.*, **256,** 9830 (1981).
175. Y. Teshima and S. Kakiuchi, *Biochem. Biophys. Res. Comm.*, **56,** 489 (1974).
176. Y. M. Lin, Y. P. Liu, and W. Y. Cheung, *FEBS Lett.*, **49,** 356 (1975).
177. W. Y. Cheung, *J. Biol. Chem.*, **246,** 2859 (1971).
178. T. S. Teo, T. H. Wang, and J. H. Wang, *J. Biol. Chem.*, **248,** 588 (1973).
179. T. E. Donnelly, *Biochem. Biophys. Acta*, **480,** 194 (1976).
180. C. H. Keller, D. C. LaPorte, W. A. Toscano, Jr., D. H. Storm, and K. R. Westcott, *Ann. NY Acad. Sci.*, **356,** 205 (1980).
181. M. A. Brostrom, C. O. Brostrom, B. McL. Breckenridge, and D. J. Wolff, *Adv. Cyclic Nucleotide Res.*, **9,** 85 (1978).
182. J. D. Potter, M. T. Piascik, P. L. Wisler, S. P. Robertson, and C. L. Johnson, *Ann. NY Acad. Sci.*, **356,** 220 (1980).
183. P. Cohen, *Nature*, **296,** 613 (1982).
184. T. Hayakawa, J. P. Perkins, and E. G. Krebs, *Biochemistry*, **12,** 574 (1973).
185. P. Cohen, *Eur. J. Biochem.*, **34,** 1 (1973).
186. P. Cohen, *Eur. J. Biochem.*, **111,** 564 (1980).
187. P. Cohen, A. Burchell, J. G. Foulkes, P. T. W. Cohen, T. C. Vanaman, and A. C. Nairn, *FEBS Lett.*, **92,** 287 (1978).
188. L. M. G. Heilmeyer, F. Meyer, R. H. Haschke, and E. H. Fischer, *J. Biol. Chem.*, **245,** 6649 (1970).
189. C. O. Brostrom, F. L. Hunkeler, and E. G. Krebs, *J. Biol. Chem.*, **246,** 1961 (1971).
190. J. R. Skuster, K. F. Jesse Chan, and D. J. Graves, *J. Biol. Chem.*, **255,** 2203 (1980).
191. K. F. Jesse Chan and D. J. Graves, *J. Biol. Chem.*, **257,** 5956 (1982).
192. C. Picton, C. B. Klee, and P. Cohen, *Eur. J. Biochem.*, **111,** 553 (1980).
193. C. Picton, C. B. Klee, and P. Cohen, *Cell Calcium*, **2,** 281 (1981).
194. S. Shenolikar, P. T. W. Cohen, P. Cohen, A. C. Nairn, and S. V. Perry, *Eur. J. Biochem.*, **100,** 329 (1979).
195. P. Cohen, C. Picton, and C. B. Klee, *FEBS Lett.*, **104,** 25 (1979).
196. P. Cohen, C. B. Klee, C. Picton, and S. Shenolikar, *Ann. NY Acad. Sci.*, **356,** 151 (1980).
197. P. J. Roach, A. A. DePaoli-Roach, and J. Larner, *J. Cyclic Nucleotide Res.*, **4,** 245 (1978).
198. N. Embi, D. B. Rylatt, and P. Cohen., *Eur. J. Biochem.*, **100,** 339 (1979).
199. A. A. DePaoli-Roach, P. J. Roach, and J. Larner, *J. Biol. Chem.*, **254,** 4212 (1979).
200. T. R. Soderling, A. K. Srivastava, M. A. Bass, and B. S. Khatra, *Proc. Natl. Acad. Sci., USA*, **76,** 2536 (1979).
201. K. Y. Walsh, D. M. Millikin, K. K. Schlender, and E. M. Reimann, *J. Biol. Chem.*, **254,** 6611 (1979).
202. T. R. Soderling, J. P. Hickenbottom, E. M. Reimann, F. L. Hunkder, D. A. Walsh, and E. G. Krebs, *J. Biol. Chem.*, **245,** 6317 (1970).

203. K. K. Schlender, S. H. Wei, and C. Villar-Palasi, *Biochim. Biophys. Acta*, **191,** 272 (1969).

204. N. Embi, P. J. Parker, and P. Cohen, *Eur. J. Biochem.*, **115,** 405 (1981).

205. R. Dabrowska, J. M. F. Sherry, D. K. Aromatoria, and D. J. Hartshorne, *Biochemistry*, **17,** 253 (1978).

206. K. Yagi, M. Yazawa, S. Kakiuchi, M. Ohshima, and K. Uenishi, *J. Biol. Chem.*, **253,** 1338 (1978).

207. B. Barylko, J. Kuznicki, and W. Drabikowski, *FEBS Lett.*, **90,** 301 (1978).

208. R. Dabrowska and D. J. Hartshorne, *Biochem. Biophys. Res. Commun.*, **85,** 1352 (1978).

209. D. R. Hathaway and R. S. Adelstein, *Proc. Natl. Acad. Sci., USA*, **76,** 1653 (1979).

210. D. R. Hathaway, R. S. Adelstein, and C. B. Klee, *J. Biol. Chem.*, **256,** 8183 (1981).

211. M. J. Yerna, R. Dabrowska, D. J. Hartshorne, and R. D. Goldman, *Proc. Natl. Acad. Sci., USA*, **76,** 184 (1979).

212. M. P. Walsh, B. Vallet, F. Autric, and J. G. Demaille, *J. Biol. Chem.*, **254,** 12136 (1979).

213. H. Wolf and F. Hofmann, *Proc. Natl. Acad. Sci., USA*, **77,** 5852 (1980).

214. R. Jakes, F. Northrop, and J. Kendrick-Jones, *FEBS Lett.*, **70,** 229 (1976).

215. W. T. Perrie, L. B. Smille, and S. V. Perry, *Biochem. J.* **135,** 151 (1973).

216. S. Ebashi, M. Endo, and I. Ohtsuki, *Q. Rev. Biophys.*, **2,** 351 (1969).

217. A. Weber and J. M. Murray, *Physiol. Rev.*, **53,** 612 (1973).

218. M. Morgan, S. V. Perry, and J. Ottway, *Biochem. J.*, **157,** 687 (1976).

219. S. Chacko, M. A. Conti, and R. S. Adelstein, *Proc. Natl. Acad. Sci., USA*, **74,** 129 (1977).

220. J. T. Barron, M. Bárány, and K. Bárány, *J. Biol. Chem.*, **254,** 4954 (1979).

221. A. Sobieszek, "Vertebrate Smooth Muscle Myosin, Enzymatic and Structural Properties," in, *The Biochemistry of Smooth Muscle*, N. L. Stevens, Ed., University Park Press, Baltimore, 1977, p. 413.

222. U. Mrwa and D. J. Hartshorne, *Fed. Proc.* **39,** 1564 (1980).

223. M. P. Walsh, R. Bridenbaugh, D. J. Hartshorne, and W. G. L. Kerrick, *J. Biol. Chem.*, **257,** 5987 (1982).

224. R. S. Adelstein, M. A. Conti, D. R. Hathaway, and C. B. Klee, *J. Biol. Chem.*, **253,** 8347 (1978).

225. M. A. Conti and R. S. Adelstein, *J. Biol. Chem.*, **256,** 3178 (1981).

226. B. Vallet, A. Molla, and J. G. Demaille, *Biochem. Biophys. Acta*, **674,** 256 (1981).

227. D. R. Hathaway, C. R. Eaton, and R. S. Adelstein, *Nature (Lond.)*, **291,** 252 (1981).

228. D. R. Manning and J. T. Stull, *Biochem. Biophys. Res. Commun.*, **90,** 164 (1979).

229. M. Hirata, J. Mikawa, Y. Nonomura, and S. Ebashi, *J. Biochem.*, **82,** 1793 (1977).

230. T. Mikawa, Y. Nonomura, and S. Ebashi, *J. Biochem.*, **82,** 1789 (1977).

231. T. Mikawa, Y. Nonomura, M. Hirata, S. Ebashi, and S. Kakiuchi, *J. Biochem.*, **84,** 1633 (1978).

232. P. Rubin, *Pharmacol. Rev.*, **22,** 389 (1972).

233. H. Schulman and P. Greengard, *Nature*, **271,** 478 (1978).

234. H. Schulman and P. Greengard, *Proc. Natl. Acad. Sci., USA*, **75,** 5432 (1978).

235. R. J. DeLorenzo, S. D. Freedman, W. B. Yohe, and S. C. Maurer, *Proc. Natl. Acad. Sci., USA*, **76,** 1838 (1979).

236. T. Ueda and P. Greengard, *J. Biol. Chem.*, **252,** 5155 (1977).

237. U. Walter, S. M. Lohmann, W. Sieghart, and P. Greengard, *J. Biol. Chem.*, **254,** 12235 (1979).
238. W. Sieghart, J. Forn, and P. Greengard, *Proc. Natl. Acad. Sci., USA*, **76,** 2475 (1979).
239. W. B. Huttner and P. Greengard, *Proc. Natl. Acad. Sci., USA*, **76,** 5402 (1979).
240. M. B. Kennedy and P. Greengard, *Proc. Natl. Acad. Sci., USA*, **78,** 1293 (1981).
241. T. Yamauchi and H. Fujisawa, *Arch. Biochem. Biophys.*, **198,** 219 (1979).
242. T. Yamauchi and H. Fujisawa, *Biochem. Biophys. Res. Commun.*, **90,** 28 (1979).
243. T. Yamauchi and H. Fujisawa, *Biochem. Int.*, **1,** 98 (1980).
244. D. M. Kuhn, J. P. O'Callaghan, J. Juskevich, and W. Lovenberg, *Biochemistry*, **77,** 4688 (1980).
245. T. Yamauchi, H. Nakata, and H. Fujisawa, *J. Biol. Chem.*, **256,** 5404 (1981).
246. D. J. Grab, K. Berzins, R. S. Cohen, and P. Siekevitz, *J. Biol. Chem.*, **254,** 8690 (1979).
247. D. J. Grab, R. K. Carlin, and P. Siekevitz, *J. Cell. Biol.*, **89,** 440 (1981).
248. R. K. Carlin, D. J. Grab, and P. Siekevitz, *J. Cell. Biol.*, **89,** 449 (1981).
249. J. G. Wood, R. W. Wallace, J. N. Whitaker, and W. Y. Cheung, *J. Cell. Biol.*, **84,** 66 (1980).
250. R. W. Wallace, E. A. Tallant, and W. Y. Cheung, *Biochemistry*, **19,** 1831 (1980).
251. R. K. Sharma, R. Desai, D. M. Waisman, and J. H. Wang, *J. Biol. Chem.*, **254,** 4276 (1979).
252. A. A. Stewart, T. S. Ingebritsen, A. Manalan, C. B. Klee, and P. Cohen, *FEBS Lett.*, **137,** 80 (1982).
253. M. E. Payne and T. R. Soderling, *J. Biol. Chem.*, **255,** 8054 (1980).
254. A. K. Verma and J. T. Penniston, *J. Biol. Chem.*, **256,** 1269 (1981).
255. H. A. Pershadsingh and J. M. McDonald, *Biochem. Int.*, **2,** 243 (1981).
256. H. A. Pershadsingh, M. L. McDaniel, M. Landt, C. Bry, P. E. Lacy, and J. M. McDonald, *Nature (Lond.)*, **288,** 492 (1980).
257. H. N. Nellans and J. E. Popovitch, *J. Biol. Chem.*, **256,** 9932 (1981).
258. A. H. Lichtman, B. G. Segel, and M. A. Lichtman, *J. Biol. Chem.*, **256,** 6148 (1981).
259. E. T. Dunham and I. M. Glynn, *J. Physiol. (Lond.)*, **156,** 274 (1961).
260. H. J. Schatzmann, *Exper. (Basel)*, **22,** 364 (1966).
261. V. Niggli, J. T. Penniston, and E. Carafoli, *J. Biol. Chem.*, **254,** 9955 (1979).
262. V. Niggli, P. Ronner, E. Carafoli, and J. T. Penniston, *Arch. Biochem. Biophys.*, **198,** 124 (1979).
263. K. Gietzen, M. Tejcka, and H. U. Wolf, *Biochem. J.*, **189,** 81 (1980).
264. T. R. Hinds and T. J. Andreasen, *J. Biol. Chem.*, **256,** 7877 (1981).
265. J. T. Penniston, E. Graf, V. Niggli, A. K. Verma, and E. Carafoli, "The Plasma Membrane Calcium ATPase," in *Calcium Binding Proteins and Calcium Function*, F. L. Siegel, E. Carafoli, R. H. Kretsinger, D. H. MacLennan and R. H. Wasserman, Eds., Elsevier/North Holland, New York, 1980, p. 23.
266. E. Graf and J. T. Penniston, *J. Biol. Chem.*, **256,** 1587 (1981).
267. V. Niggli, E. S. Adunyah, J. T. Penniston, and E. Carafoli, *J. Biol. Chem.*, **256,** 395 (1981).
268. R. M. Gopinath and F. F. Vincenzi, *Biochem. Biophys. Res. Commun.*, **77,** 1203 (1977).
269. H. W. Jarrett and J. T. Penniston, *Biochem. Biophys. Res. Commun.*, **77,** 1210 (1977).
270. G. H. Bond and D. L. Clough, *Biochim. Biophys. Acta*, **323,** 592 (1973).

271. M. G. Luthra, G. R. Hildenbrandt, and D. J. Hanahan, *Biochim. Biophys. Acta*, **419,** 164 (1976).
272. F. F. Vincenzi and M. L. Farrance, *J. Supramol. Struct.*, **7,** 301 (1977).
273. S. Katz, B. D. Roufogalis, A. D. Landman, and L. Ho, *J. Supramol. Struct.*, **10,** 215 (1979).
274. T. J. Lynch and W. Y. Cheung, *Arch. Biochem. Biophys.*, **194,** 165 (1979).
275. V. Niggli, E. S. Adunyah, and E. Carafoli, *J. Biol. Chem*, **256,** 8588 (1981).
276. S. Lotersztajn, J. Hanoune, and F. Pecker, *J. Biol. Chem.*, **256,** 11209 (1981).
277. P. Caroni and E. Carafoli, *J. Biol. Chem.*, **256,** 3263 (1981).
278. M. Tada, T. Yamamoto, and Y. Tonomura, *Physiol. Rev.*, **58,** 1 (1978).
279. M. A. Kirchberger, M. Tada, D. I. Repke, and A. M. Katz, *J. Mol. Cell. Cardiol.*, **4,** 673 (1972).
280. M. A. Kirchberger, M. Tada, and A. M. Katz, *J. Biol. Chem.*, **249,** 6166 (1974).
281. M. Tada, M. A. Kirchberger, and A. M. Katz, *J. Biol. Chem.*, **250,** 2640 (1975).
282. M. Tada and M. A. Kirchberger, *Acta. Cardiol.*, **30,** 231 (1975).
283. H. L. Wray, R. R. Gray, and R. A. Olsson, *J. Biol. Chem.*, **248,** 1496 (1973).
284. S. Katz and M. A. Remtulla, *Biochem. Biophys. Res. Commun.*, **83,** 1373 (1978).
285. C. J. Lepeuch, J. Haiech, and J. G. Demaille, *Biochemistry*, **18,** 5150 (1979).
286. G. Lopaschuk, B. Richter, and S. Katz, *Biochemistry*, **19,** 5603 (1980).
287. S. Katz, *Ann. NY Acad. Sci.*, **356,** 267 (1980).
288. M. A. Kirchberger and T. Antonetz, *J. Biol. Chem.*, **257,** 5685 (1982).
289. J. M. Anderson and M. J. Cormier, *Biochem. Biophys. Res. Commun.*, **84,** 595 (1978).
290. D. Epel, C. Patton, R. W. Wallace, and W. Y. Cheung, *Cell,* **23,** 543 (1981).
291. R. Steinhardt, R. Zucker, and G. Schatten, *Dev. Biol.*, **58,** 185 (1977).
292. D. Epel, *Curr. Top. Dev. Biol.*, **12,** 186 (1978).
293. R. A. Steinhardt and M. M. Winkler, "The Ionic Hypothesis of Cell Activation at Fertilization," in *Molecular Basis of Immune Cell Function*, J. G. Kaplan, Ed., Elsevier/North Holland, New York, 1980. p. 11.
294. E. Nishida and H. Kumagai, *J. Biochem.*, **87,** 143 (1980).
295. P. Y.-K. Wong and W. Y. Cheung, *Biochem. Biophys. Res. Commun.*, **90,** 473 (1979).
296. R. Franson and M. Waite, *Biochemistry*, **17,** 4029 (1978).
297. F. Hirata, *J. Biol. Chem.*, **256,** 7730 (1981).
298. N. Moskowitz, W. Schook, and S. Puszkin, *Science,* **216,** 305 (1982).
299. J. J. Blum and M. Hines, *Q. Rev. Biophys.*, **12,** 103 (1979).
300. B. J. Byrne and B. C. Byrne, *CRC Crit. Rev. Microbiol.*, **6,** 53 (1978).
301. J. J. Blum, A. Hayes, G. A. Jamieson, Jr., and T. C. Vanaman, *J. Cell Biol.*, **87,** 386 (1980).
302. H. P. Jones, R. W. Lenz, B. A. Palevitz, and M. J. Cormier, *Proc. Natl. Acad. Sci., USA*, **77,** 2772 (1980).
303. J. Feinberg, J. Weinman, S. Weinman, M. P. Walsh, M. C. Harricane, J. Gabrion, and J. G. Demaille, *Biochim. Biophys. Acta*, **673,** 303 (1981).
304. R. C. Weisenberg, *Science*, **177,** 1104 (1972).
305. G. G. Borisy and J. B. Olmsted, *Science*, **177,** 1196 (1972).
306. J. B. Olmsted and G. G. Borisy, *Biochemistry*, **12,** 4282 (1973).

307. J. B. Olmsted and G. G. Borisy, *Biochemistry*, **14,** 2996 (1975).
308. A. C. Rosenfeld, R. V. Zackroff, and R. C. Weisenberg, *FEBS Lett.*, **65,** 144 (1976).
309. T. Haga, T. Abe, and M. Kurokawa, *FEBS Lett.*, **39,** 291 (1974).
310. D. P. Keihart and S. Inoué, *J. Cell Biol.*, **70,** 230 (1976).
311. G. M. Fuller and B. R. Brinkley, *J. Supramol. Struct.*, **5,** 497 (1976).
312. E. D. Salmon and R. Jenkins, *J. Cell Biol.*, **75,** 295 (1977).
313. F. Soloman, *Biochemistry*, **16,** 358 (1977).
314. M. Hayashi and F. Matsumura, *FEBS Lett.*, **58,** 222 (1975).
315. S. A. Berkowitz and J. Wolff, *J. Biol. Chem.*, **256,** 11216 (1981).
316. E. Nishida and H. Sakai, *J. Biochem. (Tokyo)*, **82,** 303 (1977).
317. Y. C. Lee and J. Wolff, *J. Biol. Chem.*, **257,** 6306 (1982).
318. J. M. Marcum, J. R. Dedman, B. R. Brinkley, and A. R. Means, *Proc. Natl. Acad. Sci., USA*, **75,** 3771 (1978).
319. K. Sobue, M. Fujita, Y. Muramoto, and S. Kakiuchi, *FEBS Lett.*, **132,** 137 (1981).
320. M. J. Welsh, J. R. Dedman, B. R. Brinkley, and A. R. Means, *Proc. Natl. Acad. Sci., USA*, **75,** 1867 (1978).
321. B. Anderson, M. Osborn, and K. Weber, *Eur. J. Cell. Biol.*, **17,** 354 (1978).
322. M. J. Welsh, J. R. Dedman, B. R. Brinkley, and A. R. Means, *J. Cell. Biol.*, **81,** 624 (1979).
323. M. J. Cormier, J. M. Anderson, H. Charbonneau, H. P. Jones, and R. O. McCann, "Plant and Fungal Calmodulin and the Regulation of NAD Kinase," in *Calcium and Cell Function*, Vol. 1, W. Y. Cheung, Ed., Academic Press, New York, 1980, p. 201.
324. J. A. Cox, C. Ferraz, J. G. Demaille, R. O. Perez, D. Van Tuinen, and D. Marmé, *J. Biol. Chem.*, **257,** 10694 (1982).
325. M. J. Cormier, H. W. Jarrett and H. Charbonneau, "Role of Ca^{2+}-Calmodulin in Metabolic Regulation in Plants," in *Calmodulin and Intracellular Ca^{2+} Receptors*, S. Kakiuchi, H. Hidaka, and A. R. Means, Eds., Plenum Publishing Corp., New York, 1982, p. 125.
326. R. J. A. Grand, A. C. Nairn, and S. V. Perry, *Biochem. J.*, **185,** 755 (1980).
327. S. E. Gitelman and G. B. Witman, *J. Cell. Biol.*, **87,** 764 (1980).
328. R. E. Williamson and C. C. Ashley, *Nature*, **296,** 647 (1982).
329. T. Hayama, T. Shimmen, and M. Tazawa, *Protoplasma*, **99,** 305 (1979).
330. A. B. Hope and N. A. Walker, *The Physiology of Giant Algal Cells*, Cambridge University Press, Cambridge, 1975.
331. E. M. Dreyer and M. H. Weisenseel, *Planta*, **146,** 31 (1979).
332. C. C. Hale and S. J. Roux, *Plant Physiol.*, **65,** 658 (1980).
333. S. J. Roux, K. McEntire, R. D. Slocum, T. E. Cedel, and C. C. Hale, *Proc. Natl. Acad. Sci., USA*, **78,** 283 (1981).
334. W. Haupt, *Planta*, **53,** 484 (1959).
335. M. H. Weisenseel and H. K. Ruppert, *Planta*, **137,** 225 (1977).
336. L. A. Jaffe, M. H. Weisenseel, and L. F. Jaffe, *J. Cell. Biol.*, **67,** 488 (1975).
337. W. Herth, *Protoplasma*, **96,** 275 (1978).
338. H.-D. Reiss and W. Herth, *Protoplasma*, **97,** 373 (1978).
339. H.-D. Reiss and W. Herth, *Planta*, **145,** 225 (1979).

340. E. G. Brown, T. Al-Najafi, and R. P. Newton, *Phytochemistry*, **16,** 1333 (1977).
341. S. A. Charles and B. Halliwell, *Biochem. J.*, **188,** 755 (1980).
342. R. Kindt, E. Pahlich, and I. Rasched, *Eur. J. Biochem.*, **112,** 533 (1980).
343. M. J. Saunders and P. K. Hepler, *Science*, **217,** 943 (1982).
344. S. Muto and S. Miyachi, *Plant Physiol.*, **59,** 55 (1977).
345. H. W. Jarrett, T. DaSilva, and M. J. Cormier, "Calmodulin Activation of NAD Kinase and its Role in the Metabolic Regulation of Plants," in *Uptake and Utilization of Metals by Plants*, D. A. Robb and W. S. Pierpoint, Eds., Academic Press, London, in press.
346. P. Dieter and D. Marmé, *Cell Calcium*, **1,** 279 (1980).
347. T. Oh-Hama and S. Miyachi, *Biochim. Biophys. Acta*, **34,** 202 (1959).
348. W. L. Ogren and D. W. Krogmann, *J. Biol. Chem.*, **240,** 4603 (1965).
349. U. W. Heber and K. A. Santarius, *Biochim. Biophys. Acta*, **109,** 390 (1965).
350. S. Muto, S. Miyachi, H. Usuda, G. E. Edwards, and J. A. Bassham, *Plant Physiol.*, **68,** 324 (1981).
351. S. Muto, S. Izawa, and S. Miyachi, *FEBS Lett.*, **139,** 250 (1982).
352. H. W. Jarrett, C. J. Brown, C. C. Black, and M. J. Cormier, *J. Biol. Chem.*, **257,** 13795 (1982).
353. C. Cseke, A. N. Nishizawa, and B. B. Buchanan, *Plant Physiol.*, **70,** 658 (1982).
354. A. R. Ashton, T. Brennan, and L. E. Anderson, *Plant Physiol.*, **66,** 605 (1980).
355. R. Scheibe and L. E. Anderson, *Biochim. Biophys. Acta*, **636,** 58 (1981).
356. G. J. Kelly and E. Latzko, *Plant Physiol.*, **60,** 295 (1977).
357. P. Dieter and D. Marmé, *Proc. Natl. Acad. Sci., USA*, **77,** 7311 (1980).
358. P. Dieter and D. Marme, *FEBS Lett.*, **125,** 245 (1981).
359. A. Trewavas, *Ann. Rev. Plant Physiol.*, **27,** 349 (1976).
360. J. Bennett, K. E. Steinback, and C. J. Arntzen, *Proc. Natl. Acad. Sci., USA*, **77,** 5253 (1980).
361. A. Hetherington and A. Trewavas, *FEBS Lett.*, **145,** 67 (1982).
362. W. H. Burgess, *J. Biol. Chem.*, **257,** 1800 (1982).
363. K. P. Cambell and D. H. MacLennan, *J. Biol. Chem.*, **257,** 1238 (1982).
364. M. A. Kirchberger and M. Tada, *J. Biol. Chem.*, **251,** 725 (1976).
365. L. Meijer and P. Guerrier, *Biochim. Biophys. Acta*, **702,** 143 (1982).
366. P. M. Moore and J. R. Dedman, *J. Biol. Chem.*, **257,** 9663 (1982).
367. R. Barr, K. S. Troxel, and F. L. Crane, *Biochem. Biophys. Res. Commun.*, **104,** 1182 (1982).
368. H. Tada, M. Yazawa, K. Kondo, T. Homma, K. Narita, and K. Yagi, *J. Biochem.*, **90,** 1493 (1981).
369. C-L. A. Wang, R. R. Aquaron, P. C. Leavis, and J. Gergely, *Eur. J. Biochem.*, **124,** 7 (1982).
370. R. A. Steinhardt and J. M. Alderton, *Nature*, **295,** 154 (1982).
371. S. Muto, *FEBS Lett.*, **147,** 161 (1982).
372. J. A. Cox, M. Comte, and E. A. Stein, *Proc. Natl. Acad. Sci., USA*, **79,** 4265 (1982).
373. A. Harmon and M. J. Cormier, unpublished data.
374. C. Evans, A. Harmon, and M. J. Cormier, unpublished data.

CHAPTER 3

Calcium in Muscle

C. C. ASHLEY

University Laboratory of Physiology
Oxford, United Kingdom

CONTENTS

1 INTRODUCTION

The central role played by calcium in muscle became apparent from experiments where the ion was injected directly into isolated muscle fibers from frog skeletal muscle. Most of this early work was initiated by Heilbrunn and his colleagues in Philadelphia (1). There is now a considerable amount of literature that implicates a change in intracellular free calcium (Ca_i^{2+}) in physiological processes as diverse as muscle contraction, gating of ionic channels, and fertilization of oocytes and as a secondary messenger following hormone action on the cell membrane as well as in cytoplasmic streaming in plant cells. Not only do changes in free Ca^{2+} appear central to these diverse processes, but molecules which may act apparently as freely diffusible receptors for this calcium are also known. These molecules, or calcium-dependent regulator proteins (CDR), are of relatively low molecular weight ($\sim$16 to 17K daltons) and can form the calcium-binding subunits of intracellularly located calcium-activated enzymes (2, 3). These molecules may be ubiquitous; however, they are not present in muscle cells at such high concentrations as in many nonmuscle cells. There are present in most muscle cells other calcium binding proteins whose precise function is at present uncertain, but which exist in high concentration in the sarcoplasm; these are muscle calcium-binding parvalbumins (MCBP). Certainly the structures of CDR and MCBP are highly conserved and represent a close but diffusible relative of the actin-based calcium regulator protein subunit, troponin-C (4, 5).

In this article, some of the experiments that have led to our current ideas as to the way free calcium is regulated in muscle cells are reviewed. Much of this is slanted toward the work on invertebrate muscle fibers, since it is from this work that the initial most clear-cut experiments have been achieved, mainly through the relatively easy access to the internal environment of these large fibers.

2 THE NATURE OF TRANSPORT PROCESSES ACROSS THE CELL MEMBRANE

Muscle cells have an external calcium concentration, which is millimolar, and a membrane potential (E_m), which is minus 0.1 V. These two factors would predict that if calcium were distributed across the cell membrane according to the electrochemical gradient, then the free Ca^{2+} in the cytoplasm should be in the region of molar concentration. However, experimental evidence indicates that the free calcium concentration in muscle cells, and in fact most other cells so far examined, lies in the range of 0.1 to 0.2 μM (6, 7). This result implies a steep electrochemical gradient for calcium into the cell; therefore, that calcium must be well-sequestered and buffered by organelles within the cytoplasm of the cell and, additionally, must be ultimately actively extruded across the cell membrane. The nature of this active extrusion system is one of the areas of investigation in muscle and other cells. Two basic mechanisms have been proposed to power the active extrusion of calcium from the cell. One scheme relies entirely upon the presence in the cell membrane of an ATP-driven Ca^{2+} transport system, where the energy of hydrolysis of ATP is applied directly to the system, such as is the case with Na^{+}-K^{+} exchange. Examples of this type of mechanism are well-documented for calcium uptake into the lumina of the sarcoplasmic reticulum system of muscle, as well as across the membrane of the red cell.

The alternative mechanism relies on the energy for extrusion being derived from the movement of another ion down its own electrochemical gradient into the cell, thus providing the energy for the extrusion of calcium via a linked system. The ion which can theoretically provide this energy is sodium, so that the concept of a sodium-calcium exchange mechanism operating across the membrane of many cells has been established and has provided much impetus for fruitful lines of investigation.

These two types of extrusion system are illustrated diagrammatically in Figure 1. It is quite possible that both of these systems are able to operate independently of each other; but there is certainly strong evidence from work on isolated nerve axons, where internal perfusion is relatively easy, that ATP may participate in some way in the operation of an Na^{+}-Ca^{2+} exchange system and that this ATP must be hydrolyzable. The fact that a sodium-calcium exchange system is operating to regulate the free calcium in muscle was initially suggested by experiments on cardiac muscle where the effect upon ^{45}Ca efflux of removing external Ca^{2+} and Na^{+} was investigated (8).

This type of experiment is illustrated in Figure 2, where components of the efflux, sensitive to both $Na_o^{+}$and Ca_o^{2+} are clearly present (9). It has not been possible so far in muscle to perform the type of definitive ATP and phosphagen depletion experiments so elegantly performed on nerve axons (10), and which implicated an additional requirement for ATP in the extrusion mechanism. However, it has recently been demonstrated that the ion vanadate, which affects ATP-dependent extrusion systems, also inhibits the

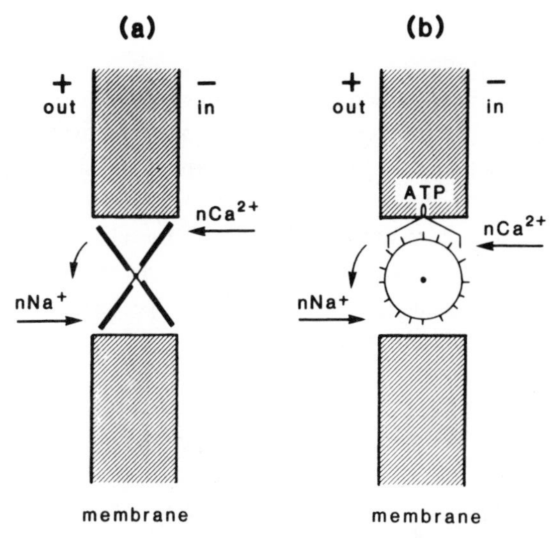

Figure 1 Schematic diagram of the functioning of the Na^{+}/Ca^{2+} exchange system in (*a*) where only the membrane potential (E_m) and the ionic concentration gradients for sodium and calcium are factors, and (*b*) where additionally an ATP involvement in an "escapement" system can be envisaged. In both (*a*) and (*b*) the "transporter" (windmill) is considered to be uncharged.

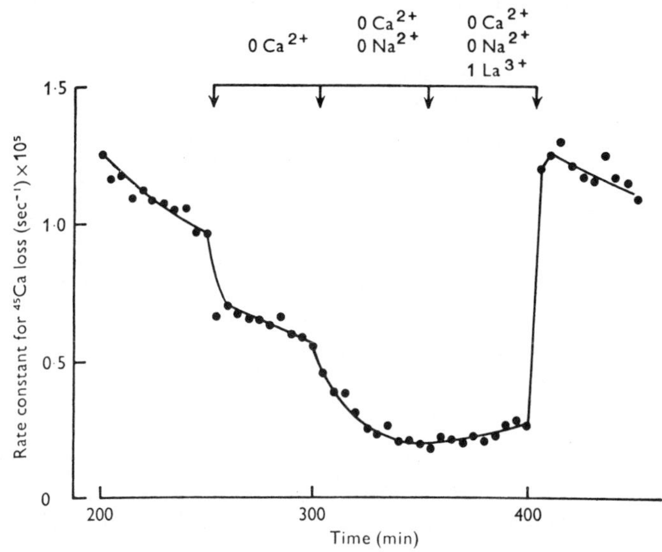

Figure 2 The effect of replacing external Ca^{2+} and Na^{+}, and the addition of 1 m*M* La^{3+} on the efflux of ^{45}Ca from a *Balanus* muscle fiber. The resting potential is −48 mV. The fiber is in physiological saline initially. (From ref. 9).

Ca^{2+} efflux system of crustacean muscle fibers (11). This lends support to the idea that the ATP-dependent Na^+-Ca^{2+} extrusion mechanism may also be operating across the cell membranes of these crustacean muscle cells, as well as nerve axons. The precise way that ATP may be involved in the extrusion system, and whether it is required only to maintain a *very* low free Ca^{2+} concentration internally, has yet to be determined. However, if a Na^+-Ca^{2+} exchange system operates to extrude Ca^{2+} from the cell, it should be possible to reverse the Na^+ gradient across the cell membrane and stimulate Ca^{2+} entry. Thus, much experimental work has been performed examining the nature of the Ca^{2+} influx into muscle.

3 USE OF ^{45}Ca TO EXAMINE CALCIUM INFLUX

One advantage of using large muscle fibers to study the ways in which calcium is transported and balanced within the cell is the fact that the isotope can be injected directly into the cell. In this case, the unidirectional efflux can be assessed and, in the ideal case, be free from problems of the extracellular space component. Similar problems confront the experimenter wishing to investigate unidirectional influx, particularly when the influx of the ion in question is relatively slow. Generally, for influx determinations, an aggregate technique has been used in which several fibers are sacrificed at each time point to obtain a statistically satisfactory influx. In addition, with influx measurements, in order to overcome or attempt to overcome the problems of extracellular space, an extracellular space marker, such as radioactively labeled inulin, is included with the isotope of interest.

One method, particularly for use with large muscle fibers (and axons), is the glass scintillator probe method (12), which permits the influx of low energy β-emitting isotopes to be determined. Although the technique is also able to detect the influx of the labeled material into the extracellular compartment, this can be corrected for in the final calculation.

Another method that is particularly, although not exclusively, applicable to the large muscle fiber preparation, is that of internal perfusion. This elegant technique, developed initially for work with large nerve axons, has been applied to examine the influx of calcium into large muscle fibers. Initially, the method used a glass capillary having porous properties, but now employs cellulose acetate tubing permeable to molecular species of <600 daltons (13).

The experimental arrangement for detecting isotope influx using the scintillator probe is illustrated in Figure 3, and the result of a typical ^{45}Ca influx experiment into an isolated muscle fiber is shown in Figure 4*a*. There is an initial rapid increase in the rate of the apparent probe counts over the first 60 to 90 min, followed by a phase in which the probe count rate increases roughly linearly with time. The method for calculating the influx rate from the probe count is indicated in the legend to Figure 4. The slope of the linear phase of calcium influx predicts an influx rate of ~ 10 pmol cm^{-2} sec^{-1}. This value is considerably higher than the value calculated for the same fiber

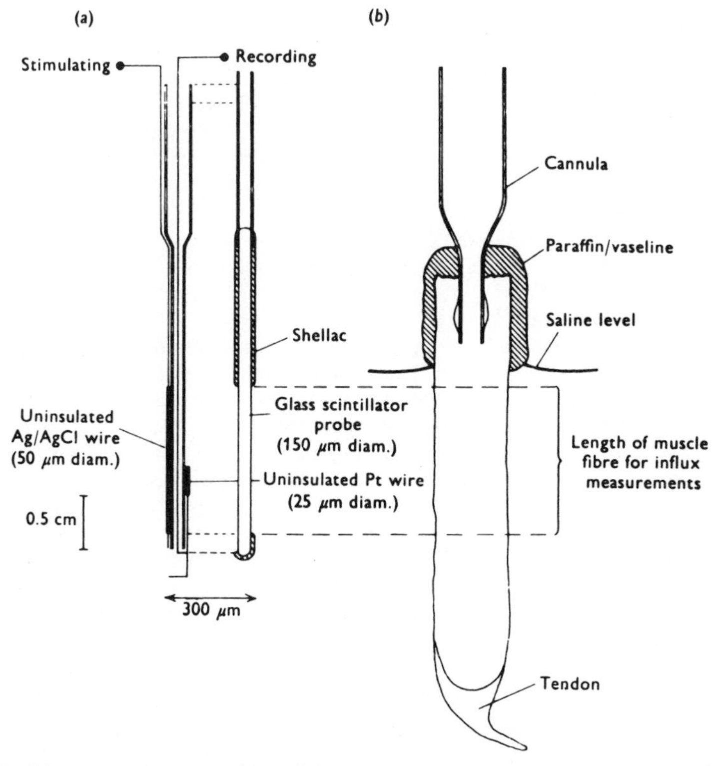

Figure 3 Diagrammatic representation of the apparatus for recording the glass scintillator probe responses. A dual axial stimulating-recording electrode can be attached to the probe (*a*) and the assembly inserted into the cannulated muscle fiber (*b*), when the effects of electrical stimulation are to be investigated. The photomultiplier tube (2 in. diameter, EMI 9324) was operated at + 1600 V and the output was anode-decoupled and fed to a Panax modular counting system. The fiber was contained in a glass cuvette (volume = 2.5 ml), whose optical face was coupled (Dow-Corning coupling compound 20-057) to the faceplate of the photomultiplier tube (12, 25).

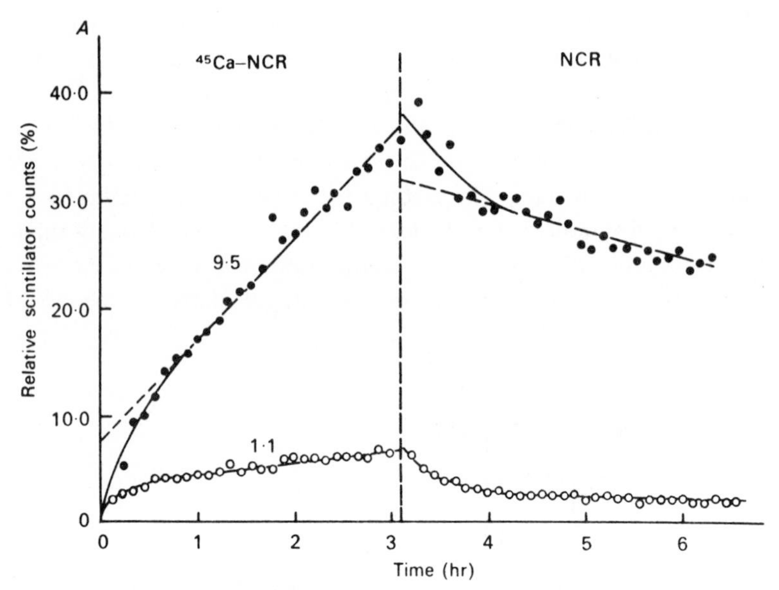

Figure 4 (a) The influx and efflux of ^{45}Ca in an uninjected muscle fiber (●) and an EGTA-injected fiber (○) (final EGTA concentration, 4.2 mmol/kg) as measured with the internal scintillator probe. The relative scintillator counts (Y_m/Y_s) are plotted against time, which can be converted to apparent influx (pmol cm^{-2} sec^{-1}) = $Y_m/Y_s \cdot M_s \cdot d \times 10^9)/4t$, where Y_m is count/unit time of probe in fiber compared to external solution (Y_s), M_s is the ionic concentration externally, d is the fiber diameter, and t is immersion time (sec). After 3 hr, the external saline (NCR containing 100 μCi ^{45}Ca/ml) was replaced by ^{45}Ca-free saline; numbers near the linear phases of the influx curves (as in subsequent figures) are values for Ca^{2+} influx (pmol cm^{-2} sec^{-1}), calculated from the slope; resting membrane potentials (E_m, mean): -50 mV (●) and -52 mV (○); the uninterrupted curves in this figure and others have been drawn by eye unless otherwise specified. (*b*) An enlarged version of the curve in (a) for the EGTA-injected fiber. The uninterrupted curve has been calculated from a two-parallel compartments model, in which the time constants are 0.046 and 0.0019 min^{-1}, and the corresponding compartment sizes are 3.7 and 9.8%, respectively (expressed in relative scintillator counts). The dashed line shows how extrapolation of the slow influx phase to zero time gives an estimate of the size of the rapidly exchanging compartment. (From ref. 12, 25).

using more conventional methods, for example, by simply sacrificing the fiber at the end of the experiment and assessing directly the radioactive calcium content. The extent of the *influx* assessed by this latter method was similar to the values of the residual *efflux* determined by axial microinjection of the isotope, that is, in the range of 1 to 2 pmol cm^{-2} sec^{-1} at 20°C.

Further experimentation revealed that if the calcium-binding agent EGTA (ethylene glycol *bis*-(β-amino-ethyl ether) *N,N'*-tetraacetic acid; as the potassium salt, pH 7.1) was initially injected into the cell, the apparent influx rate

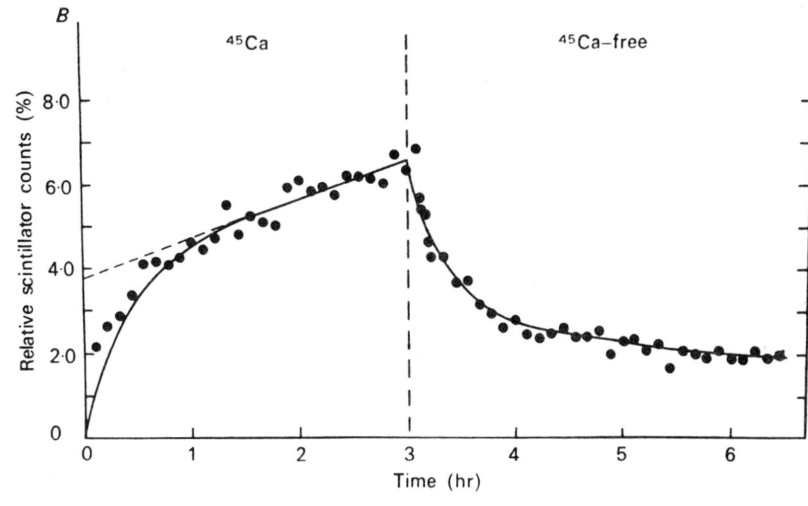

Figure 4 Continued.

assessed by the scintillator probe was reduced by some ten times to values similar to conventional efflux values, and to those resulting from total fiber assay (Fig. 4*b*). A number of reasons have been put forward to account for the effectiveness of this calcium-binding agent in reducing the apparent probe flux for ^{45}Ca, since with other isotopes, such as ^{14}C and ^{3}H, the probe gives influx values closely correlated to the conventional influx values as determined by total fiber assay. One likely explanation is that local damage to the extensive extracellular space of these fibers (the cleft system) occurs upon probe insertion, producing a locally high calcium influx rate into the SR lying within the detection range of the scintillator. This range, which depends upon the energy of the isotope, would be, for calcium (mean = 0.08 MeV), within an annulus of radius 45 μm around the surface of the probe. The presence of EGTA, however, would produce a more uniform distribution of the entering ^{45}Ca, so that when distributed throughout the cross section of the whole cell, the extra calcium entry would not significantly affect either the final apparent influx determined from the rate of the probe counts, or the ^{45}Ca content estimated by final liquid scintillation assay (see Table 1) (12).

It is also possible to assess the size of the extracellular space in these fibers from the probe experiments. Extrapolation of the linear phase of the probe counts to zero time, the time of application of the bulk phase ^{45}Ca, gave a value in the range of 4 to 10% of the volume of the fiber. This estimate is sim-

Table 1 ^{45}Ca **influx into single muscle fibers**[a]

		Calcium Influx [(pmol cm^{-2} sec^{-1}) ± S.E., 20°C]		
$(EGTA)_i$ (mM)	Number of Fibers	Probe	Liquid Scintillation Assay	Liquid Scintillation Assay With 5% Cleft Correction
0	5	13.8 ± 2.3	1.3 ± 0.3	0.6 ± 0.3
3.3–7.0 (mean 4.8)	6	1.31 ± 0.12	2.87 ± 0.41	1.53 ± 0.44

$(EGTA)_i$ = intracellular concentration of EGTA allowing for fiber dilution.
[a]Data from Ref: 12.

ilar to that determined either directly with the probe, by employing a nonpermeant solute, or conventionally, by using markers such as the radioactively labeled polysaccharide inulin.

When the fibers are immersed in ^{45}Ca containing saline, the initial phase (representing filling of the extracellular space) occurs in 60 to 90 min, with a $T_{1/2}$ of some 15 to 20 min. When the fibers were subsequently placed in isotope-free saline, the rate of the probe counts decreased rapidly in the first 60 to 90 min, and then more slowly; the rate constant for this latter phase was 1.5 to 1.9 $\times$ 10^{-3} min^{-1}, which was similar to that determined from the residual phase of efflux observed after axial injection of these fibers (14) (Fig. 4*b*).

The probe method has been used to tackle two basic problems in these fibers: (a) that of the nature of the calcium transport across the cell membrane at rest, and (b) the extent of calcium entry across the cell membrane during electrical stimulation. It is also possible to correlate the extent of calcium entry at rest, with the change in the free Ca_i^{2+} within the sarcoplasm of the cell determined by the photoprotein technique.

The result of such an experiment is shown in Figure 5*a*, where upon replacing the external sodium with choline, the calcium influx measured with the probe increases markedly. The influx rate from the probe increases from a resting value of 1.7 pmol cm^{-2} sec^{-1} to a value in the range of 6.2 to 7.1 pmol cm^{-2} sec^{-1} in $0Na_o$. In these fibers no contraction was observed, as would normally be expected, as they were injected with some 3 to 6 mmol/kg of the calcium-binding agent EGTA. The interesting observation is that the influx of calcium continues to increase linearly with time over periods as long

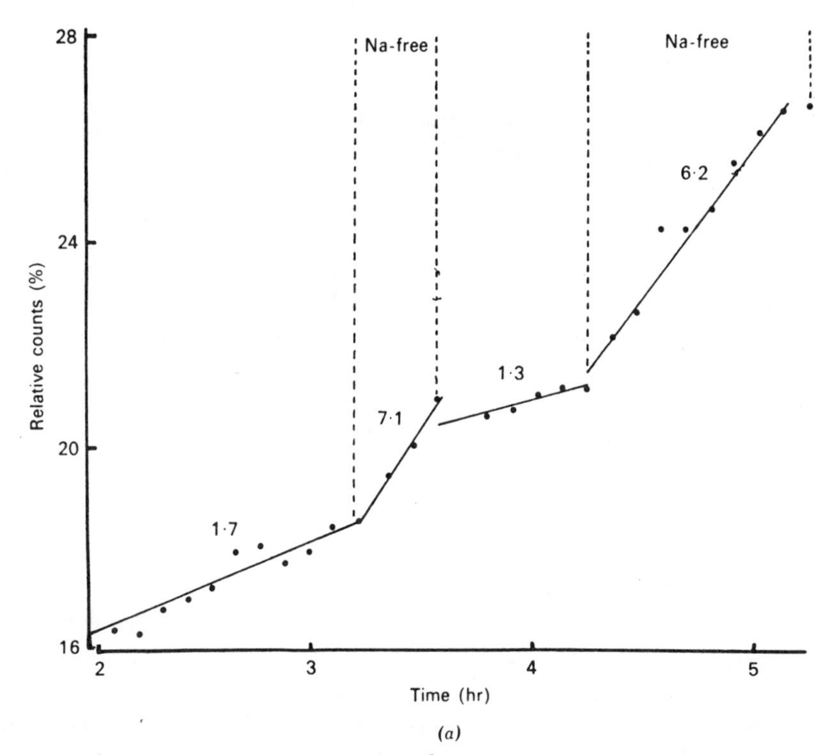

(*a*)

Figure 5 (a) The effect of sodium replaced [0Na (Li)] salines upon the time course of the relative scintillator counts for a single *Balanus* muscle fiber. The normal and 0Na salines were matched so that they had virtually identical specific activities. The mean fiber resting potential is −43.0 mV and the fiber diameter is 1.2 mm (from ref. 12, 25). The initial fast phase was omitted and the EGTA-injected fiber reached 6 mmol kg^{-1}. Values give the apparent influx values in pmol cm^{-2} sec^{-1}. The temperature is 20°C. (*b*) The effect of $0Na_o$ (Li) saline (↑) on the light emission from an aequorin-injected crustacean muscle fiber. The effect was reversible (↓) and was not observed on application of $0Na_o$, $0Ca_o$ (Li) salines. The vertical calibration is 20 nA; the horizontal is 10 sec; Ca_o in Figures 5*a* and *b* is 11.8 m*M*. (From ref. 9).

as 1 hr, with no sign of a decline or "fade." This result is to be contrasted with those from a similar experiment where the free calcium concentration is determined using the aequorin method (see discussion below). Here the free calcium concentration, as indicated by the aequorin light emission, rises to a new steady value within some 30 sec of the application of the $0Na_o$ saline (Fig. 5*b*), and is blocked by La^{3+} (1 m*M*) (12).

It seems that although the *total* calcium influx measured with the probe continues to increase, the *free* calcium is buffered and reacts readily to the

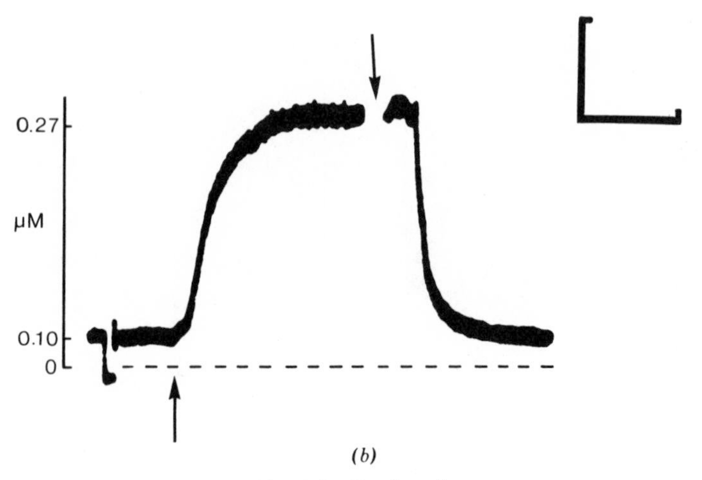

(*b*)

Figure 5 Continued.

imposed calcium load by rapidly achieving a new steady-state value. Calculations suggest that, under these conditions, the free calcium increases some 2.7 times, assuming that the relation between free calcium and aequorin light emission follows a 2.5 power relation (15, 6). The influx and efflux values measured with the probe method were higher than the values recorded with the intracellular dialysis technique (i.e., 0.11 to 0.38 pmol cm^{-2} sec^{-1}) (16, 17, 11) and these lower values may be attributable to dialysis artifacts (17).

The conclusions drawn from these experiments are that both the influx and efflux of Ca^{2+} from the cell, across the surface membrane, is sensitive to the external sodium concentration (Na_o), and hence to the electrochemical gradient for this ion.

The effects of electrical stimulation of the muscle fibers upon the total calcium influx are illustrated in Figure 6*a*. The fibers were initially injected with EGTA for the reasons already outlined, but at a concentration of at least 4 mmol kg^{-1}. In this case, the fiber produced all-or-none spike potentials as originally described by Hagiwara and Naka (18), whereas in the non-EGTA injected fiber these potentials were either graded or absent altogether. Calculations indicate that the *free* calcium concentration is reduced from ~200 to ~2 n*M*, assuming that all the fiber calcium is able to come into equilibration with the injected K_2EGTA which, after fiber dilution, is in the region of 20 m*M*. In addition, careful examination of the total fiber magnesium concentration suggests that this too will be appreciably altered by injected EGTA

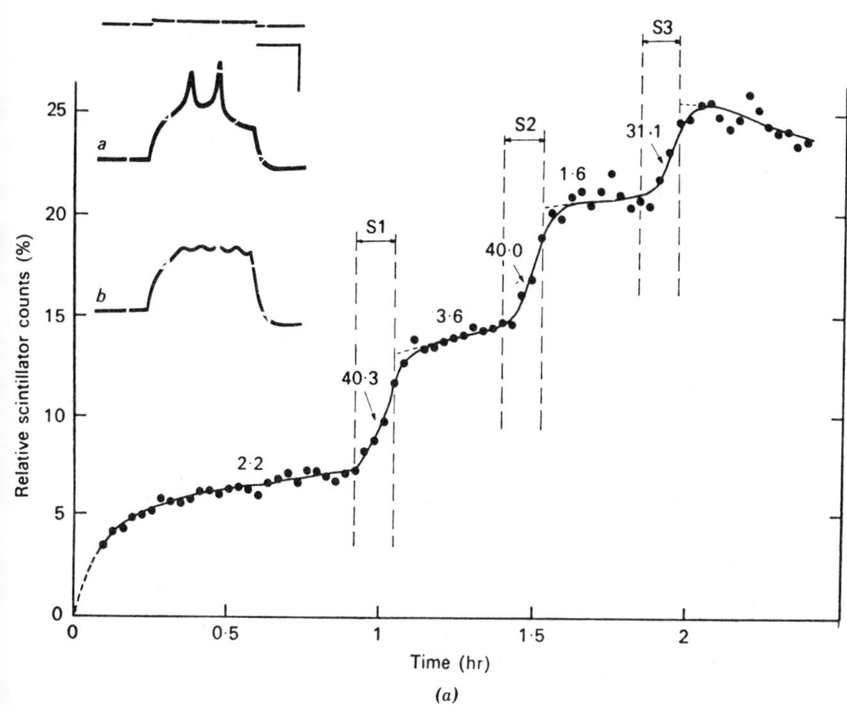

Figure 6 (a) The effects of membrane depolarization on the slow Ca^{2+} influx measured by a scintillator probe in a fiber injected with EGTA (final concentration 16.0 mmol kg^{-1}). During the periods marked S1, S2, and S3, the fiber was stimulated with constant amplitude, rectangular current pulses (*inset* top trace), 250 msec in duration and at a frequency of 1.0 sec^{-1}, to give depolarizing membrane responses shown in the inset *(a)* at the start and *(b)* at the finish of each period. Mean $E_m = -54$ mV. Inset calibrations for *(a)* and *(b)*: vertical, top 300 μA, 20 mV; horizontal, 100 msec. (12, 25, 144). Values give the apparent influx values in pmol cm^{-2} sec^{-1}. The temperature is 20°C. (*b*) Relation between (a) aequorin light (■); (b) isometric force (▲), and (c) inward membrane current (I_m) from Keynes et al. (141) as the result of a single voltage clamp pulse (holding potential −40 mV; temperature 11°C). Membrane current derived from a non-EGTA, TEA-perfused fiber, measured at 50 msec and considered to represent mainly I_{Ca}. (From unpublished observations).

(see ref. 6). The free magnesium concentration (Mg_i^{2+}), estimated in these fibers as ~5 mmol kg^{-1} (19), would be decreased to between 2 and 3 mmol kg^{-1} by the injection of EGTA. This alteration in free Mg^{2+} concentration would produce a change not only in the sensitivity of the pCa:tension curve of isolated myofibrils (20, 21) but also in the sensitivity of the photoprotein aequorin to calcium (6, 15, 22). Certainly, there is little information concern-

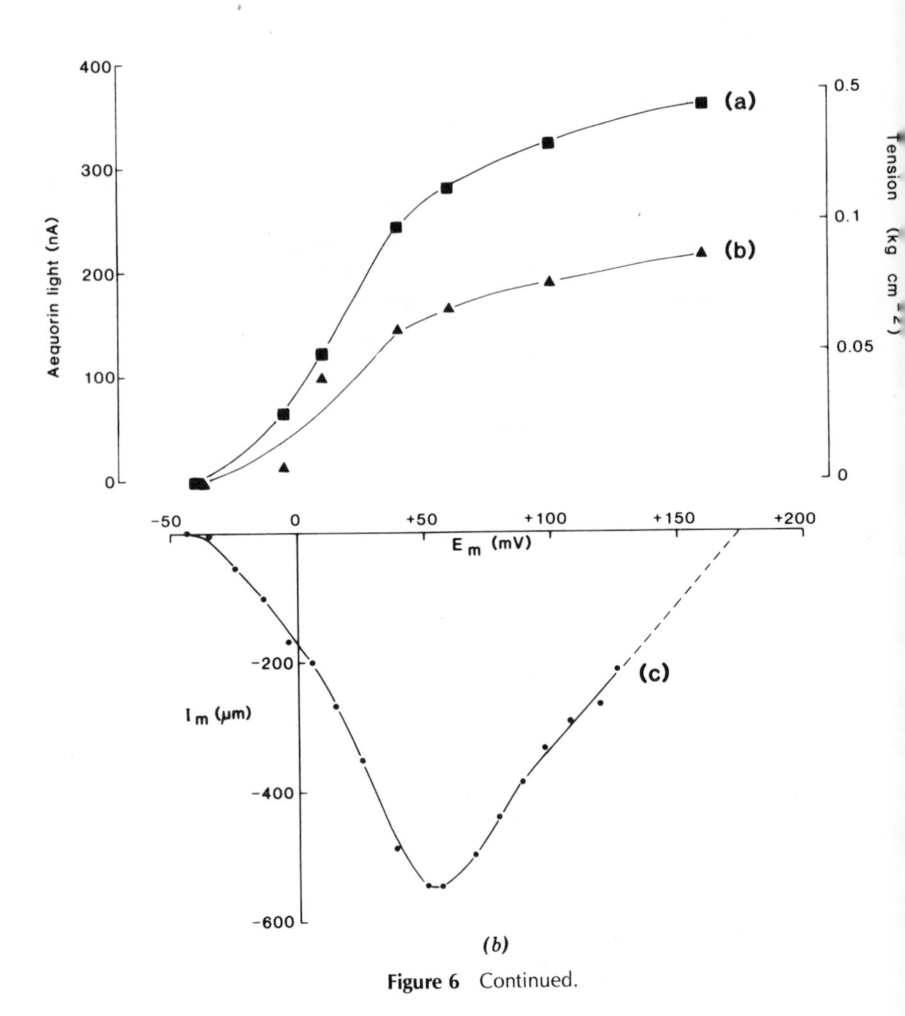

(b)

Figure 6 Continued.

ing the way that membrane conductance channels are influenced by changes in the internal concentration of free Mg^{2+}, as a result of chelator injection. Depolarization of the fiber for a period of 8 min, at a frequency of 1 sec^{-1}, with a pulse length of 250 msec, produced a marked increase in the probe count rate to a value some 20 times the resting value; this returned to resting value soon after the stimulation ceased. These effects were readily repeatable in the same fiber. Calculations indicate that, after making allowance for the resting influx rate, the extra calcium influx per pulse was 37.4 pmol cm^{-2} (Fig. 6*a*) and varied between 31 and 95 pmol cm^{-2} for similar duration pulses with a mean value of 55 $\pm$ 22 pmol cm^{-2}. This increase in influx was only observed when the action potential or graded electrical activity was observed, whereas subthreshold stimuli caused little or no increase in the influx. Even at lower concentrations of EGTA (in the range of 4 to 11 mmol kg^{-1}) where the estimated free Ca_i^{2+} was some 60 nM as a result of the injection, the fiber membrane produced only oscillatory membrane responses; but there was still an appreciable ^{45}Ca influx associated with such pulses. The range of Ca_i^{2+} values in which oscillatory membrane responses are obtained were different from those reported by Hagiwara and Nakajima (23), even when more recent values for the apparent affinity of calcium for EGTA were taken into consideration.

In addition, the pharmacological agent D600 (24, 146, 147), known to block the inward calcium currents in many muscle fiber preparations, particularly where inward calcium currents play an important part in the action potential mechanisms, abolished both the spike potential as well as the extra influx of calcium associated with electrical stimulation (25, 26).

In summary, it has been possible to use the probe method to assess the entry of calcium in an isolated single muscle fiber as a result of the removal of external sodium, and to compare this with the effectiveness of the internal buffering at regulating the free calcium ion concentration as a result of the imposed calcium load. This method has also given important information as to the extent of calcium entry during electrical excitation, so that an assessment can be made of the relative importance of surface membrane calcium movement compared to the amount of calcium required for activation of the contractile system derived solely from internal calcium compartments.

4 IMPORTANCE OF SURFACE MEMBRANE CALCIUM MOVEMENTS

The probe method discussed in the section above can be used to determine the concentration of calcium entering the cell as a result of electrical activation. The estimated influx values indicate that the total calcium within the cell increases to a level in the range of 1 to 4.1 μmol kg^{-1} per pulse (ΔV +50 mV). It is important to assess whether this amount of entering calcium would produce a significant tension response by interacting with the myofibrillar calcium-binding proteins. In these fibers, of course, tension development is considerably suppressed by the presence within the sarcoplasm of millimolar concentrations of the calcium-binding agent EGTA, so that the question is theoretical. Nevertheless, in the absence of EGTA within the cell, a depolarization of +50 mV and 226 msec duration would produce about 1 kg cm^{-2}, which is about 20% P_o, where P_o, the maximum achievable tension, is 4.9 kg cm^{-2} (27). The amount of force expected from this calcium (taking 4.1 μM per pulse) entering across the cell membrane would be 1.2% P_o (calculated assuming that at least both the calcium-specific binding sites on each troponin molecule need to be 100% occupied), or some 6% of the force developed at this level of membrane depolarization (28). This has been calculated on the basis of the model proposed by Ashley and Moisescu (29) (see also refs. 22 and 30), where two calciums are considered to bind independently to spatially different sites in the same functional unit for tension; the details are given later in this article.

This result strongly suggests that the amount of calcium entering the cell under these conditions is insufficient to activate directly the contractile system at the level required. A similar, if more qualitative conclusion has been reached from the experiments of Caputo and DiPolo (31) and of Ashley and Thomas (32), using voltage clamp methods. One of these experiments is illustrated in Figure 6*b*, where the effect of voltage clamp pulses of increasing intensity upon the amplitude of the aequorin light emission and the isometric tension response is shown. If the aequorin response and tension were strongly influenced and dependent upon transsarcolemmal flux of calcium, both the light response and tension, as a result of the transient voltage clamp step, should be strongly dependent upon the inward calcium current. Theoretically, the shape of this curve is such that, as the depolarization approaches the calcium equilibrium potential E_{Ca}, the inward calcium current declines to zero and beyond E_{Ca} would reverse. Experimentally, as E_{Ca} is approached there is certainly a decline in the inward calcium current, but no decline in the aequorin light response or tension. This result indicates that the aequorin light response represents internal calcium movements between compartments and is not dominated by transsarcolemmal fluxes. It also supports the results obtained with the scintillator probe, which indicate that upon electrical excitation the amount of calcium entering the cell from the outside represents only the minor fraction of the calcium required for tension development in this muscle fiber preparation. If, however, a total of 4 Ca^{2+} is required per functional unit, then the force developed is even smaller, only 0.06% P_o, calculated assuming all 4 Ca^{2+} sites are independent in their Ca^{2+} binding ability (33–37).

It is difficult to relate the change in surface membrane flux to events occurring within the muscle cell if the associated change in free Ca_i^{2+} is not known with any certainty. As the schematic diagram in Figure 7*a* indicates, the changes in calcium transport across the cell membrane are certainly reflected in changes in the free Ca_i^{2+} within the cell cytoplasm, but without a direct measure of this there must be considerable doubt about their magnitude and time course, dependent upon the complexity of the internal calcium sequestering systems. In many cells, the free Ca_i^{2+} response once measured, even with some degree of precision, is often difficult to interpret, since it is not always clear whether it represents the total calcium change (i.e., the system is weakly buffered) or simply a concentration change reflecting relatively large movements of calcium (i.e., the system is well buffered for calcium) (Fig. 7*b*). The extent of calcium buffering within muscle depends upon the presence not only of organelles, such as the SR, the mitochondria, and the calcium-binding components of the contractile proteins, but also of the diffusible calcium-binding components, such as calmodulins (2) and the parvalbumins (4, 35, 38, 66).

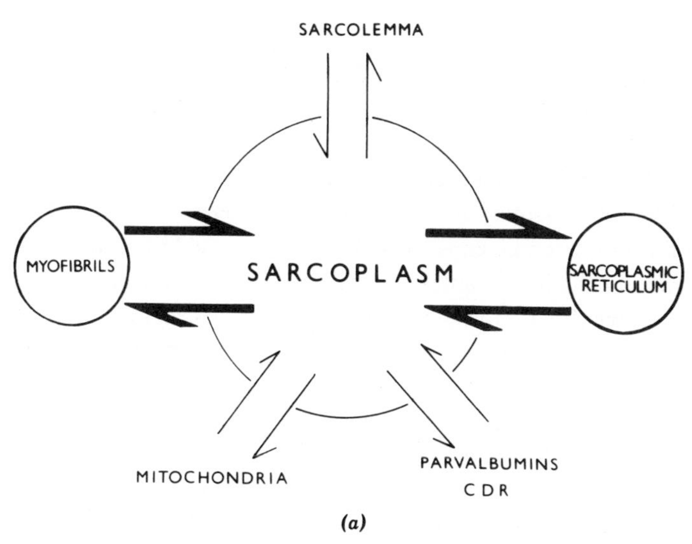

(a)

Figure 7 Schematic diagram of the flux pathways for calcium through a muscle cell. The heavy arrows represent the dominant pathways in muscles possessing a low content of mitochondria and parvalbumins (from unpublished observations). (*b*) Schematic diagram of the effect of strongly Ca^{2+}-buffered intracellular environment upon the free Ca_i^{2+} concentration change following an increased calcium flux. The intracellular indicator is located in compartment (2) and responds with a relatively small concentration change ($Ca_I^{2+} \rightarrow Ca_{II}^{2+}$) in free Ca^{2+} following a release of Ca^{2+} from compartment (1), as a result of effective and possibly cooperative Ca^{2+} buffering at the target site (3) (from ref. 6).

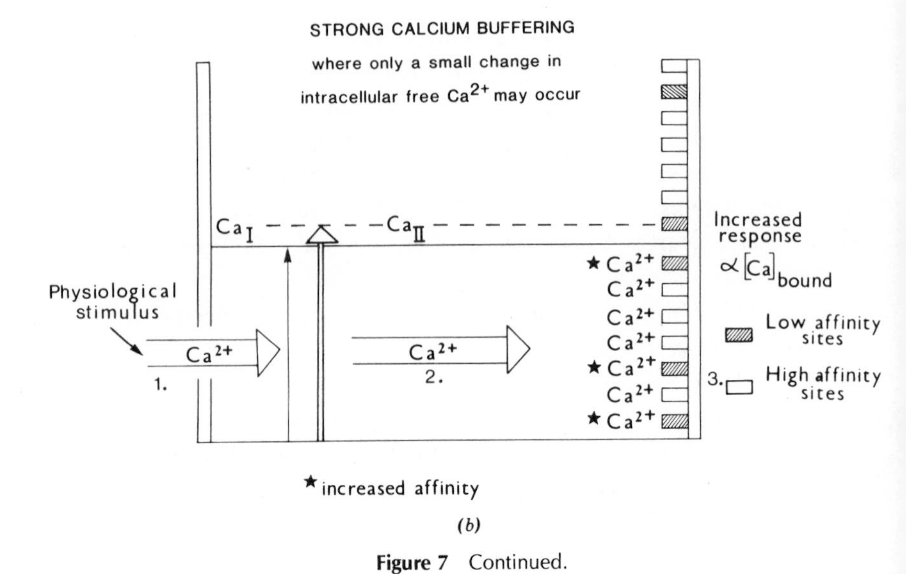

(*b*)

Figure 7 Continued.

5 DETECTION OF FREE CALCIUM CHANGES DURING ACTIVATION USING CALCIUM-SENSITIVE INDICATORS

Until the development of optical methods for measuring free calcium changes within cells, the microinjection experiments of Heilbrunn and Wierczinski (1) and the iontophoretic experiments of Niedergerke (39) indicated only that an elevation of the intracellular free calcium initiated a mechanical response from the muscle fibers. Little was known of the precise value of the free calcium concentration within the cell (Ca_i^{2+}) at rest, apart from the observation that "isolated vesicles of the sarcoplasmic reticulum, the 'erschlaffungs-grana' " were shown to be able to lower the free calcium within *in vitro* solutions to below 0.1 μM (40, 41). In addition, a technique using the properties of the highly specific calcium chelating molecules having high affinities for Ca^{2+} (Table 2) was used to determine the upper limits of the resting cell calcium (Ca_i^{2+}) (42) as well as to assess, in single fibers, the total calcium released during contraction (43). These methods gave neither information as to the time course of the relations between the intracellular free calcium change and the mechanical event, nor a value for the precise Ca_i^{2+} at rest.

Optical methods were also employed fairly widely to determine free Ca^{2+} changes in both intact cells and across the membranes of cell fractions, using

Table 2 Properties of the Most Commonly Used Ca^{2+} Indicators and Chelators[a]

	Murexide	Arsenazo III	Antipyrylazo III	Quin 2	EGTA	BAPTA	HDTA
Molecular weight	284	776	746	541	380	500	348
Solubility (mM)	>20	>50	>20	>100	>100 pH 7.0	>100 pH 7.0	>100 pH 7.0
Predominant stoichiometry	CaL	CaL_2	CaL_2	CaL	CaL	CaL	CaL
Approximate K'_D under physiological conditions	~3 mM	10^{-9}–$2\times 10^{-9}\ M^2$	$\sim 2\times 10^{-8}\ M^2$	$\sim 1\times 10^{-7}\ M$	$1\times 10^{-7}\ M$ $Mg^{2+}=0$	$1\times 10^{-7}\ M$ $Mg^{2+}=0$	$0.15\ M$ $Mg^{2+}=0$
Maximum value of $\Delta\epsilon$ per mole of indicator	6000	25,000	7000	30,000 (in UV) plus fivefold fluorescence increase	—	—	—
Suggested measurement wavelength pairs (nm)	540–507	660–690	670–690 or 720–790	520 fluorescence (340 excitation)	—	250–300 nm	—
K'_D of MgL complex under physiological conditions (mM)	125	7–10	~3	~2	~21 (pH 7.1)	~20 (pH 7.1)	125 (pH 7.1)
pH interference	Slight	Strong	Strong	Very slight	Strong	Slight	Slight
Relaxation time, 50 μM L	<2 μs	>2.8 ms	180 μs	~100 μs	$k_{on}\ 2\times 10^6\ \cdot M^{-1}s^{-1}$ (pH 7.0 25°C)	$k_{on}\ 10^8\ M^{-1}\cdot s^{-1}$ (unconfirmed)	—

[a]Data from references 50, 61, and 97.

the metallochromic indicator (dye), murexide (ammonium purpurate). Early uses examined Ca^{2+} transport in vesicles isolated from the sarcoplasmic reticulum (44), mitochondria (45), and a number of nonmuscle cells (see ref. 46 for review). It is important that the murexide technique was the first used to give some indication as to the temporal relations between free calcium and force in skeletal muscle fibers. These experiments used whole toad sartorius muscle which had been loaded with the indicator by external permeation (47). They indicated the relation between free calcium and force was not likely to be a simple one. More recently, murexide has been used to follow free Ca^{2+} changes within smooth muscle cells (48). However, the most successful use of metallochromic indicators has come from the development of dyes having a higher affinity than murexide, together with the use of time-shared, multiwavelength microspectrophotometry. One of the most widely used of these "second generation" dyes has been arsenazo III (ASIII) (49, 46, 50, and 7; Table 2). Initially, these dyes appeared to offer many advantages over the use of photoproteins for free Ca_i^{2+} detection (see discussion following), not the least being their ready availability. Unfortunately, however, there is still much uncertainty about their exact Ca^{2+} stoichiometry, the problems associated with possible binding to intracellular components (51), and their exact response time to rapid changes in free Ca^{2+}. In addition, they must be used inside cells at concentrations that undoubtedly cause appreciable buffering of the internal calcium (52, 53), and their presence would have to be considered in the quantitative assessment of relations between free Ca^{2+} changes and force development. This latter point may not be too serious a drawback in cells where the calcium buffering is high (see discussion following), such as is likely in muscle (29); where the calcium buffering is weak, however, it is certain that an appreciable amount of the *available* cell calcium will be complexed to the dye rather than to the natural target.

In motile cells there is an even more serious drawback to the use of the dyes, which became apparent in early murexide experiments. This is because there is a large light-scattering event associated with the mechanical response, which in itself may provide important information as to the molecular events underlying the contractile process. Nevertheless, it is usually much larger in magnitude than the wavelength specific event associated with the dye response, but from which it has to be clearly distinguished. To overcome some of these problems, the muscle fibers are stretched to a mean sarcomere length at which there is either no overlap between the myofilaments, or a considerably reduced amount, and consequently reduced force development (54, 55). Alternatively, the fibers are treated with hypertonic salines that reduce movement but do not apparently interfere with the process of calcium release. In invertebrate fibers, hypertonic salines do reduce the calcium release process, as indicated by the reduced size of the calcium transient response to an electrical pulse (56). This lack of response could not be readily attributed to a change in the responsiveness of aequorin, as the fibers still showed well-characterized effects when challenged with caffeine or with zero sodium salines (57). Unfortunately, changes in sarcomere length not only alter force but also affect Ca^{2+} release (58); thus, it is not an ideal parameter to use for this purpose.

More recently, the use of multiwavelength time-shared microspectrophotometry (Fig. 8) has made it possible to attain a correlation between the time course of the Ca^{2+} specific absorbance change and that of isometric tension, although an accurate assessment of the resting Ca_i^{2+} is still not readily achievable with the method. By careful positioning of the light path and adjustment of the mechanical arrangement of the fiber, the light-scattering event assessed at the isosbestic point of the dye can be reduced to a minimum, leav-

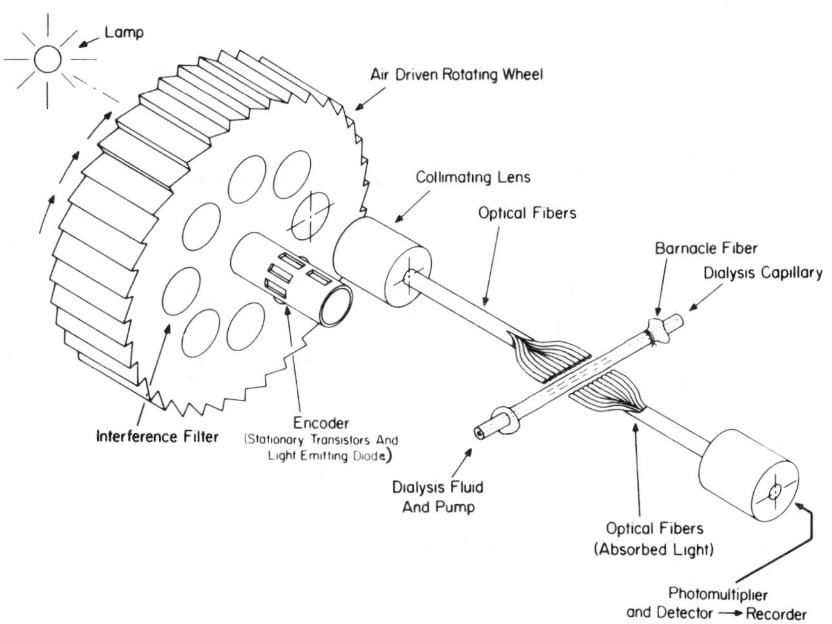

Figure 8 Diagram of a time-shared multiwavelength micro spectrophotometer. The eight-filter version allows the measurement of several wavelength pairs in a single large cell (from ref. 46).

ing the absorbance at the Ca^{2+} specific wavelength unaffected (59, 32). Both wavelengths can be determined within a few milliseconds of each other with the time-sharing method. A typical response from an ASIII-injected barnacle muscle fiber is illustrated in Figure 9*a*. The major point is that the time course of the absorbance change at this wavelength pair, which can be attributed mainly to the change in the free Ca_i^{2+}, largely precedes the time course of the isometric force response. The peak of the absorbance change leads the peak of the force response by some 200 msec. The decline in the calcium specific absorbance change is slower than the decline in the aequorin light transient (see discussion following), and is elevated above the baseline at a time when the force response is only 10 to 20% of its peak value. It is also clear that, in double pulse experiments, the smaller free Ca^{2+} event first observed with aequorin (60) following a second pulse is not directly attributable to an indicator utilization phenomenon (32,142; Fig. 9*b*). Either the release of calcium from the SR is less effective as a result of the second pulse, or the free calcium is better buffered by the cell.

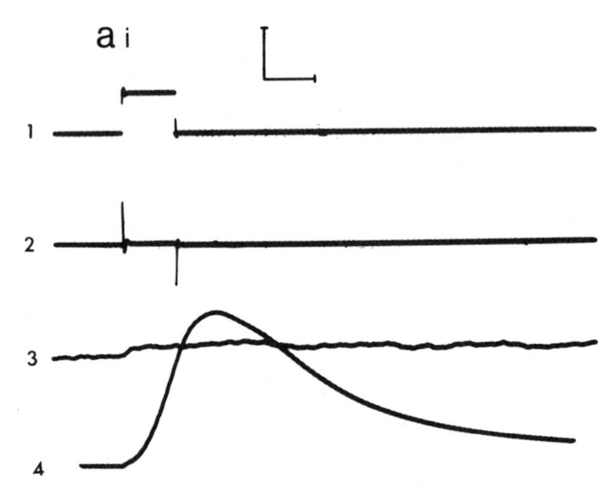

Figure 9 Result of voltage-clamp stimulation of a single barnacle muscle fiber microinjected with arsenazo III (ASIII). Trace 1, membrane voltage; trace 2, membrane current; trace 3, absorbance; trace 4, isometric force (temperature 16 to 17°C). (*a i*) Absorbance at isosbestic wavelength pair λ720 to λ750. (*a ii*) At wavelength pair λ660 to λ720 (mainly Ca^{2+}) (vertical: (1) 20 mV, (2) 0.5 mA, (3) 0.01 ΔA, (4) 0.014 kg/cm^2; horizontal: 200 msec; holding potential: −30 mV; ASIII concentration in fiber: 100 μM; pulse duration: 200 msec; $\Delta V = +15$ mV). (*b*) Double-pulse experiment in the same fiber at the same vertical gains shown in (a) (horizontal, 500 msec; temperature, 10 to 11°C; pulse length, 200 msec; $\Delta V = +18$ mV) (from ref. 32 and unpublished observations).

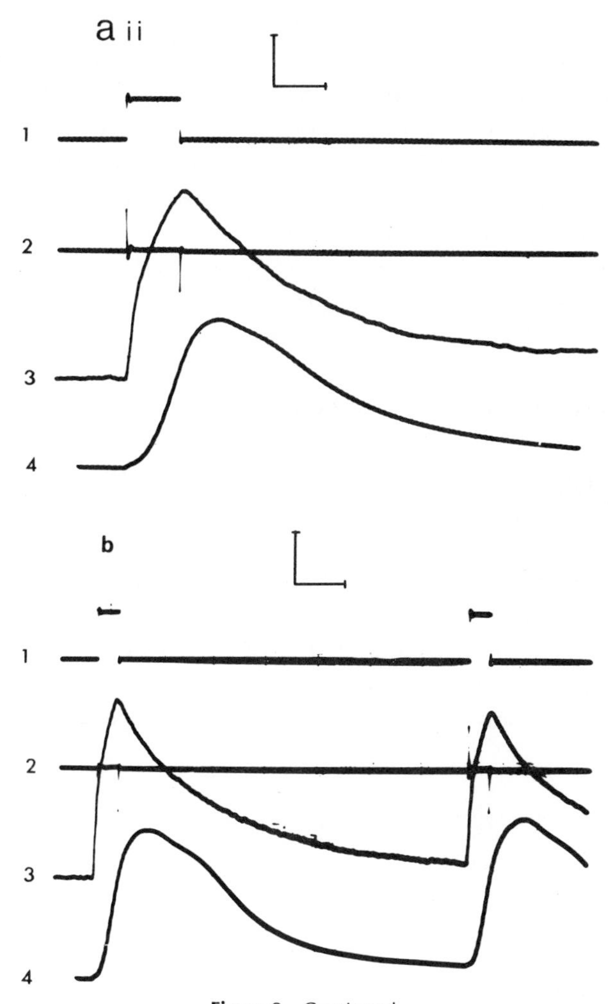

Figure 9 Continued.

6 USE OF CALCIUM-SENSITIVE PHOTOPROTEINS TO DETERMINE THE FREE Ca_i^{2+} CHANGE

As implied in the last section, the use of metallochromic indicators does not permit a ready detection and quantitative assessment of the resting Ca_i^{2+}. The newer generation of fluorescent indicators (such as Quin 2) have, however, been used to assess resting Ca_i^{2+}, (61, 50, 145), but only recently in muscle. With photoproteins, such as aequorin, particularly in large cells, the resting light emission after injection is easily measured and its calibration in terms of free Ca^{2+} is not too complex (62–64, 22, 6). This can be achieved by matching the resting light emission (with an *in vitro* system containing the same concentration of photoprotein) together with a solution having an ionic composition similar to that existing within a muscle cell (Fig. 10). The result indicates three important points: (1) that the resting free calcium concentration in these nonperfused muscle cells is close to 250 n*M*; (2) that the relation between aequorin light emission and free calcium concentration is a power function; and (3) that there is apparently a free Ca^{2+} below which the photoprotein system is insensitive to changes in free calcium concentration (64, 65). However, the resting free calcium concentration in these and frog muscle cells lies above the value at which the aequorin apparently becomes insensitive to free Ca^{2+} (about 5 n*M*) (6, 7); and over the range of free Ca^{2+}, where transient changes in calcium occur (~250 n*M*), the stoichiometry of light emission and free calcium concentration is constant.

By employing the photoprotein technique in conjunction with simultaneous calcium influx measurements, it has been possible to obtain information as to the efficiency of the calcium buffering system within the muscle cell. This type of experiment is discussed in Section 3, where the change in free calcium detected with aequorin, and the total calcium influx measured with the glass

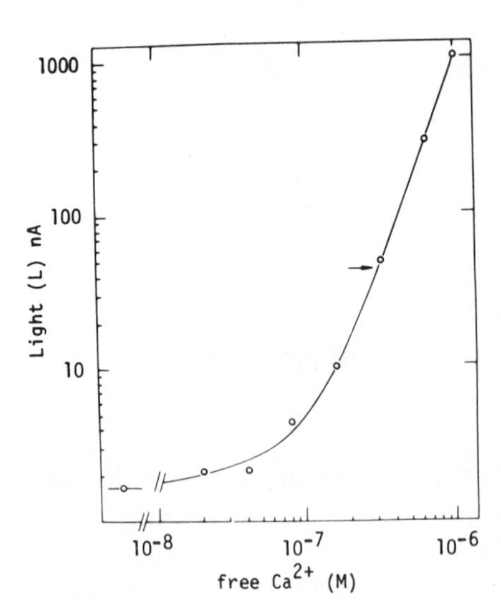

Figure 10. Relation between aequorin light emission and free Ca^{2+} in an assay solution of composition (V = 50 μl) similar to that of a muscle fiber (m*M*) 180 K^+, 20 Na^+, 4.2 free Mg^{2+}, 50 $EGTA_{total}$; each (○) represents the mean of two determinations. The arrow points to the mean value (n = 11) of resting light from a muscle that was fiber injected with the same concentration of aequorin (temperature, 20°C) (from ref. 6).

scintillator probe, are compared (Fig. 5*a* & *b*). It is clear that when the system is perturbed, in this case by reversing the $Na^+ - Ca^{2+}$ exchange system, the calcium influx continues steadily and the cell gains calcium linearly over a period of at least an hour. The *free* Ca^{2+}, on the other hand, rises rapidly to a new value and stabilizes at apparently a new steady state, which is achieved within 5 to 10 sec (25, 66).

Other interventions, which are able to perturb the steady state as reflected in the resting free Ca^{2+}, are the effects of acidification. In experiments where the effect of acidification upon the contractility of fibers was investigated (see later section), there was a marked increase in the resting free calcium concentration associated with an increase in the resting force of the muscle if permeant species such as CO_2 or phthalate were used (67, 78). This effect is illustrated in Figure 11*a*, and has been demonstrated in cardiac muscle, as well as directly and indirectly in other muscle and nonmuscle cells (68–70). The opposite effect, a decrease in the internal free calcium concentration, was observed upon application of ammonium chloride (Fig. 11*b*).

It was not clear from these experiments performed on intact muscle fibers whether the increase in free Ca^{2+} was brought about, particularly in the CO_2 experiments, by the internal change in sarcoplasmic pH (pH_i), which is of the order of 1 pH unit when the CO_2 concentration is 100% (71, 72), or by an effect upon an internal organelle. One way of investigating this problem was to use skinned muscle fibers (73). Here the surface membrane of the muscle cell is removed under mineral oil, and bundles of myofibrils and associated SR are exposed directly to a solution resembling the internal environment of the cell (20, 21, 74–77). It was possible, by increasing the concentration of bicarbonate, to maintain the pH of the medium surrounding the myofibrils at a constant level and, at the same time, increase the CO_2 concentration to 100%. In bundles of myofibrils exposed to this protocol, CO_2 increased the free calcium concentration of the medium, indicated by an increase in aequorin light (Fig. 12*a*) and produced a force response from the myofibrillar bundle preparation in crustacean (79) as well as frog skinned fibers (80).

The increase in free calcium appears to be the result of an effect of CO_2/HCO_3^- upon the SR, and is blocked by agents that inhibit the calcium-induced calcium release process, such as procaine (see Fig. 12*b*). This result, together with the finding that the increase in free calcium is also obstructed by weak bases, such as amantidine, has suggested that the mechanism may be a release process initiated by a change in the pH within the lumen of the SR: a proton-mediated calcium-release process (79, 80, 143). A proton-

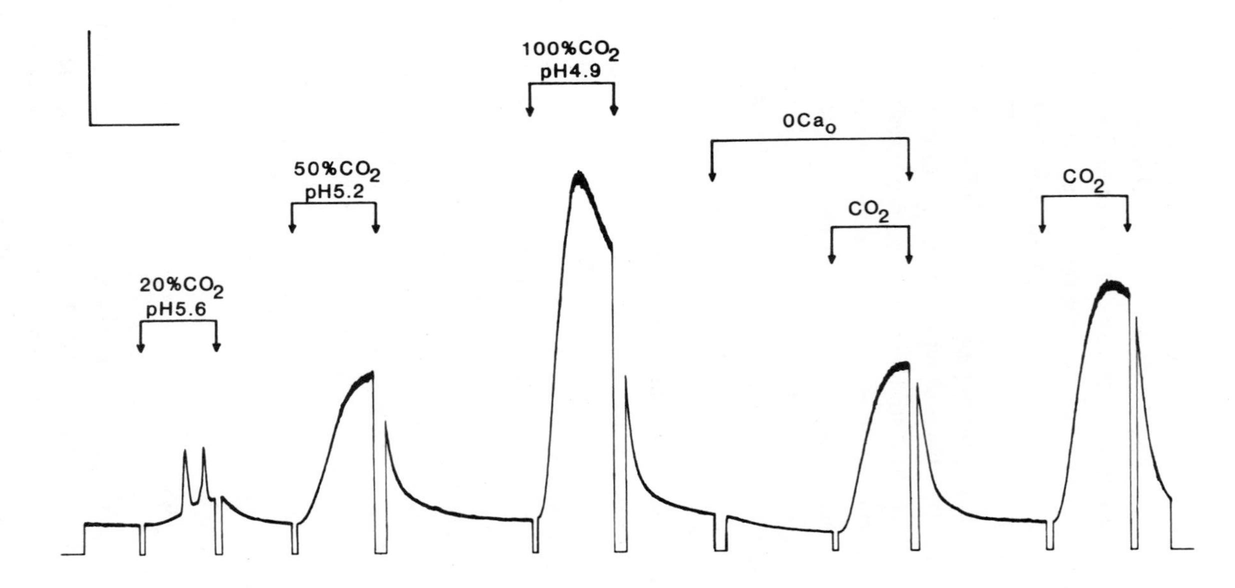

(a)

Figure 11 (*a*) The effect of externally applied salines containing different CO_2 concentrations upon the aequorin light emission from single *Balanus* muscle fiber. Here $0Ca_o$ salines reduce, but do not completely abolish, the CO_2 response. $E_m = -50$ mV. Photomultiplier switched off during saline changes. (Calibration: vertical, 100 nA; horizontal, 15 min.) (From ref. 78 and unpublished observations.) (*b*) The effect of externally applied salines containing CO_2 or NH_4Cl upon the aequorin light emission from a single barnacle muscle fiber. $E_m = -50$ mV. (Calibration: vertical, 200 nA; horizontal, 15 min.) (From unpublished observations.)

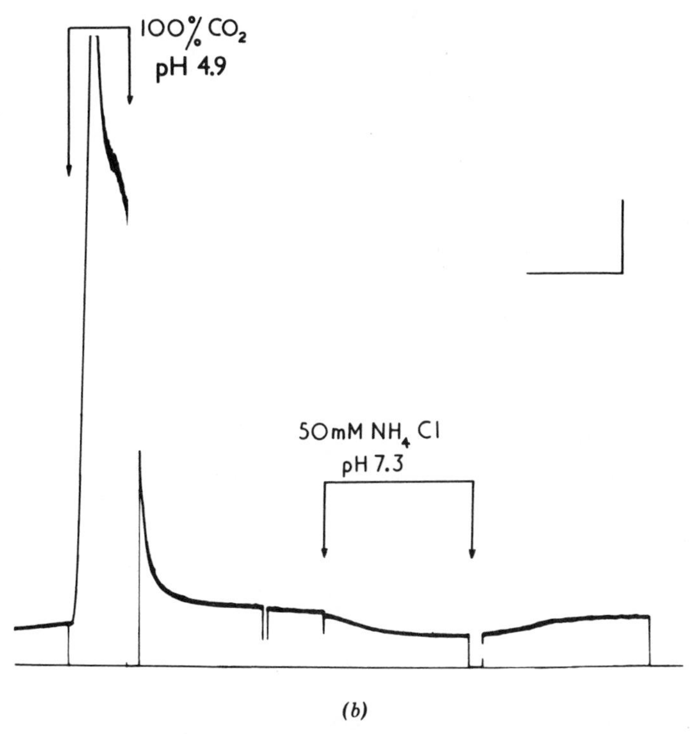

(b)

Figure 11 Continued.

ophore-induced (FCCP) release of calcium has also been reported from skinned rabbit skeletal muscle (81).

In summary, perturbations of the calcium transporting system, whether across the cell membrane (25, 9, 12), or involving a disturbance in the steady state of an intracellular organelle by inhibiting the uptake or stimulating the release of calcium, will be reflected by changes in the free Ca^{2+}. These ideas are summarized in Fig. 7*a* & *b*; the results emphasize the importance of the balance of calcium fluxes throughout the muscle cell and indicate that in the resting state the steady state free Ca^{2+} is finely controlled by the careful balancing of the resting fluxes. This latter point is indicated by the result in Figure 13, which demonstrates the effect of altering the external calcium concentration upon the resting free calcium concentration. It seems clear that, upon lowering the external calcium concentration, there is a decrease in the surface calcium flux, which is reflected in a decrease in the resting calcium

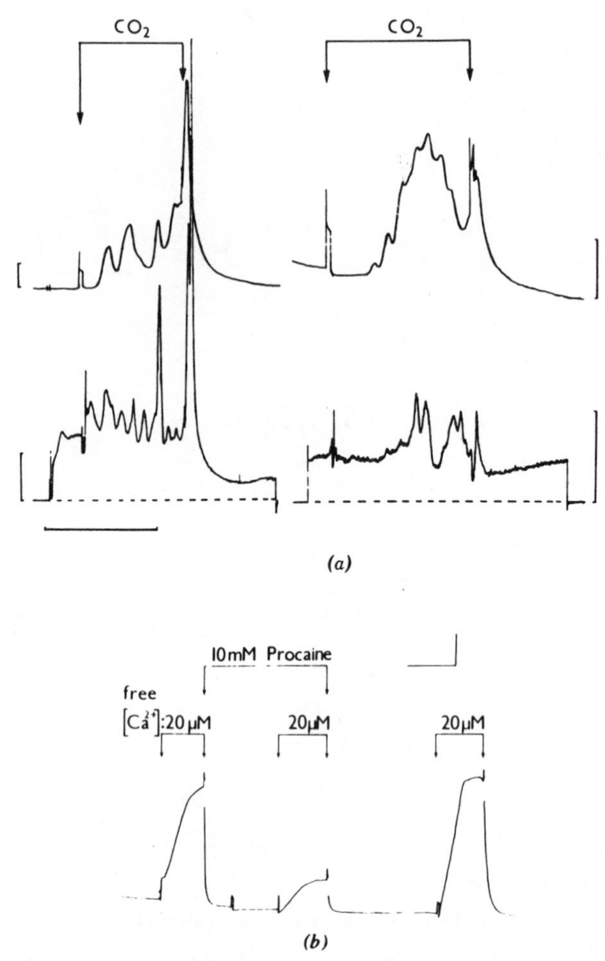

Figure 12 (a) Isometric tension (upper trace) and aequorin light emission (lower trace) from a crustacean (*Maia*) myofibrillar bundle preparation with intact SR. Note the change in calibrations after the first CO_2 challenge. (Calibration: vertical, upper trace, 0.1 mN; lower trace, 10 nA; horizontal, 5 min.) (From ref. 79.) (*b*) Ca^{2+}-induced calcium release from a myofibrillar bundle preparation with intact SR (*Balanus*) activated by 20 μM free Ca^{2+} and its inhibition by procaine. Independent experiments indicate that this concentration of procaine is ineffective upon the tension:pCa relation. (Calibration: vertical, 0.1 mN; horizontal, 4 min.) (From unpublished observations.)

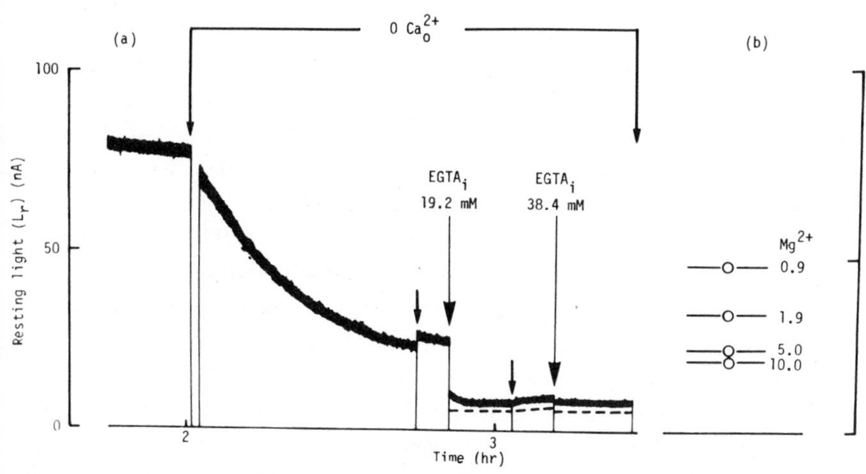

Figure 13 (a) Effect of 0Ca$_o^{2+}$ saline and intracellular injections of EGTA-K_2, pH 7.1 upon the aequorin-light emission from a single crustacean muscle fiber. Final EGTA concentrations after fiber dilution are given in the figure. (*b*) Results of an *in vitro* assay containing the same aequorin concentration at different free Mg^{2+} concentrations. The results in (*b*) suggest that *in vivo* (a), the EGTA-irreducible light is suppressed by some four times *in vivo* compared to *in vitro* (from ref. 6).

concentration. Under these conditions, the SR is presumably able to maintain a lower free Ca^{2+} in the sarcoplasm. Finally, it is not obvious as to what part calcium-binding proteins, present in the sarcoplasm of the muscle cell, play in the balance of the resting fluxes and in the determination of the resting free Ca_i^{2+}. This is partly complicated by the fact that some of the calcium-binding proteins can bind mainly calcium, whereas others bind both calcium and magnesium. These latter proteins, the parvalbumins (Table 3), at the free Mg^{2+} concentration present in muscle cells (19, 84–87) will be virtually satu-

Table 3 Kinetic parameters for Calcium-Binding Proteins[a]

	Ca^{2+}			Mg^{2+}		
	K_d (M)	k_{off} (sec^{-1})	k_{on} ($M^{-1}sec^{-1}$)	K_d (M)	k_{off} (sec^{-1})	k_{on} ($M^{-1}sec^{-1}$)
T-sites	$10^{-6.5}$	$10^{1.5}$	$10^{8.0}$	$10^{-2.3}$	$10^{2.3}$	$10^{4.6}$
P-sites	$10^{-8.0}$	$10^{0.0}$	$10^{8.0}$	$10^{-4.1}$	$10^{0.5}$	$10^{4.6}$

[a]Concentrations ($10^{-3}M$): T-sites—from troponin: 0.14; P-sites—from troponin: 0.14; P-sites—from parvalbumin: 0.70 (frog, carp), 1.20 (hake, pike), and 3.00 (swimbladder muscle). (For further details see ref. 35.)

rated with this ion. The displacement of magnesium upon elevation of the free Ca_i^{2+} is a relatively slow process [~1 sec^{-1} (88)], so that depending upon both the extent to which and the length of time that the free calcium is elevated above its resting value, there will be a gradual displacement of bound magnesium and sequestration of calcium to these diffusible calcium-binding proteins. The exact role for the parvalbumins is obscure, although a role in the process of relaxation has been postulated (35, 66). The four calcium-binding sites on the other class of proteins, the calmodulins, and which are competed for by magnesium ions, so that calcium, when elevated above its resting level in the cell, may need to displace Mg^{2+} from these sites. It is also not clear, as yet, whether a calmodulin molecule needs to have all four sites occupied by calcium to be in its activated conformation. The molecule exists in at least three conformational states, depending upon the extent of its saturation with Mg^{2+} (Fig. 14) and, hence, upon the Ca_i^{2+} to which it is exposed (89). The calcium saturated conformation is affected by the binding of anti-psychotic drugs of the phenothiazine type. In rabbit skeletal muscle, 35 to 40% of the calmodulin is thought to be bound as the subunit of phosphorylase kinase, and much of the remainder is likely to be bound to other calmodulin-dependent enzymes, such as myosin light chain kinase (90–93) and activation of these enzyme complexes by calcium will presumably occur by free calcium diffusing to the enzyme-calmodulin complex. Thus, the concentration of free calmodulin within the skeletal muscle cell is likely to be small.

In summary therefore, it seems that a large fraction of the muscle calmodulin is already bound to the receptor site, awaiting a change in the resting free Ca^{2+}. What fraction of these sites, if all are saturated, whether they are par-

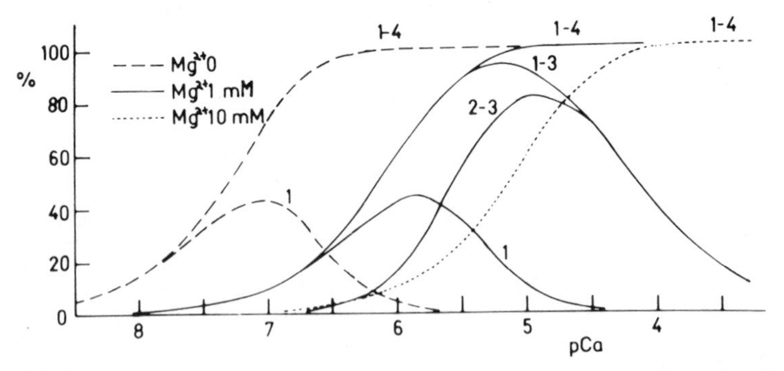

Figure 14 The variation in the Ca^{2+} complexed form of calmodulin (from CaL to Ca_4L form) at different free Mg^{2+} concentrations (from ref. 3).

tially occupied by magnesium, and whether all four need to be occupied by calcium, still need to be determined. Finally, the answer to many of these questions depends upon the value of the free calcium in the muscle cell, and the range of the transient changes in free calcium experienced when the muscle cell is activated.

7 TRANSIENT CHANGES IN FREE Ca^{2+} DETECTED WITH PHOTOPROTEINS

Although the potentially useful indicator properties of the photoprotein aequorin were reported by Johnson and Shimomura in 1962 (94), it was not until 1967 that this molecule was used successfully within a biological system, a striated muscle fiber, to detect a transient free Ca^{2+} response as a result of electrical activation. Of equal importance was the fact that it was also possible to relate this transient phenomenon to the force response, which was recorded simultaneously (95). The initial observations made with this method indicated that the relations between free calcium and force were not simple ones. Thus, the free calcium and force did not rise and fall concomitantly; there was an appreciable delay between the maximum of the aequorin light emission, and hence free calcium in the cell (see discussion following), and peak force output. This rather unexpected finding led to the immediate suggestion that free calcium changes within the muscle cell may be controlling the rate of force development, so that the reactions in which calcium was involved to produce force were rate controlling steps and were out of equilibrium with the free calcium concentration, at least during a short duration response (95, 96). That is to say, for relatively short periods of time after the onset of calcium release, and for relatively high calcium concentrations compared to resting values, the reactions in which calcium is involved to produce force can be considered unidirectional (29). Hence, the forward reaction step may be considered more important than the reverse reaction. This implies that the reactions involving calcium and resulting in force production equilibrate relatively slowly, so that the end product, force, lags behind the changes in the free calcium concentration (22, 29, 83).

The delay between the free calcium concentration change and force in a short duration transient is well-demonstrated in the experimental record shown in Figure 15*a*, as the result of a single voltage-clamp pulse applied to an isolated muscle fiber.

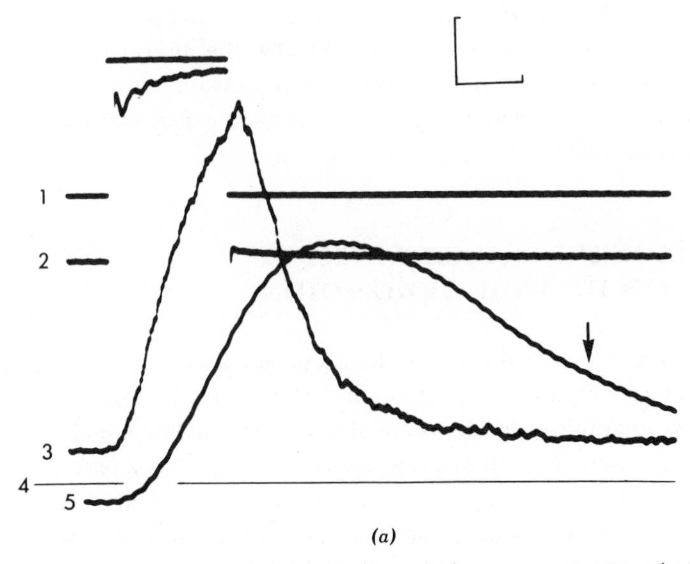

(a)

Figure 15 (a) Results of a single voltage-clamp pulse applied to an aequorin-injected crustacean muscle fiber. Trace 1, membrane voltage; trace 2, membrane current; trace 3, aequorin-light emission; trace 4, baseline for trace 3; trace 5, isometric force. Clamp pulse: 200 msec duration; $\Delta V = +45$ mV; holding potential $= -47$ mV; temperature, 18°C; time constant for trace 3, 10 msec. Gains: vertical (1) 20 mV; (2) 100 μA; (3) 50 nA; (4) 0.021 kg cm^{-2}; horizontal, 100 msec. The arrow indicates the light emission at time of half relaxation of force. (*b*) Deconvolution of the aequorin light response from (a). Trace 1, membrane voltage; trace 2, membrane current; trace 3, calculated free Ca^{2+}, from a deconvolution and normalization of the light response [$V(L)$] according to $Ca^n \rightarrow ACa \rightarrow L + dL/dt \cdot 1/k \rightarrow V + RC\ (dV/dt) + (d/kdt)\ [V + RC\ (dv/dt)]$, where RC = 10 msec, $1/k$ = 11 msec, and n = 2.8 (modified from refs. 29, 102, and unpublished results); trace 4, aequorin-light response [$V(L)$]; trace 5, indicates decay of calcium if exponential; trace 6, isometric force—$r(L)$ is resting light, $a(L)$ is zero light. Gains are the same as for (a). Arrows (1) and (2) indicate the half-time of the decay of light [$V(L)$] and calculated free Ca^{2+}. The arrow on the force trace indicates the position of half-relaxation of force. (c) Result of a short (I) and a long (II) voltage-clamp pulse applied to an aequorin injected crustacean muscle fiber where light was recorded at high and low gain. In both: trace 1, membrane current; trace 2, aequorin-light at high gain; trace 3, aequorin light at low gain; trace 4, isometric force. Solid horizontal lines illustrate the respective baseline for the two light records. Gains—vertical: (1) 50 μA, (2) 20 nA, (3) 50 nA, (4) 0.08 kg cm^{-2}, horizontal: (I) 100 msec, (II) 1 sec. (I): 200 msec pulse, $\Delta V = +20$ mV, (II) 5 sec pulse; $\Delta V = +12$ mV; holding potential, -53 mV; time constant of light trace, 10 msec. The transient pulse toward the end of II is artifactual. (From ref. 60 and unpublished results.)

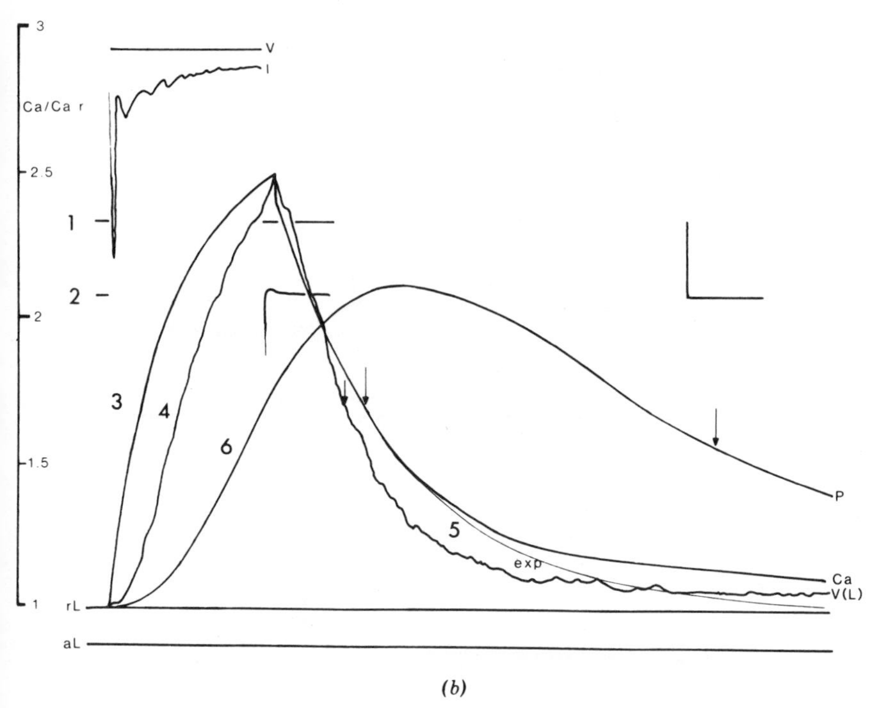

(b)

Figure 15 Continued.

Three main points arise from this record: (1) the comparison of the time course of the free calcium response with that of the force response which has already been mentioned; (2) the precise location of the calcium in the cell during the relaxation phase of the force response; and (3) the magnitude of the transient free Ca^{2+} event compared to the total number of calcium-binding sites in the cell, and in particular to those located upon the contractile proteins and associated with force development. In vertebrate striated muscle, these are thought to be mainly located upon the calcium-binding sites on the calcium-binding subunit of troponin, troponin-C.

If the reactions for force in which calcium is involved equilibrate slowly, at least for the time course of the pulses used in Figure 15*a*, it should be possible to prolong the free Ca^{2+} event in order to allow both the free calcium and force to come into a new equilibrium. This can certainly be achieved by prolonging the duration of the voltage-clamp pulse, and the result is illustrated in Figure 15*c*. In this case, force equilibrates in about 10 sec at this free Ca^{2+}.

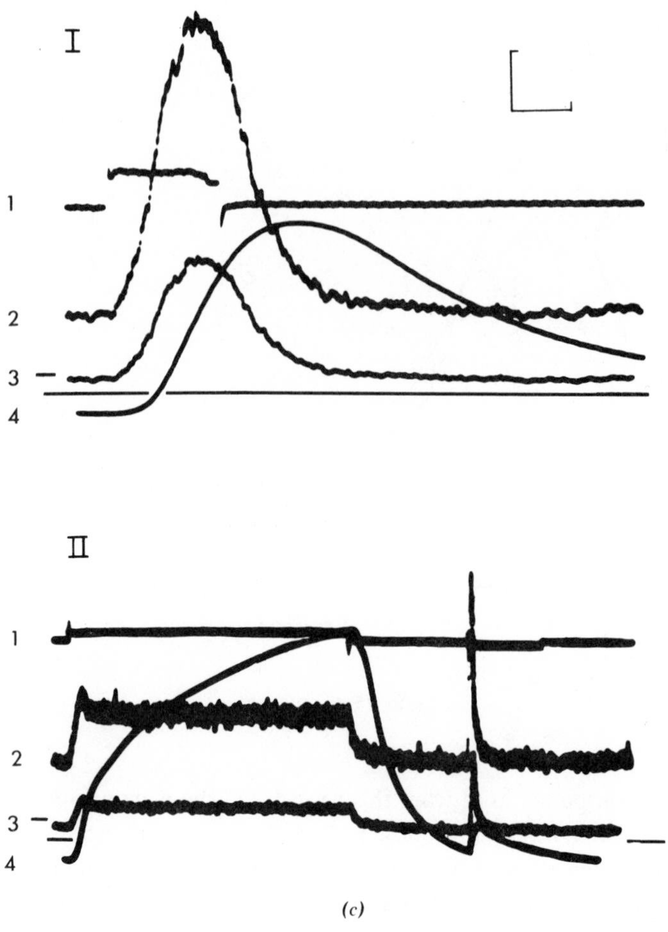

(c)

Figure 15 Continued.

It is noticeable that, at a low free calcium concentration, force develops relatively slowly, and yet eventually produces as much force as a clamp pulse of greater intensity but shorter duration. Similar responses can be observed in isolated bundles of myofibrils activated with the pCa clamp method (see Section 8; 30, 82, 83, 97). Experiments of this type support the notion that, at least for short duration pulses, the calcium-dependent reactions for force are not in equilibrium, and that calcium-dependent reactions are important steps in force development (22, 98, 99).

The second point concerns the falling phase of the calcium transient and, in particular, the relation between the free calcium concentration and force during the relaxation phase of the mechanical event. At first sight, it would appear that the free calcium concentration, as indexed by the aequorin light emission (Fig. 15*a*), has returned to its resting value at, or just before, the onset of relaxation. This finding might suggest that the free calcium concentration in the cell has returned to its resting value, possibly indicating that the calcium bound to the contractile proteins has returned to the SR. The process of relaxation would, therefore, involve processes other than calcium removal from the calcium-binding sites on the contractile filaments. However, if the relation between aequorin light and free calcium concentration is nonlinear, small but highly important changes in free calcium concentration may be neglected; since if these occur at values close to the resting free calcium concentration, they will have less effect on the light emission than a similar change in free calcium concentration at a more elevated calcium value (see Fig. 10). Therefore, to determine the free calcium concentration at any particular instant during the course of the transient response, and particularly during the later phases of the relaxation process, it is important that the light signal be deconvoluted (6, 22, 29). The details of this method are given later in this section. As a result of deconvolution of the aequorin light signal, it is evident that there is an appreciable and elevated free calcium concentration during the whole phase of force relaxation Fig. 15*b*. Ashley and Ridgway (56) proposed that calcium was still attached to the contractile system during the later phase of the relaxation process, and returned to the SR at a low, but elevated value and at high flux rates. This suggestion formed the basis of models developed by this group (29, 83, 101) and subsequently adopted by others (34, 35). The fact that deconvolution of high gain aequorin light records (Fig. 16*a*) obtained during the later phases of relaxation (60, 102) indicates an appreciable elevation of the free Ca^{2+} at this time, supports this notion. It would, of course, be possible that this is calcium initially chelated by soluble calcium-binding proteins, such as parvalbumins, and then returned to the SR during the later stages of relaxation (103). This is not likely in these particular crustacean fibers, as the concentration of parvalbumins and related molecules is very low (less than 60 mg/kg wet wt) (146). In addition, an elevation of the free calcium during the process of relaxation has also been observed in frog muscle fibers (104), but in this muscle millimolar concentrations of parvalbumins do exist (38) (Table 3). The fact that calcium is still attached to the contractile system, and either detaches fairly slowly or possibly recycles, is

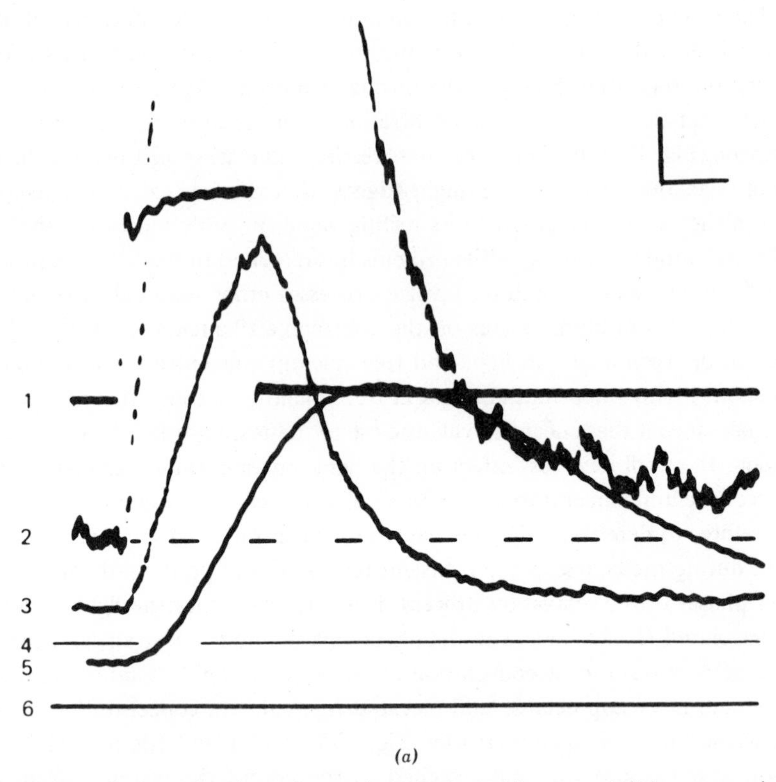

(a)

Figure 16 Result of a single voltage-clamp pulse (as in Fig. 15*a*) but aequorin light was recorded at both high and low gain to demonstrate the "tail" of the light response. Trace 1, membrane current; trace 2, high gain light signal, trace 3, low gain light signal; trace 4, baseline for trace 3; trace 5, isometric force; trace 6, baseline for trace 2. Pulse duration, 200 msec; $\Delta V = +45$ mV; holding potential = -47 mV; vertical gains as for Fig. 15*a*; horizontal, 200 msec; temperature, 16 to 18°C. (From ref. 60 and unpublished observations.) (*b*) Double pulse experiment using identical voltage-clamp stimuli applied to an aequorin injected crustacean muscle fiber. Aequorin light recorded at both high and low gain. Trace 1, membrane current; trace 2, high gain aequorin light response; trace 3, baseline for trace 2; trace 4, low gain aequorin light response; trace 5, baseline for trace 4; trace 6, isometric force. Both clamp pulses: 200 msec duration; $\Delta V = +28$ mV; spacing 600 msec, holding potential, -51 mV. Time constant for light = 10 msec. Vertical: (1) 50 μA; (2) 5 nA; (3) 50 nA; (5) 0.03 kg/cm^2; horizontal: 200 msec; temperature: 18°C (from ref. 105 and unpublished results).

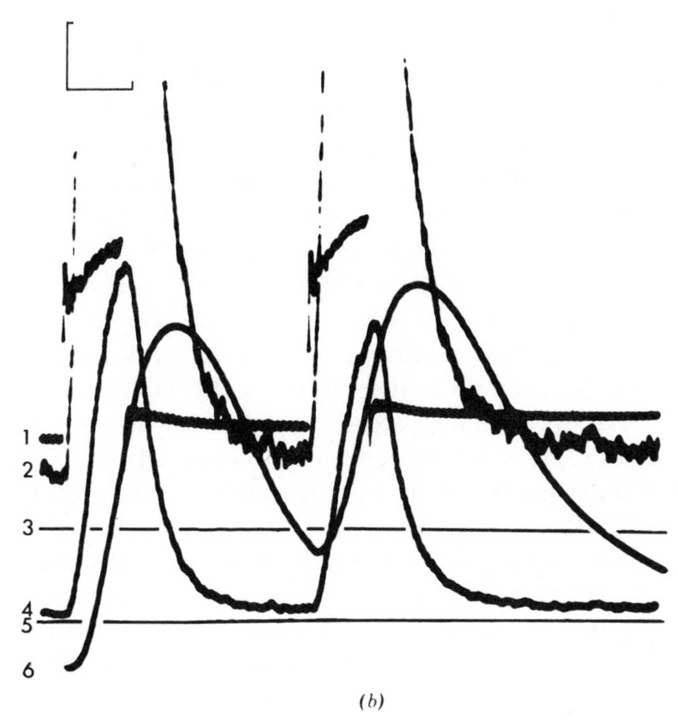

(*b*)

Figure 16 Continued.

suggested also by the experiment in Figure 16*b*. Here two identical voltage clamp pulses are applied to the same muscle fiber. The second tension response is potentiated, providing that the second pulse is applied at a time when the resting light is still elevated above its resting value (60, 105). The fact that the second calcium transient is also smaller emphasizes the potentiation of the force record. Its explanation could have three simple interpretations: (1) an aequorin utilization phenomenon (this is made less likely as a result of the experiment performed with nonutilized indicators, such as ASIII) (see Fig. 9*b*); (2) the process of calcium release shows some degree of refractoriness after the first pulse, up to periods of several seconds later; or (3) the buffering capacity of the cell has increased so that the flux of calcium is higher and the size of the free calcium ion transient may consequently be smaller. This latter explanation could be the case if already formed cross-bridges increase the affinity of the troponin for calcium, and even perhaps increase the effective number of troponin-binding sites (106).

The final point concerns the calibration of the calcium response in terms of the absolute free calcium ion concentration. Calibration of the aequorin light emission during a transient response using a square law relation (62) indicated that during tetanic stimulation the light increased some 100 to 200 times above the resting value, which itself was determined by a "null point" method (6). This indicated that the free calcium increased by some 10 to 11 times, so that the free Ca^{2+} at peak was in the region of 0.7 to 2.2 μM, and increased to a lesser extent when the degree of activation was smaller (see also ref. 56). Recent estimates of the free calcium concentration in these fibers at rest, based partly upon the more recent value for the free ionized magnesium concentration (19), place the resting calcium at closer to 250 nM. Also, the observation that the relation between free calcium and aequorin light emission is closer to a 2.5 power function than a square law relation (6, 15, 22, 65) implies that the transient Ca^{2+} changes are somewhat smaller. Nevertheless, the implications for the model are essentially unaltered, that is, the free calcium change during maximum activation is considerably less than the total troponin concentration in the muscle cell, which is estimated to be in the range of 70 to 100 μM in crustacean muscle fibers (107) and closer to 100 μM for vertebrate muscle fibers. A similar discrepancy between the free Ca^{2+} change and the total troponin is observed in frog skeletal muscle fibers where the change in free calcium concentration has been estimated to be about 2 μM at maximum (64). If all the troponin sites on the thin filament (either two or four per molecule) must be occupied with calcium for it to be fully activated, and the extent of cooperativity involves only the seven G actins and one tropomyosin (108), then at maximum force (P_o) either 140 μM or 400 μM calcium needs to be bound to the thin filament in crustacean muscle. This is certainly in the range of estimates determined from equilibrium binding experiments on glycerol extracted rabbit psoas muscle (37). The discrepancy between the change in *free* calcium observed with indicators [values from ASIII experiments also suggest μM concentrations at peak (52, 59)] and the *total* calcium change, where dye experiments suggest that between 100 and 220 μM total Ca^{2+} is released in a twitch (109, 53), a value similar to that estimated from EGTA injection experiments (43). The photoproteins are only, it seems, recording changes in free concentration, and the flux of calcium through the sarcoplasm is very much greater. In other words, the system is strongly calcium-buffered, by both the myofibrils and the SR (see Fig. 7*b*) (22, 29, 83, 101, 102; Section 8).

This flux model formed the basis of the calculations in which the transient free calcium response observed during activation was used to predict the time course of relative force recorded simultaneously (29), and also to reconstruct the free Ca^{2+} event (83).

To predict force quantitatively by the model, a precise knowledge of both the time course and magnitude of the transient free calcium change is required. The method adapted for the calibration of the magnitude of the free calcium change has already been discussed. The precise method for deconvolution of the light record, from which the time course of the free Ca^{2+} concentration change is derived, is based upon the kinetic scheme given below. This was determined from experiments in which the photoprotein was mixed rapidly (~1 msec), in stopped-flow experiments, with calcium *in vitro*, and where the rise time of the resulting light emission response was observed (110). This scheme has been modified to accommodate 3Ca (15, 65).

$$3\text{Ca} + \text{Aeq} \underset{k''_{-1}}{\overset{k''_1}{\rightleftharpoons}} 3\text{Ca} - \text{Aeq} \xrightarrow{k''_2} X \xrightarrow{k''_3} Y^* \xrightarrow{\text{fast}} Y + h\nu \qquad (1)$$

For simplicity, k''_1 and k''_{-1} are taken to be the same per site. The rise time of the light emission, which was independent of both the calcium and magnesium concentration at a pCa < 7, predicts an apparent rate constant of 100 $\sec^{-1}$ at 20°C (80 $\sec^{-1}$ at 10°C) and is virtually the value of k''_3 in eq. (1), since the first step is extremely rapid (110). The steady-state decay of light is relatively slow in comparison and is governed by the product of k''_2 (~1.3 $\sec^{-1}$ at 20°C) together with the relative concentration of the component (3Ca − Aeq). The correction based upon this relatively rapid rise time of the light emission results in a free calcium transient *in vivo*, which is translated in front of the aequorin light emission trace by some 12–13 msec at 10–12°C, the temperature of the experiments.

A calcium dependency of the rate of rise of light emission from aequorin has been reported (111), using rapid-mixing continuous flow techniques, but it predicts a rate constant that was not less than 100 $\sec^{-1}$ at 25°C. If this scheme is inferred together with a simple square-law relation (15, 62, 63), then the error in determining the Ca^{2+} within the muscle fiber is ±5 msec, assuming that, for the free calcium changes involved (some five times, see Fig. 17*a*) the rate of rise is not modified by more than two times (111). This produces only a relatively small uncertainty in the calculated relative tension response in these slowly responding striated muscle fibers. More recent analysis of photoprotein kinetics does not, however, support a calcium dependency of the rate of rise of light (112, 113).

To determine the kinetic relationship, it was important to know both the number of the slowly equilibrating calcium-dependent reactions that lead to force and the stoichiometry of these reactions as regards calcium. In these calculations a basic functional unit has been considered and, presumably, this partly involves the calcium-regulatory protein, troponin, located periodically along the actin filaments. The functional unit, in the absence of calcium, has a repressing action on force development and myofibrillar ATPase, but the binding of calcium in the reactions to be discussed derepresses its activity. The nature of the functional unit is not predicted from kinetic studies of this type, but it is all-or-nothing in its response. However, if the complete population of functional units is considered, their resultant activity must of course be graded, not all being activated simultaneously. Thus, only the active (derepressed) functional units contribute to the development of force.

It is a relatively straightforward procedure, based upon the overall reversibility of the calcium reactions, to determine both the number of slow reactions and the calcium stoichiometry of the functional unit directly from the *in vivo* experiments using aequorin (56). Analysis in the transient state of the rate of development of force (dP/dt) for different experimental traces, compared to the calculated value of the free calcium, indicated that two calciums were involved in the functional unit to produce force in these muscle fibers. A similar analysis to determine the number of slowly equilibrating reactions indicated that this was greater than one and was most probably two (21, 29, 97). This analysis was based upon a Ca–Aequorin stoichiometry of 2:1, but the more recent values of 3:1 (6) would automatically imply a greater calcium:functional unit stoichiometry for force (114).

There are three basic ways that two calciums can be combined with two slow reactions in the same functional unit. It is possible that the calciums are bound simultaneously in time in a single slow reaction, followed by the second slow reaction in which free calcium is not directly involved. The development of this kinetic scheme does not provide a particularly good reconstruction of either the recorded isometric force response based solely upon the free calcium event in the transient state, or the relation between ATPase and calcium bound in the steady state. The two kinetic schemes that provide the best construction of tension in the transient state, as well as accurate predictions of steady-state ATPase and calcium bound, both involve the cooperative interaction of two calciums. Here calcium is bound separately and at different points in time; there is, however, interaction between the two calcium-binding sites within the same functional unit to produce force. They both provide a good fit to the experimentally recorded responses under circumstances where they are mathematically identical. In one case, the *independent* scheme, the two sites (M_1, M_2) in the functional unit are considered to be independent in their calcium-binding ability, but only when both are occupied with calcium can the component ($CaM_1 \infty CaM_2$), which is proportional to the product of the concentrations of CaM_1 and CaM_2, give rise to force. This scheme can be expressed as

$$M_1^* \infty M_2^* + Ca \overset{\text{slow}}{\rightleftharpoons} \left\{ \begin{matrix} CaM_1 \infty M_2^* \\ \text{or} \\ CaM_2 \infty M_1^* \end{matrix} \right\}$$

$$\rightleftharpoons CaM_1 \infty M_2Ca \sim \text{force (ATPase)} \tag{2}$$

Here an asterisk represents an active calcium binding site. The concentration of $M_1^* \infty M_2^* + CaM_1 \infty M_2^* + CaM_2 \infty M_1^* + CaM_1 \infty M_2Ca$ is a constant and is the concentration of the functional unit. In the alternative scheme, a *consecutive* reaction, the second calcium can only be bound after the first and this is indicated as

$$M_1^* \infty M_2 + Ca \overset{\text{slow}}{\rightleftharpoons} CaM_1 \infty M_2^* + Ca$$

$$\overset{\text{slow}}{\rightleftharpoons} CaM_1 \infty M_2Ca \sim \text{force(ATPase)} \tag{3}$$

where $M_1^* \infty M_2 + CaM_1 \infty M_2^* + CaM_1 \infty M_2Ca$ is a constant and is the concentration of the functional unit.

If the *independent* scheme is developed in more detail for the case where calcium is varying in time, as is the situation in the muscle fiber experiments, then for a site i, where $i = 1, 2$ in this case

$$Ca + M_i \underset{k_{-i}}{\overset{k_i}{\rightleftharpoons}} CaM_i \tag{4}$$

where $M_i + CaM_i = C_i$, and C_i is the concentration of the calcium-binding sites that lead to force production. It can be shown that, in the transient state

$$\mathrm{CaM}_i(t) = C_i\left(1 - \exp k_{-i}[-K_i\mathrm{Ca}_r I(t) - t] \times \left\{k_{-i}\int_0^t \exp k_{-i}[K_i\mathrm{Ca}_r I(t')]\,dt' + (1 + K_i\mathrm{Ca}_r)^{-1}\right\}\right) \quad (5)$$

where $K_i = k_i/k_{-i}$; $I(t) = \int_0^t \mathrm{Ca}(t')/\mathrm{Ca}_r dt'$; and Ca_r is the resting fiber calcium. Therefore, from the model, force is given as

$$\mathrm{force(ATPase)} \sim \prod_i \mathrm{CaM}_i \quad (6)$$

and

$$\text{calcium bound} \sim \sum_i \mathrm{CaM}_i \quad (7)$$

The results from eqs. (4) to (6) are indicated in Figure 17*a*, where the value of $\mathrm{Ca}/\mathrm{Ca}_r$ was determined from the experimentally recorded aequorin light emission. The value of $K_i\mathrm{Ca}_r$ was here taken as 0.15, using the original resting calcium concentration values for these fibers, which range from 0.07 to 0.1 μM (62). The value of K_i was determined from steady-state force versus calcium concentration relations for this preparation (21), depending upon the free magnesium concentration. The value of k_{-i}, the reverse rate constant, was estimated from the falling phase of tension as about 1 sec^{-1}, and a value of 1.29 sec^{-1} was used at +10°C. Other details are mentioned in the legend. Although the equations for 2Ca per functional unit have been developed here, the case involving 3Ca is discussed elsewhere (6).

In more detailed calculations it has been possible to take the process of calcium diffusion, from the site of release to its site of binding on the functional unit, into consideration for the kinetic scheme. The myofibrils are considered to be a series of cylinders of radius *a* surrounded uniformly by SR. The radius of each myofibrillar cylinder was subdivided into a large number of intervals and the free calcium and tension calculated at each point. The results from these calculations are illustrated in Figure 17*b* & *c* compared to isometric force. Although 1 to 2 hr is left for aequorin diffusion radially, it was instructive to compare the results for the condition of either aequorin uniformly located within the myofibril or located only at its surface. In the calculations, a radius of 2 μm was employed, which is close to the observed values for these fibers. Incorporating the diffusion of calcium, with uptake and release, makes no significant difference to the calculated fit to force us-

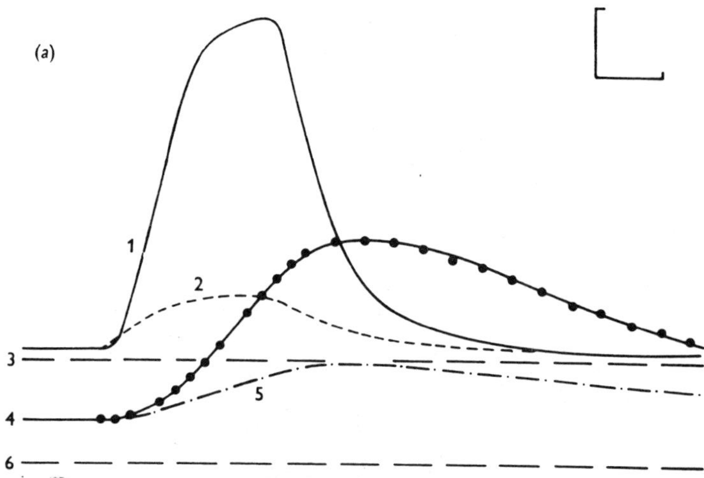

Figure 17 (a) Reconstruction of an isometric force response based upon the free Ca^{2+} change occurring in a single crustacean muscle fiber. Trace 1, aequorin light emission; trace 2, free Ca^{2+} calculated according to eq. (1); trace 3, baseline for light and free Ca^{2+}; trace 4, isometric force; trace 5, calculated Ca^{2+} bound to the force sites; trace 6, baseline for bound Ca^{2+} and force. The solid dots (●) represent force values calculated from model, Eqns 5 and 6 (see text). For eq. 2 and 4 $K_i = k_i/k_{-i} = K; K \times Ca_r^{2+} = 0.15; k_{-i} = 1.29\ sec^{-1}$. For eq. (3) ${}_1K = 2K, {}_2K = \frac{1}{2}K; K \times Ca_r^{2+} = 0.15$; ${}_{1-}k = \frac{1}{2}({}_{2-}k) = 1.29\ sec^{-1}$. For resting calcium Ca_r^{2+} is 0.7 to $1.0 \times 10^{-7} M$ and K is 1.5 to $2.0 \times 10^6\ M^{-1}$. ${}_1K$ and ${}_2K$ are the equilibrium constants for the first and second calcium binding steps; ${}_{1-}k$ and ${}_{2-}k$ are the respective reverse rate constants for these two steps for the *consecutive* reaction scheme. Horizontal: 100 msec. Vertical: (1) 100 nA; (2) $6.67 \times Ca_r^{2+}$; (4) 2.5 g wt (2.6% P_o); (5) 20% calcium bound (see text). Temperature: 10 to 12°C. (From refs. 22 and 29.) (*b*) Time course of recorded force (●) from (a). Solid line: the time course of relative isometric force calculated from the *independent* scheme (eq. 2) incorporating the process of calcium diffusion for aequorin *uniformly* distributed; Dashed line: calculated relative isometric force, aequorin considered at the *surface* of the myofibril (see text and ref. 29). The parameters for the force calculation are as in (a). $D\,[(\partial^2 Ca(r,t)/\partial r^2) + 1/r\,(\partial Ca(r,t)/\partial r)] = (\partial Ca(r,t)/\partial t) + k\,Ca(r,t)\,M(r,t) - k\,CaM\,(r,t)$, and where force $P(t)$ is given by $P(t) \sim \int_0^a CaM^2\,(r,t)r\,dr$, r is the distance from the axis of the myofibril, M is the concentration of free calcium binding sites and CaM the concentration of occupied calcium binding sites. The boundary condition is the following for $Ca(r,t)$: case (1) aequorin uniformly distributed within the cell, light intensity $I(t) = \int_0^a Ca^2\,(a,t)r\,dr$; case (2) aequorin located at surface of myofibril (only), light intensity $I(t) \sim Ca^2\,(a,t)$, where a, the radius of the myofibril, is 2 μm and D, the apparent diffusion coefficient for calcium, is $6 \times 10^{-6}\ cm^2\ sec^{-1}$ (from refs. 22 and 83). (c) Three-dimensional reconstruction of the free calcium transient across the diameter of the myofibril calculated for the condition of aequorin uniformly distributed [see (*b*)]. The complete transient is shown only at the surface of the myofibril. The lines joining the transients are at points of equal time and illustrate the effect of diffusion upon the time course and magnitude of the free calcium: *o*, axis of myofibril; *a*, surface of myofibril (from refs. 22 and 83).

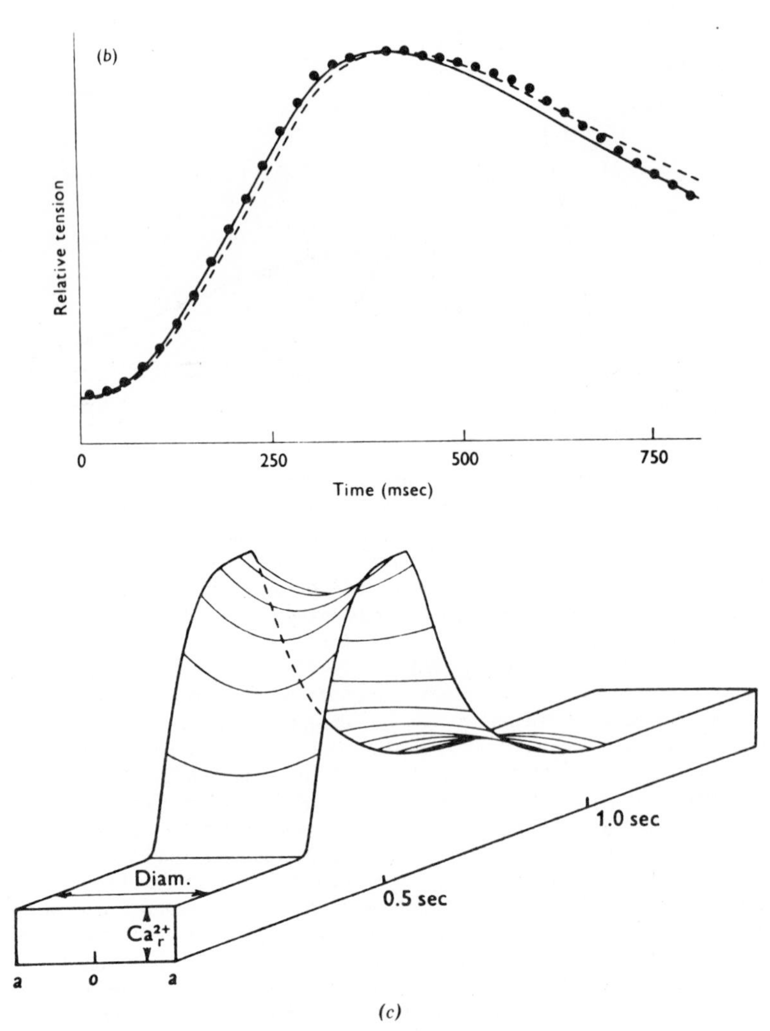

(c)

Figure 17 Continued.

ing the same constants as in Figure 17*a*, when the aequorin is considered uniformly distributed. Even if aequorin were located only on the outside of the myofibril, the change in the time course of force is very small and can readily be accommodated by a slight change in the reverse rate constant. Even if larger values of *a* are used in the calculations, the results are virtually the same. It seems, therefore, that the relatively slow velocities observed for the calcium reactions in this muscle cannot be readily attributed to a local limitation of the available Ca^{2+} (83).

It is also quite possible to assume that each calcium is initially rapidly bound by an electrostatic mechanism with a binding constant of 10^3 to 10^4 M^{-1}, followed by a relatively slow reaction with an equilibrium constant of 10^3 to 10^2, giving an overall binding constant of 10^6 to 10^7 M^{-1}. For the small range of calcium concentrations examined in these *in vivo* experiments, the kinetic model expressed by eqs. (2) and (5) is practically unaffected (29).

8 THE EFFECT OF CALCIUM ON THE CONTRACTILITY OF MYOFIBRILS

A method has been developed by which the contractile proteins can be exposed directly to activating solutions while retaining their basic geometric array. This has enabled physiologists to investigate more directly the influences of ions upon the contractile system, as well as upon the stages in the excitation-contraction (E-C) coupling process, free from the influence of the surface membrane (73, 74, 76, 115–120). In many cases the SR can also be retained in an intact and functional state (121).

Manual mechanical removal of the cell membrane under mineral oil, as first described by Natori (73), is the most effective method if the surface membrane is not required. Other methods available for the complete or partial removal of the surface membrane include treatment with glycerol as well as chemical skinning with agents such as the detergents Brij, Triton, deoxycholate (DOC), and saponin.

Chemical treatment may involve simply making the surface membrane more permeant to ions and solutes but retaining the overall structure, as in glycerol extraction methods. This procedure has been used extensively, and often for systems where the availability of fresh muscle is poor, as it enables material to be kept in a functioning state for several months. Detergents such as Brij (122) will remove both the surface and SR membranes; saponin is more selective, as it renders the surface membrane relatively more permeant to ions than the membranes of the SR that remain functional (123). Large muscle cells are certainly amenable to the mechanical skinning methods, whereas for smaller cells, such as those from vertebrate smooth muscle, saponin has been used (124). Alternatively, mechanical separation of the cells by homogenization, followed by manual skinning, has also been successfully applied (119, 125).

With skinned muscle fibers or bundles of myofibrils isolated from skinned fibers, not only are the qualitative responses of the contractile proteins to imposed ionic changes detectable either by the naked eye or under the microscope (73, 126), but it is also possible to produce a quantitative record of these responses either by recording the extent of shortening or, if external shortening is prevented, by force production. Attaching the bundles of myofibrils to the recording apparatus is certainly made easier if the original tendons of the intact fibers are used, that is, if the individual single fiber is first mounted *in situ* and connected to the recording system, and then has its surface membrane removed by mechanical skinning. If the skinning occurs beforehand, the preparation has to be connected either by clamping, tying, or glueing the myofibrils to the recording system, which may involve some damage.

One of the puzzling observations noted when skinned frog fibers were activated by calcium was the relatively slow development of isometric force (74) even when the calcium concentrations used were in the range encountered in intact cells (56, 62, 127, 128). In examining the possible reasons for this effect, one must take into account those components still functionally intact, existing within the myofibrillar preparation, and capable of accumulating calcium from the intermyofilament space. These components have already been discussed (see Fig. 7*a*). Hellam and Podolsky (74) were the first to examine the influence of the SR upon the rate of development of tension in frog skinned preparation when challenged with an activating solution containing a high free Ca^{2+} concentration (that is a low pCa value, $pCa = -\log[Ca^{2+}]$). The rate of development of force, in preparations where the elements of SR remained intact, was compared with those that had subsequently been treated with detergent to remove, or at least inactivate, the SR and mitochondria, both of which are capable of accumulating calcium ions. In these experiments, although the half-time for force development was considerably decreased compared to the situation where the SR was still functioning, it was still much slower than the rates of force development observed *in vivo*. Endo (122) observed a similar slow rate of force development, and indicated that this did not appear to be related to the changes in fiber volume, and hence to the accompanying changes in lattice spacing that occurred upon skinning (129, 130). Perhaps in the process of skinning, geometrical or biochemical changes occur in the preparation that no longer permit force development to happen as rapidly or as fully as that encountered *in vivo*; diffusible factors such as the calmodulins or the enzymes involved in myosin light chain phosphorylation processes may well be lost in the skinning and subsequent equilibration in aqueous media (147). Thus, a demonstration that the skinned fiber preparation could react rapidly to changes in free Ca^{2+} was important if this preparation was to be used as a reliable guide to processes that occur within intact fibers upon activation. Previous authors employed the same concentration of $EGTA_{total}$ in the relaxing as in the activating solution, and this was in the range of 1 to 5 m*M* (74, 131). Endo (122) observed an increase in the rate of force development in skinned frog fibers when the total concentration of EGTA and, hence, the Ca–EGTA complex in the activating solution, was increased although the total EGTA concentration in the relaxing solution was also increased. The rate of force development observed under these circumstances was as fast as that witnessed when the calcium accumulating ability of the SR was reduced or destroyed by the presence of detergent (Brij-58, 0.5% w/v). Nevertheless, the rate was still far slower than that observed *in vivo*.

Ford and Podolsky (118) had also observed that, in myofibrillar preparations where the SR was intact, the delay in force development was dependent upon the concentration of the Ca–EGTA complex in the activating solution. The detailed interpretation of the force transients was complicated by the presence of the SR and of the process of regenerative release of Ca^{2+}, but the delay was attributed at least partly to the time taken to fill the elements of the SR with calcium.

After analyzing the way free calcium changes can affect isometric force in intact muscle fibers (29), it was apparent that the presence within the skinned fiber bundle of both diffusible (EGTA) and fixed calcium-binding sites (such as troponin) could well modify the rate at which the externally applied calcium buffer system reached equilibrium (Fig. 18*a*). One way to test this idea was to alter the protocol by which activation of the myofibrillar bundle was achieved in the absence of an effective SR (83, 101). The main alteration was to reduce the concentration of free EGTA in the relaxing solution to a value some 200 times smaller than the total EGTA concentration in the activating solution, while maintaining the same ionic strength in the two solutions by ionic replacement and removing or inactivating the SR. Now the low concentrations of EGTA in the relaxing solution present within the preparation should not appreciably disturb the rate at which the externally applied calcium buffer system achieves its equilibrium as it diffuses into the bundle.

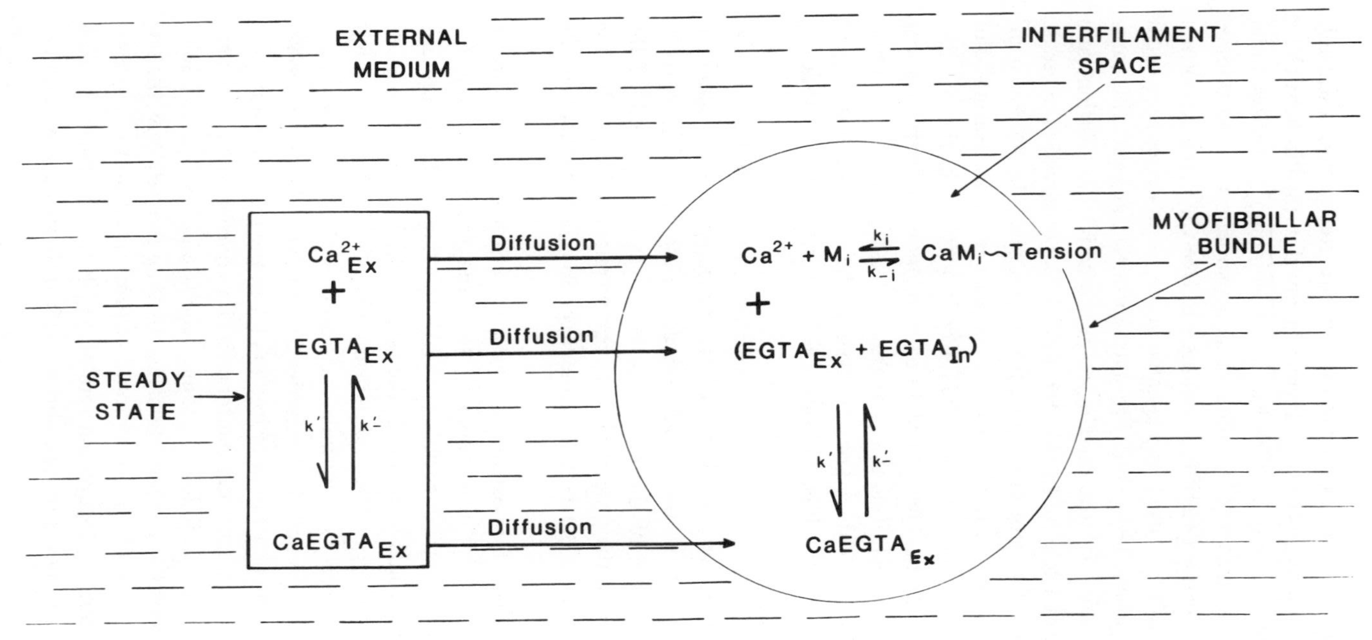

(a)

Figure 18 (a) Schematic diagram for the activation of myofibrillar bundles with EGTA. The concentration of Ca–EGTA within the bundle before activation is considered negligibly small in this scheme (unpublished observations). (*b*) Estimate of the free Ca^{2+} change at axis when $r = 0$ as a function of time in a myofibrillar bundle (dia. 175 μm). In $L \rightarrow L$ (Ca–EGTA + EGTA) = 3 m*M* in relaxing and activating solutions; in $L \longrightarrow H$ it is 0.1 m*M* in the relaxing and 20 m*M* in the activating. Free Ca^{2+} concentration in activating solution $8.7 \times 10^{-8}M$. Apparent diffusion coefficient for EGTA = 4.6×10^{-6} cm^2 sec^{-1}. Based upon Crank (132); no account taken here of the kinetics of Ca^{2+} binding to and dissociation from EGTA. Conditions are similar to Figure 20*b* (unpublished observations).

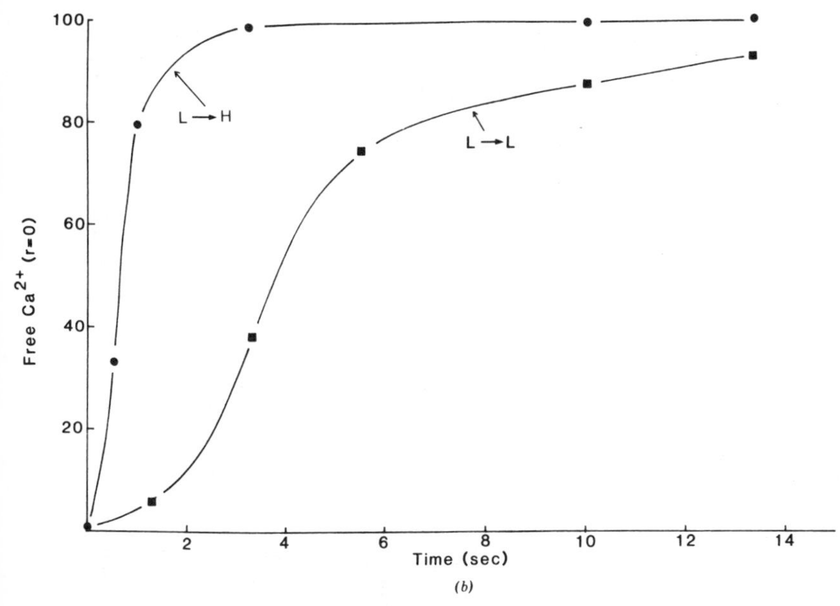

Figure 18 Continued.

Experimentally, the results of this activation protocol are illustrated in Figure 19 for a skinned frog muscle fiber. When the preparation is activated by solutions containing the same concentration of EGTA as in the relaxing solution, the rate of tension development is slow (Fig. 19*a*). This was increased by including in both the relaxing and activating solutions, DOC, to inactivate the SR (Fig. 19*b*). If, however, the newer activation protocol is employed to reduce the free EGTA in the relaxing solution without a significant change in the ionic strength between the relaxing and activating solutions, the rate of tension development is now considerably faster (Fig. 19*c*). In crustacean myofibrillar bundles the rate approaches that observed *in vivo* (83, 101) (Fig. 19*d*); similar responses were observed with frog (97, 98). The kinetic situation during activation can be approximated to the problem of the diffusion of a molecular species into an empty cylinder (see ref. 132), of radius r, containing a finite concentration of material (C_1). The predicted time course of the free calcium change as a result of two different types of activation procedure is illustrated in Fig. 18*b* for a skinned fiber bundle of radius 85 μm, employing an experimentally determined value for the apparent diffusion coefficient for EGTA of 4.0–4.6 $\times$ 10^{-6} cm^2 sec^{-1} for relaxed myofibrils (30, 98). The

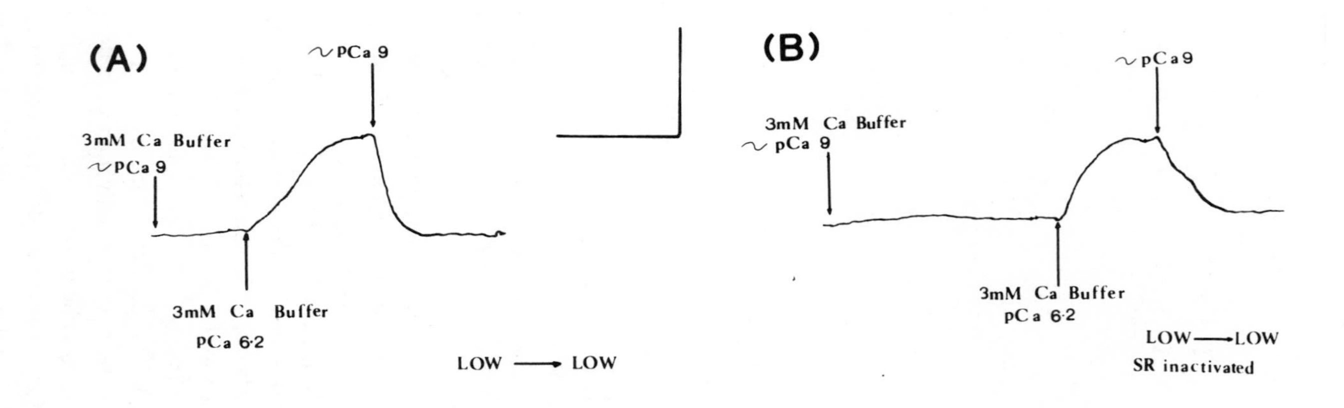
(A)
~PCa 9
3mM Ca Buffer
~PCa 9
3mM Ca Buffer
PCa 6·2
LOW ⟶ LOW
(B)
~pCa 9
3mM Ca Buffer
~pCa 9
3mM Ca Buffer
pCa 6·2
LOW ⟶ LOW
SR inactivated

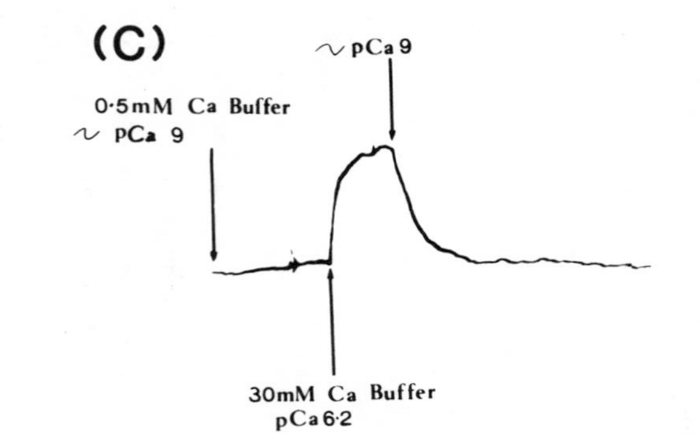
(C)
~pCa 9
0·5mM Ca Buffer
~PCa 9
30mM Ca Buffer
pCa 6·2
VERY LOW ⟶ HIGH

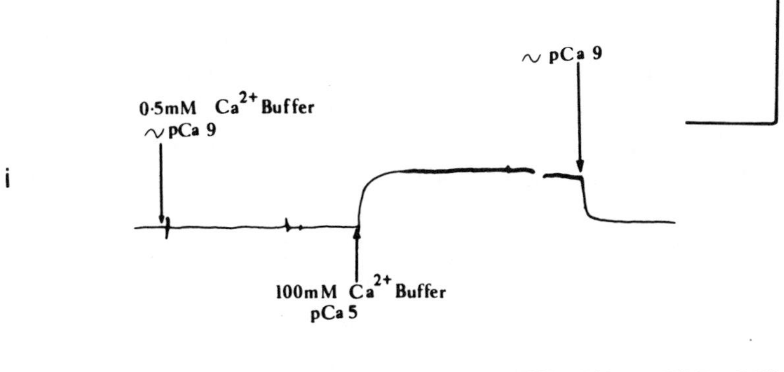

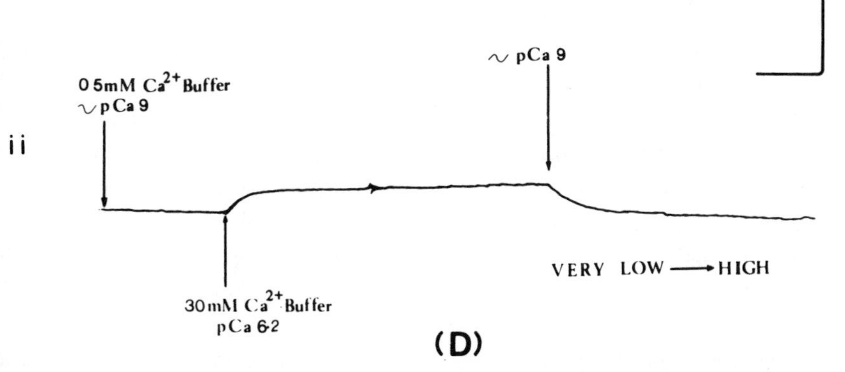

(D)

Figure 19 (*a*), (*b*), and (*c*) Isometric force records from bundles of myofibrils isolated from a mechanically skinned frog sartorius fiber and activated by an externally applied Ca–EGTA buffer solution. Myofibrils dissected under mineral oil and in a relaxing solution (pCa ∼ 9). Same bundle for A → C diameter, 82 μm; segment length, 1.5 to 2.0 mm. Temperature, 20°C. Calibration: vertical, 50 mg; horizontal, 60 sec. Solutions: all pH 7.1 ± 0.03, ATP 4 m*M*; free Mg^{2+} 0.05 m*M*. In calcium buffer solutions the concentration of (EGTA + Ca–EGTA) is 3 m*M* in A and B; it is 0.5 and 30 m*M* in C. Ionic strength throughout, 0.17 to 0.18 *M*. In B all solutions contain deoxycholate to inactivate the sarcoplasmic reticulum. (*d*) Isometric force records from bundles of myofibrils isolated from mechanically skinned crustacean muscle fiber (*Balanus nubilus*) and activated by externally applied Ca–EGTA buffer solutions. Same procedure and conditions as for (a) to (c). Diameter: (i) 85 μm; (ii) 51 μm. Solutions as for (a) to (c). Except free Mg^{2+} ∼ 1.2 mM, ionic strength (i) 0.55–0.6 M; (ii) 0.27–0.29M. Calibration: vertical (i) and (ii) 50 mg wt; horizontal (i) 10 sec, (ii) 2.5 sec (from ref. 100).

interfilament free calcium concentration, as a result of the low $EGTA_{total}$ (L) to high $EGTA_{total}$ (H) activating condition, reaches some 90% of the external concentration within 1 to 2 sec. When the activating protocol used by other authors is investigated (L to L), where the total EGTA concentration is the same in the activating as in the relaxing solution, the free calcium concentration within the bundle only approaches 90% of the external concentration after some 13 sec. In these calculations no account was taken of the fairly slow dissociation rate of the Ca–EGTA complex (74, 133), but the results for the L to H activating protocol are essentially unchanged, even when this and other factors (such as an unstirred layer surrounding the skinned preparation) are also taken into consideration (30, 83, 98).

Although the previous calculations suggest that the free Ca^{2+} concentration within the bundle resulting from the L to H activation protocol should equilibrate with that externally within 1 to 2 sec, it was important to determine whether this was actually so experimentally. The calcium-sensitive photoprotein aequorin was employed (6, 7, 94, 134), and the time course of light emission from the aequorin within the bundle was used as an index of the mean free Ca^{2+} within the myofilament space. It was initially necessary to have an accurate estimate of the apparent diffusion coefficient for aequorin within the myofibrillar preparation, so as to be certain that the major fraction of the light response observed upon activation was from aequorin actually within the bundle after loading, rather than from aequorin that had already diffused out into the external medium during the activation period.

Bundles of myofibrils were transferred for known periods of time into a loading solution containing aequorin, rinsed briefly in an aequorin-free relaxing solution, and activated in an aequorin-free solution at a higher free Ca^{2+} concentration. This protocol was repeated for the same skinned fiber bundle, but for increasing loading times. The area of each light emission response following activation represented the complete utilization of all the aequorin molecules in the preparation, and hence was directly proportional to the number of aequorin molecules that penetrated the bundle in the particular loading time. The areas reached constant values after times greater than 6 min for bundles of diameter ~ 160 μm, suggesting that diffusion at this time was virtually complete into the space available to the photoprotein. The result suggested an apparent diffusion coefficient of $\sim 1 \times 10^{-7}$ cm^2 sec^{-1} at 20°C, at an ionic strength of 0.13 M (9, 30), assuming aequorin was able to diffuse into an empty cylinder (Fig. 20*a*). Thus, for a bundle whose radius is 75 μm, 90% loading with aequorin occurs in about 5 min. This apparent diffusion coefficient for aequorin is some 15 to 20 times slower than

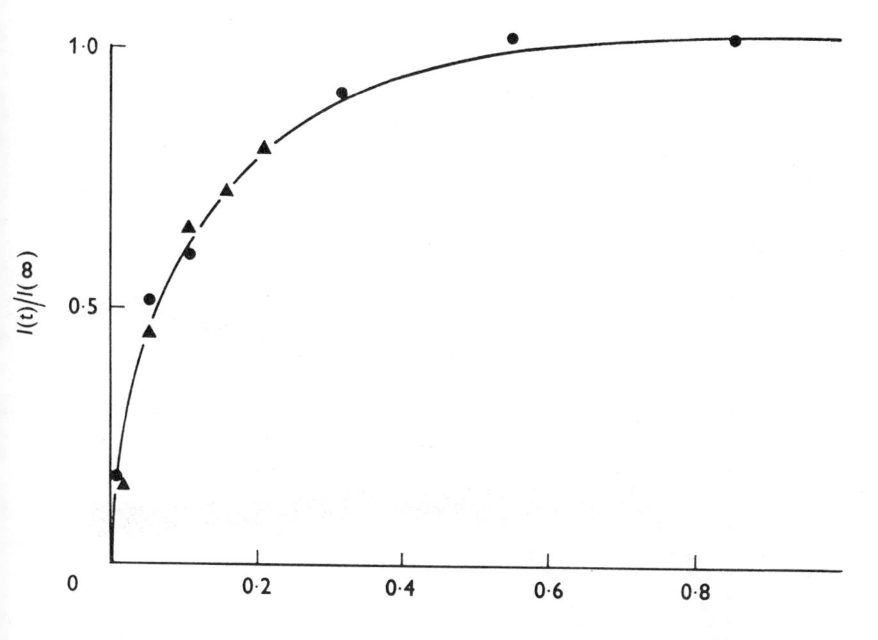

Figure 20 (a) Determination of aequorin diffusion in an isolated bundle of myofibrils (*Balanus*). Two sets of experimental results (●, ▲), both with radii, $r = 78\ \mu m$; temperature, 20°C. The line is the predicted diffusion of aequorin filled by an apparent diffusion coefficient of 1×10^{-7} cm^2 sec^{-1}. (ionic strength = 0.13 M) To convert the abscissa to time units, multiply by 10 (min). The light intensity $I(t)$ is proportional to the aequorin concentration within the bundle at time t, and $I(\infty)$ is proportional to the aequorin concentration at steady state (i.e., time ∞), so that $I(t)/I(\infty) = 1 - 4\sum_{n=1}^{\infty} 1/\mu_n^2 \exp[-\mu_n^2 (D'_{Aeq}t/r^2)]$, where D'_{Aeq} is the apparent diffusion coefficient for aequorin, r is the radius of the bundle, and μ_n is the solution of the Bessel function of order zero ($J_o(\mu) = 0$) (from ref. 30). (*b*) Activation of a myofibrillar bundle (*Balanus*) after preloading with aequorin. Aequorin-light (trace 1) and force (trace 2) responses (diameter 175 μm) activated externally with Ca–EGTA buffer solution (free $Ca^{2+} = 8.7 \times 10^{-8} M$). Initially, the preparation was in a high EGTA relaxing solution (10 min), followed by a low EGTA relaxing solution (5 min). Aequorin, in low relaxing, was applied uniformly along the segment while immersed under mineral oil (3 mm length). After 6 min the bundle was washed in low relaxing (2 to 3 sec). The "blip" on trace 1 indicates the moment of immersion in the activating solution. Insert: theoretical force points (solid dots) calculated from the following equation assuming a "step" increase in calcium (dotted line). The solid line represents the force in trace 2 (scale reduced by half). $P/P_\infty = 1 + (\gamma_2/\gamma_1 - \gamma_2)\exp(-\gamma_1 t) + (\gamma_1/\gamma_2 - \gamma_1)\exp(-\gamma_2 t)$, where P/P_∞ is the ratio between the instantaneous value of tension and its steady state value; the parameters $\gamma_1 = 0.38$ sec^{-1} and $\gamma_2 = 0.48$ sec^{-1} depend upon the rate constants and free Ca^{2+} in the *consecutive* scheme [eq. (2)] (29). Composition of solutions: high EGTA relaxing (free $Ca^{2+} < 10^{-9} M$), 20 mM EGTA; low EGTA relaxing (free $Ca^{2+} \sim 10^{-9} M$), 0.1 mM EGTA, 20 mM K_2SO_4; activating (free $Ca^{2+} = 8.7 \times 10^{-8} M$), total EGTA, 20 m$M$. All solutions contained in addition (mM): 40 KCl, 10 TES, 0.1 Mg^{2+}, 4 ATP (total), A23187 2.5 μg ml^{-1}, buffered with KOH to 7.1 ± 0.01. Ionic strength = 0.13 M, temperature 20°C. Calibration: vertical, 100 nA (1), 0.15 mN (2); horizontal, 2 sec (from ref. 30, 32).

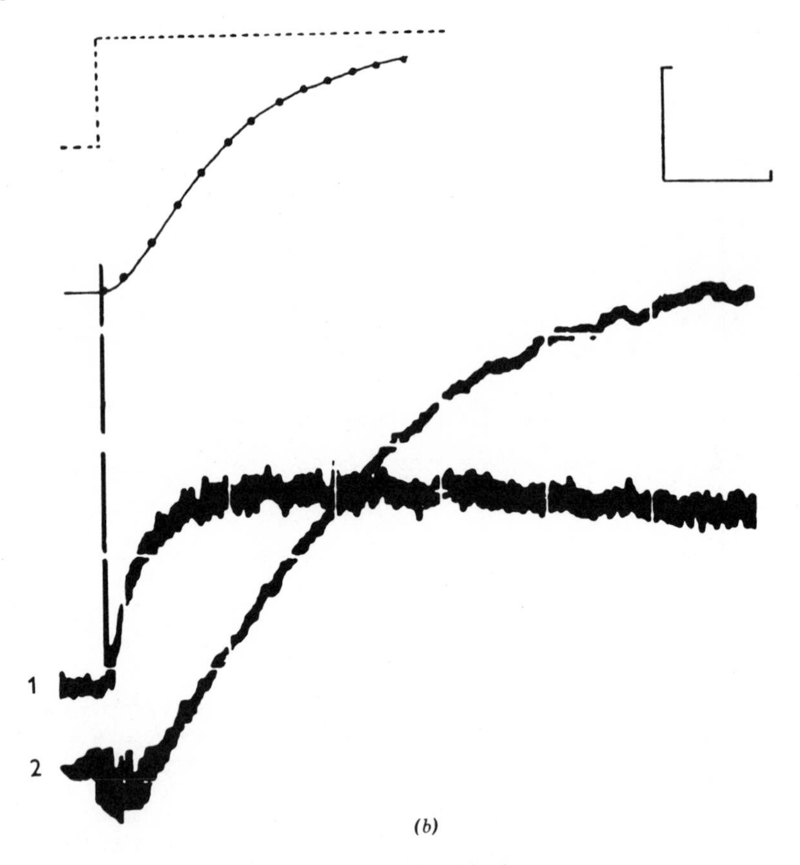

Figure 20 Continued.

the diffusion coefficient for aequorin in free solution (135, 136) and some 4 times slower than the diffusion coefficient for radial diffusion in large axons (63), but similar to the value observed for radial diffusion of this protein in large crustacean muscle fibers (9). Additionally, proteins of similar molecular weight to aequorin are accessible to the myofilament lattice space, but those of higher molecular weights are virtually excluded (137). In frog muscle, a slightly smaller value was observed for the longitudinal diffusion of aequorin of 0.5×10^{-7} cm^2 sec^{-1} at 15°C (7, 58), but the major point is that the diffusion of aequorin was some 40 times slower than that of EGTA (or Ca-EGTA). Calculations based upon this value for the apparent diffusion coefficient indicate that ~80% of the light signal would be due to aequorin still located within the myofibrillar preparation after some 10 sec in an aequorin-free solution, when the bundle was initially fully "loaded" with photoprotein (22, 30, 98).

The photoprotein, therefore, could be reasonably used as an indicator of the free Ca^{2+} within the bundle and hence give an index of the rapidity with which the Ca^{2+} achieved equilibration, compared to force development.

Bundles of both frog and barnacle myofibrils were skinned under mineral oil (73) and subjected to the following protocol, the precise times in each solution being dependent upon the diameter of the bundle. First, the bundle was transferred from mineral oil to a relaxing solution containing a high total EGTA, pCa > 9, to remove contaminating calcium. The myofibrillar bundle was then transferred to a relaxing solution with a low total EGTA content, pCa ~ 9, and then back to oil, where 20 to 40 nl of aequorin contained in a low EGTA relaxing solution was applied for a period of time adequate for ~90% loading. The preparation was removed from the oil, briefly rinsed in aequorin-free low EGTA relaxing solution, and immediately activated in a solution with a high total EGTA content and low pCa. In addition, all solutions contained the divalent ionophore A23187 (2.5 μg/ml) to equilibrate SR Ca^{2+}. The experimental light and tension records obtained are illustrated in Fig. 20*b* for barnacle and frog myofibrillar bundles, respectively. In both preparations, the light responses (trace 1), and hence the free calcium, equilibrate rapidly rising to relatively steady values within 1 to 2 sec, as predicted by the previous calculations. Force, however, did not reach a steady value at this free Ca^{2+} for either barnacle or frog until some 10 sec later, suggesting that in both preparations free Ca^{2+} equilibrates in the bundle more rapidly than subsequent reactions that lead to force production. This work, using the "pCa clamp" method, adds support to the ideas on Ca^{2+} activation derived from intact muscle fibers (see earlier section).

In a more detailed study of the kinetics of tension development in frog, Moisescu (97) and Moisescu and Thieleczek (98) employed solutions that were carefully ionically balanced between the activating and relaxing solutions by employing a substitute for EGTA, hexamethylenediamine-NNN′N′-tetraacetic acid (HDTA), rather than sulfate, which had been used previously. The compound HDTA is virtually ideal, as it has the same ionic charge and a similar molecular weight but a negligible affinity for divalent cations, compared to EGTA (Table 2).

The results of experiments using this protocol on skinned frog fibers also clearly demonstrate a Ca^{2+}-dependency of the rate of force generation (Fig. 21*a* & *b*), and provides support for the notion that, in both vertebrate and

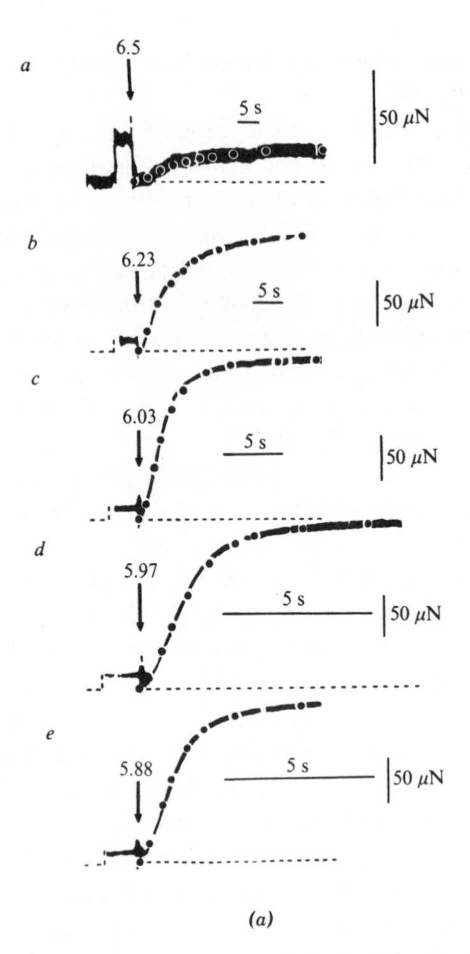

Figure 21 (a) Isometric force records from a myofibrillar bundle isolated from a mechanically skinned frog iliofibularis muscle fiber, activated by externally to applied Ca–EGTA buffer solutions of gradually increasing free Ca^{2+} concentration (*a to e*). The pCa of the activating solutions is given in caption; bundle diameter 50 μm, length 1–2 mm, sarcomere length 2.3 to 2.5 μm. The relaxing solution contained 50 m*M* $HDTA^{2-}$ (Table 2) and 0.15 m*M* free $EGTA^{2-}$ (pCa ~ 8.0), and the activating solution contained 50 m*M* ($EGTA^{2-}$ + Ca–$EGTA^{2-}$). In addition, all solutions contained (m*M*): creatine phosphate (Na_2) 10, ATP–Na_2 8, caffeine 10, TES 140, creatine kinase 10 U.ml^{-1}. pH 7.1 ± 0.01, total K^+ 137 ± 2 m*M*, free Mg^{2+} 1 m*M*, temperature 2 ± 0.5°C. (From ref. 97). The solid dots represent predicted values for relative tension (see ref. 97). (*b*) Activation of a myofibrillar bundle isolated from a frog sartorius muscle after preloading with aequorin. Trace 1, aequorin light and trace 2, force responses (diameter 80 μm) activated externally with Ca–EGTA buffer solution (free Ca^{2+} = 5×10^{-7}*M*). Same protocol and solutions as for Figure 20*b*, temperature 20°C. Calibration: vertical (1) 200 nA; (2) 0.07 mN; horizontal, 2 sec. (From ref. 30).

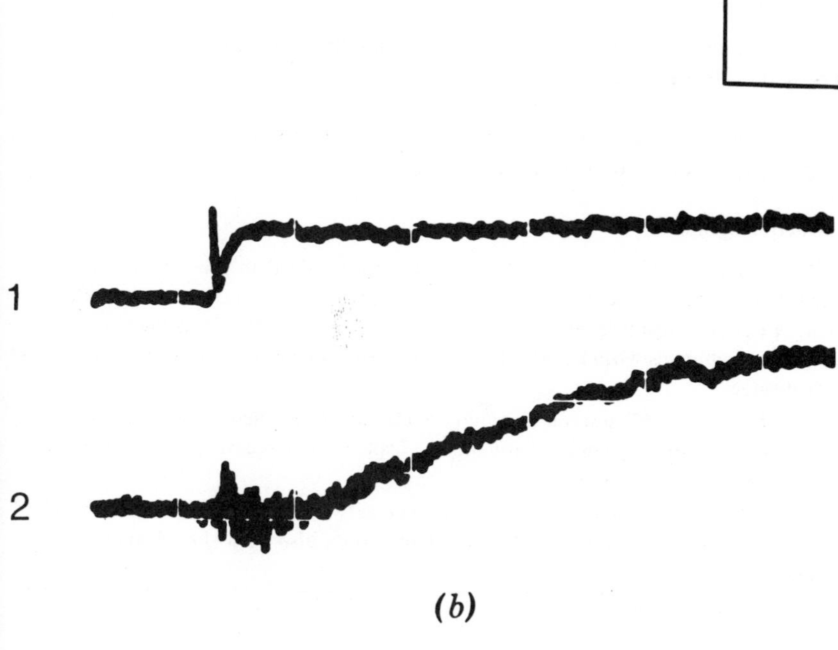

(b)

Figure 21 Continued.

invertebrate muscle, free Ca^{2+} plays a crucial role in the rate of force development.

A kinetic scheme similar to that proposed earlier from the work based upon studies of intact fibers (29) provided some interpretation of the kinetics of contraction and its dependency on free Ca^{2+} observed in frog fibers, although here a total of 6Ca per functional unit were required, but in two relatively slow reactions (97).

It is not clear at present, from work on either intact fibers or isolated myofibrils, as to the nature or exact location of the rate-limiting steps that lead to force production in skeletal muscle. In addition, rate limiting steps in the pre-steady state development of isometric tetanic force may not be rate limiting once a steady state has been achieved and is being maintained. These ideas are summarized in Table 4 (see also refs. 138 and 139).

The methods developed for the rapid activation of myofibrils, the "pCa

Table 4 Kinetic steps in force production

Intact Fibers[a]

Experimentally, the free Ca_i^{2+} rises more rapidly than does isometric force. This observation could be explained by:

1. A slow Ca^{2+} attachment rate to the functional unit for force,
2. A rapid Ca^{2+} attachment followed by slower steps.

Isolated Troponin

1. If only Ca^{2+} specific sites in troponin are involved in the functional unit for force, then the attachment rate at low pCa should be fast, and the slow force development *in vivo* may result from another step or steps. Some of the delay may involve conformational changes within the troponin-tropomyosin complex, or attachment rate of cross-bridges.

2. If Ca^{2+}/Mg^{2+} sites on troponin are also involved, then the attachment rate of Ca^{2+} to these sites is slower, via a Mg^{2+} displacement reaction. The k_{off} of Ca^{2+} from these sites is also slow.

3. If the functional unit involves myosin as well as troponin,[a] then the Ca^{2+} attachment rate to the myosin (P) light chain would also be a slow displacement reaction.

[a]See reference 29, 105.

clamp" (30, 82) and ionic balancing of the solutions (97), can also be employed to examine the details of the process of force relaxation.

Intracellular acidification of intact muscle cells by CO_2 leads to a prolongation of the relaxation phase of force (67) (Fig. 22) and similar effects are observed with the permeant anion phthalate.

This prolongation of relaxation could arise by a direct effect of H^+ upon the contractile proteins, for example, either by influencing the "off" rate for calcium from the troponin complex, or by affecting the cross-bridge and ATP turnover rates. Its most likely action is, however, a change in the Ca^{2+} accumulating activity of the SR.

The possible direct effect of H^+ upon the contractile proteins was investigated in bundles of myofibrils (80 to 100 μm diameter). Caffeine was present (20 m*M*) to deplete the SR of calcium and so bypass the activity of the SR Ca^{2+} pump. After an initial equilibrium in very low relaxing solution containing no EGTA, the preparation was activated by virtually unbuffered free calcium (weakly buffered by ATP), so that the free Ca^{2+} concentration was

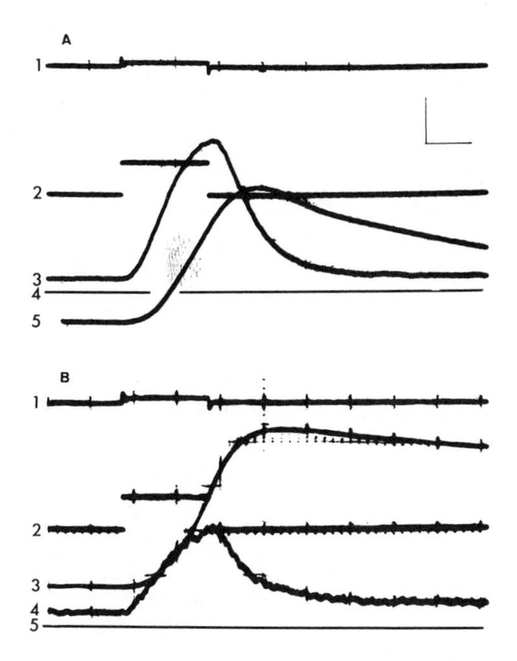

Figure 22 The effect of CO_2 application on the contractility of a single aequorin-injected crustacean (*Balanus*) muscle fiber. (A) Control. (B) 10 min after a brief (5 min) application of a CO_2 saline (pH 4.9). Voltage clamp pulse: 200 msec, $\Delta V = +38$ mV; holding potential -45 mV. Temperature 16 to 18°C. Trace 1, membrane current; trace 2, membrane voltage. (A) Trace 3, aequorin light emission; trace 4, baseline for trace 3; trace 5, isometric force. (B) Trace 3, isometric force; trace 4, aequorin light; trace 5, baseline for trace 4. Time constant for light (A) 20 msec, (B) 5 msec. Calibration: vertical; (A) and (B), (1) 0.2 mA; (2) 50 mV; (A)(3) and (B)(4); 0.5 μA; (A)(5) for 0.03; and (B)(3) 0.006 kg cm^{-2}. Horizontal: 100 msec. Time constant (A)(3) 20 msec; B(4) 5 msec. (From ref. 67 and unpublished results).

13 μM at pH 7.1. Force development occurred and was allowed to come to equilibrium with this free Ca^{2+} before application of the relaxing solution, which varied in $EGTA_{total}$ from 0.3 (low relaxing) to 50 mM (high relaxing). Rates of relaxation were examined at both pH 7.1 and 6.0 in separate bundles. The relaxing solutions caused force to decline approximately exponentially. At pH 7.1 the relaxation half-time appeared to approach a plateau minimum value as the EGTA concentration was increased. At 50 mM EGTA, the rate-limiting step is less likely to be the reduction in the free Ca^{2+} concentration in the myofibrillar space, but more probably the kinetics of Ca^{2+} release from the myofibrillar proteins.

For a given EGTA concentration the rate of relaxation was longer at pH 6.0 than at 7.1 (Fig. 23*a*). This can be explained by the fact that the apparent affinity constant of Ca^{2+} to EGTA is reduced by a factor of about 150 on reducing pH from 7.1 to 6.0, and a plot of the half-time against the free Ca^{2+} of the relaxing solution shows that pH 6.0 causes a shift to the left, suggesting that relaxation may be accelerated at the lower pH (Fig. 23*b*). This was also found with single bundles relaxed at both pH values with the same relaxing free Ca^{2+}. However, at pH 6.0 the relaxing rate was still probably determined by the rate of EGTA diffusion into the bundle, even at 100 m*M* EGTA, and further increases in the buffering power of the solution (at pH 7.1) could reduce the half-time (Fig. 23*b*) (140).

As a result of using 100 m*M* EGTA in the high relaxing solutions at pH 6.0, the ionic strength was raised from 220 to 370 m*M*; this increase did not

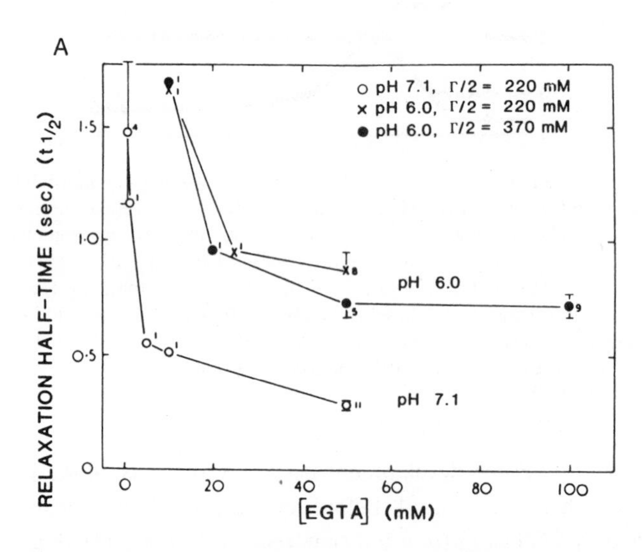

Figure 23 (*a*) Effect of EGTA concentration, and (*b*) effect of pCa, upon the relaxation of isometric force in a myofibrillar bundle preparation from crustacean muscle (*Balanus*) at different pH values. In (a) points are mean values from several myofibrillar bundles (diameter 80 to 100 μm) with number of observations alongside the points and S.E. bars shown where appropriate. Relaxing solutions at pH 6.0 had an ionic strength, (I), of either 220 (x) or 370 m*M* (●). Activating solutions at pH 6.0 contained either 11 or 67 μM free Ca^{2+}. Activation solution contained (m*M*): 131 K^+, 30 Na^+, 12 Cl^-, 18 TES, 5.0 ATP_{total}, 4.3 MgATP, 10 creatine phosphate, 20 $U.ml^{-1}$ creatine kinase, 20 caffeine, 1 to 1.5 free Mg^{2+}. (*b*) Relaxation half-time data from (a), replotted against the pCa value of the relaxing solutions (from ref. 140).

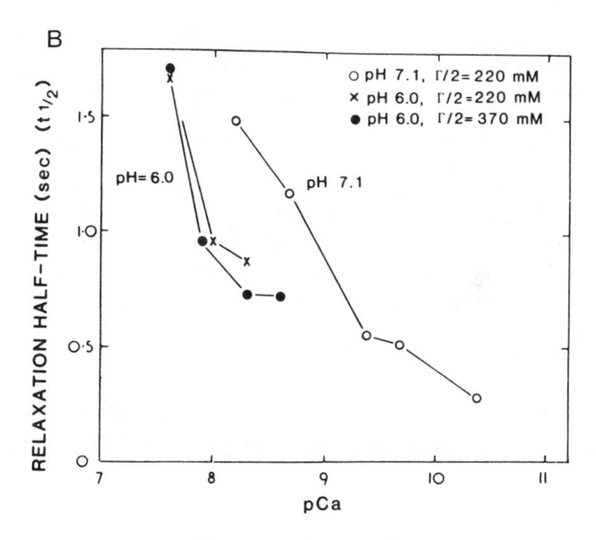

Figure 23 Continued.

affect the relaxation rate by itself. The amplitude of contraction in response to very low activating solution at pH 6.0 was smaller than that at pH 7.1 for the same pCa value, a finding previously observed in barnacle myofibrils (21). Usually, at pH 6.0 it was necessary to raise the free Ca^{2+} from 11 to 67 μM to get a maximal tension response. The half-time for relaxation at a given pH and EGTA concentration was detectably longer (by about 20%) if the bundle was activated by the higher free Ca^{2+} solution. The nature of the equilibrating solution before contraction did not affect the subsequent relaxation rate, only the rate of tension development.

Finally, the use of one of the "second generation" calcium buffers, BAPTA (61), has enabled the effect of H^+ to be examined unequivocally (Table 5), as this chelator is little affected by pH changes in the range 6.0 to 7.0 (Table 2). Relaxation was not significantly different at these two pH values (140).

Thus, the results presented here make it unlikely that the prolonged relaxation observed in intact muscle fibers exposed to 100% CO_2 (67) can be attributed to an effect of acidification on the relaxation kinetics of the myofibrillar proteins, since the range of pH values studied *in vitro* (i.e., 7.1 to 6.0) is similar to the observed decrease in intracellular pH (pH_i) of muscle fibers exposed to 100% CO_2 (71, 72). A more likely cause of the effect of CO_2 on intact

Table 5 Relaxation times for isolated myofibrils with pH[a]

Relaxing Solution Number	Calcium Chelator	pCa	pH	Relaxation Half-time (s) (±S.E.)	Number of Determinations
2/3	5 m*M* EGTA	9.4	7.1	1.18 ± 0.25	4
11	5 m*M* BAPTA	9.6	7.1	1.17 ± 0.22	3
12	17 m*M* BAPTA	9.6	6.0	1.34 ± 0.32	4

[a]Two myofibrillar bundles were used; diameters = 138 and 163 μm. As in the experiments with EGTA-containing relaxing solutions, each bundle was first equilibrated in a VLR (very low relaxing) solution for 3 min and then contracted in a VLA (very low activating) solution that contained either 13 μ*M* free Ca^{2+} (pH 7.1) or 75 μ*M* free Ca^{2+} (pH 6.0). (See ref. 140 for details.)

fibers is an impairment of the SR Ca^{2+} pump as a result of the decrease in pH_i, or possibly the increase in intracellular CO_2 and HCO_3^- concentrations.

9 SUMMARY

In this chapter, aspects of the ways in which calcium is transported and regulated within muscle cells have been considered, with particular reference to crustacean muscle fibers. The large size of these fibers permits easy access to the internal environment of the cell, allowing it to be altered by microinjection or microperfusion. At rest, calcium is not in equilibrium across the cell membrane and enters the cell down a steep electrochemical gradient. The free calcium ion concentration at rest is maintained at a value close to 100 n*M* by a combination of internal buffering systems, mainly the sarcoplasmic reticulum, mitochondria, and the fixed and diffusible calcium-binding proteins, as well as by an energy-dependent extrusion system operating across the external cell membrane. This system relies upon the inward movement of sodium down its own electrochemical gradient to provide the energy for the extrusion of calcium ions. It is not clear, however, what part, if any, the high energy phosphate compound ATP may play in this extrusion process. As a result of electrical excitation, voltage-sensitive channels for calcium are activated and permit calcium to enter the cell more rapidly than at rest. It has been possible to determine both the amount of calcium entering by this step, and what part this externally derived calcium plays in the development of force as well as in the free calcium change, which can be determined directly by calcium-sensitive indicators present in the cell sarcoplasm. A combination of techniques, allowing both the total and free calcium changes to be assessed during electrical excitation, has not only given valuable information as to how muscle cells buffer their calcium in order to regulate the extent of the change in the free calcium concentration, but it has also indicated that the entering calcium can only make a small direct contribution to the force developed by the cell. The implication here is that the major source of calcium for contraction must be derived from the internal calcium storage sites within the sarcoplasmic reticulum system.

The ability to measure the free calcium concentration changes within a single cell during activation has also provided the opportunity to analyze in detail the likely relations between free calcium and the process of force development in muscle.

The fact that the free calcium change precedes the development of force implies that there are delays in the mechanism, either at the site of calcium attachment on the myofibril, or at some later stage in the process of force development not previously anticipated. The experimental evidence also supports the notion that calcium is still attached to the contractile system during the relaxation phase of the mechanical event, returning to the sarcoplasmic reticulum at relatively low concentrations, but high rates of flux.

The article emphasizes the versatility of these large, single muscle fibers for physiological investigations. It also stresses the importance of the ideas derived from these experiments, and their applicability to other muscle cells where similar effects and, in particular, time relations between free calcium and force development have more recently been shown to occur.

REFERENCES

1. L. V. Heilbrunn and F. J. Wiercinski, *J. Cell. Comp. Physiol.*, **29,** 15 (1947).
2. D. M. Watterson and F. F. Vincenzi, *Ann. NY Acad. Sci.*, **356,** (1980).
3. O. Scharff, *Cell Calcium*, **2,** 1 (1981).
4. R. H. Kretsinger, *Ann. Rev. Biochem.*, **45,** 239 (1976).
5. R. H. Kretsinger, *Ann. NY Acad. Sci.*, **356,** 14 (1980).
6. C. C. Ashley and A. K. Campbell, *The Detection and Measurement of Free Ca^{2+} in Cells.* Elsevier: North Holland, New York, 1979.
7. J. R. Blinks, W. G. Wier, P. Hess, and F. G. Prendergast, *Prog. Biophys. Mol. Biol.*, **40,** 1 (1982).
8. H. Reuter and N. Seitz, *J. Physiol.*, **195,** 451 (1968).
9. C. C. Ashley, J. C. Ellory, and K. Hainaut, *J. Physiol.*, **242,** 255 (1974).

10. R. DiPolo, *Nature (Lond.)*, **274,** 390 (1978).
11. M. T. Nelson and M. P. Blaustein, *Nature*, **289,** 314 (1981).
12. C. C. Ashley and T. J. Lea, *J. Physiol.*, **282,** 307 (1978).
13. F. J. Brinley and L. Mullins, *J. Gen. Physiol.*, **50,** 2303 (1967).
14. C. C. Ashley, P. C. Caldwell, and A. G. Lowe, *J. Physiol.*, **223,** 735 (1972).
15. D. G. Moisescu and C. C. Ashley, *Biochim. Biophys. Acta*, **460,** 189 (1977).
16. R. DiPolo, *Biochim. Biophys. Acta*, **298,** 279 (1973).
17. J. M. Russell and M. P. Blaustein, *J. Gen. Physiol.*, **63,** 144 (1974).
18. S. Hagiwara and K-I. Naka, *J. Gen. Physiol.*, **48,** 141 (1964).
19. C. C. Ashley and J. C. Ellory, *J. Physiol.*, **226,** 653 (1972).
20. C. C. Ashley and D. G. Moisescu, *J. Physiol.*, **239,** 112*P.* (1974).
21. C. C. Ashley and D. G. Moisescu, *J. Physiol.*, **270,** 627 (1977).
22. C. C. Ashley, *Ann. NY Acad. Sci.*, **307,** 308 (1978).
23. S. Hagiwara and S. Nakajima, *J. Gen. Physiol.*, **49,** 807 (1966).
24. A. Fleckenstein, K. Nakayama, G. Fleckenstein-Grün, and Y. K. Byon, "Interactions of vasoactive ions and drugs with Ca-dependent excitation-contraction coupling of vascular smooth muscle," in *Calcium Transport in Contraction and Secretion*, E. Carafoli, et al., Eds., Elsevier/North Holland, Amsterdam, 1975, pp. 555–566.
25. C. C. Ashley and T. J. Lea, "Photoproteins and glass scintillators in assessing transport properties," in *Membranous Elements and Movement of Molecules*, Vol. 6, E. Reid, Ed., Ellis Horwood, Chichester, U.K., 1977, pp. 79–96.
26. R. A. Sjodin, *Transport in Skeletal Muscle.* Wiley, New York, 1982.
27. G. Hoyle and T. Smyth, *Comp. Biochem. Physiol.* **10,** 291 (1963).
28. S. Hagiwara, K. Takahashi, and D. Junge, *J. Gen. Physiol.*, **51,** 157 (1968).
29. C. C. Ashley and D. G. Moisescu, *Nature, New Biol.*, **237,** 208 (1972).
30. C. C. Ashley and D. G. Moisescu, "The part played by Ca^{2+} in the contraction of isolated bundles of myofibrils," in *Calcium Transport in Contraction and Secretion*, E. Carafoli, F. Clementi, W. Drabikowski, and A. Margreth, Eds., North Holland, Amsterdam, 1975, pp. 517–525.
31. C. Caputo and R. DiPolo, *J. Gen. Physiol.*, **71,** 467 (1978).
32. C. C. Ashley and M. V. Thomas, *J. Muscle Res. Cell Motil.*, **3,** 505 (1982).
33. J. M. Potter and J. Gergely, *J. Biol. Chem.*, **250,** 4628 (1974).
34. S. P. Robertson, J. D. Johnson, and J. D. Potter, *Biophys. J.*, **34,** 559 (1981).
35. J-M. Gillis, D. Thomason, J. Lefevre, and R. H. Kretsinger, *J. Muscle Res. Cell Motil.*, **3,** 377 (1982).
36. F. Fuchs, *Biochim. Biophys. Acta*, **585,** 477 (1979).
37. F. Fuchs and B. Black, *Biochim. Biophys. Acta*, **622,** 52 (1980).
38. J. Demaille, E. Dutruge, J. P. Capony, and J-P. Pechère, "Muscular parvalbumins: a family of homologous calcium-binding proteins. Their relation to the calcium-binding troponin component," in *Calcium Binding Proteins*, W. Drabikowski et al., Eds., Elsevier, Amsterdam, 1974, pp. 643–678.
39. R. Niedergerke, *J. Physiol.*, **128,** 12*P.* (1955).
40. T. Nagai, M. Makinose, and W. Hasselbach, *Biochim. Biophys. Acta*, **43,** 223 (1960).

41. S. Ebashi and F. Lipmann, *J. Cell Biol.*, **14,** 389 (1962).
42. H. Portzehl, P. C. Caldwell, and J. C. Rüegg, *Biochim. Biophys. Acta*, **79,** 581 (1964).
43. C. C. Ashley, P. C. Caldwell, A. G. Lowe, C. D. Richards, and H. Schirmer, *J. Physiol.*, **179,** 32*P.* (1965).
44. T. Ohnishi and S. Ebashi, *J. Biochem. (Tokyo)*, **55,** 599 (1964).
45. L. Mela and B. Chance, *Biochemistry, NY*, **7,** 4059 (1968).
46. A. Scarpa, "Measurement of calcium ion concentrations with metallochromic indicators," in *The Detection and Measurement of Free* Ca^{2+} *in Cells*, C. C. Ashley and A. K. Campbell, Eds., North Holland, Amsterdam, 1979, pp. 85–115.
47. F. F. Jöbsis and M. J. O'Connor, *Biochem. Biophys. Res. Comm.*, **25,** 246 (1966).
48. K. Kometani and H. Sugi, *Experientia*, **34,** 1469 (1978).
49. A. Scarpa, F. J. Brinley, T. Tiffert, and G. R. Dubyak, *Ann. NY Acad. Sci.*, **307,** 86 (1978).
50. M. V. Thomas, *Techniques in Calcium Research*, Academic Press, New York, 1982.
51. T. J. Beeler, R. H. Farmen, and A. N. Martonosi, *J. Memb. Biol.*, **62,** 133 (1980).
52. S. M. Baylor, W. K. Chandler, and M. W. Marshall, *J. Physiol.*, **331,** 139 (1982).
53. R. Miledi, I. Parker, and P. H. Zhu, *J. Physiol. Lond.*, **333,** 655 (1982).
54. A. M. Gordon, A. F. Huxley, and F. J. Julian, *J. Physiol. Lond.*, **184,** 170 (1966).
55. F. Julian, M. R. Sollins, and R. L. Moss, *Proc. Roy. Soc. B*, **200,** 109 (1978).
56. C. C. Ashley and E. B. Ridgway, *J. Physiol.*, **209,** 105 (1970).
57. P. J. Griffiths, *Pflug. Arch.*, **368,** R22 (1977).
58. J. R. Blinks, R. Rüdel, and S. R. Taylor, *J. Physiol.*, **277,** 291 (1978).
59. G. R. Dubyak and A. Scarpa, *J. Muscle Res. Cell Motil.*, **3,** 87 (1982).
60. C. C. Ashley and J. Lignon, *J. Physiol.*, **318,** 10*P.* (1981).
61. R. Y. Tsien, *Biochemistry, NY*, **19,** 2396 (1980).
62. C. C. Ashley, *J. Physiol.*, **210,** 133*P.* (1970).
63. P. F. Baker, A. L. Hodgkin, and E. B. Ridgway, *J. Physiol.*, **218,** 709 (1971).
64. D. Allen and J. R. Blinks, "The interpretation of light signals from aequorin-injected skeletal and cardiac muscle cells: a new method of calibration," in *Detection and Measurement of Free* Ca^{2+} *in Cells*, C. C. Ashley and A. K. Campbell, Eds., North Holland, Amsterdam, 1979, pp. 159–174.
65. D. Allen, J. R. Blinks, and F. G. Prendergast, *Science*, **195,** 996 (1977).
66. C. C. Ashley and P. J. Griffiths, *J. Physiol.*, **344** (1983), in press.
67. C. C. Ashley, F. Franciolini, T. J. Lea, and J. Lignon, *J. Physiol.*, **296,** 71*P.* (1979).
68. B. Rose and R. Rick, *J. Membrane Biol.*, **44,** 377 (1978).
69. O. H. Petersen, R. C. Collins, and I. Findlay, *Pflug. Arch.*, **392,** 163 (1981).
70. D. M. Bers and D. Ellis, *Pflug. Arch.*, **393,** 171 (1982).
71. P. C. Caldwell, *J. Physiol.*, **142,** 22 (1958).
72. C. Aickin and R. C. Thomas, *J. Physiol.*, **252,** 803 (1975).
73. R. Natori, *Jik. Med. J.*, **1,** 119 (1954).
74. D. C. Hellam and R. J. Podolsky, *J. Physiol.*, **200,** 807 (1969).
75. L. E. Ford and R. J. Podolsky, *Science, NY*, **167,** 58 (1970).

76. M. Endo, M. Tanaka, and Y. Ogawa, *Nature*, **228,** 34 (1970).
77. A. Fabiato and F. Fabiato, *Circ. Res.*, **31,** 293 (1972).
78. T. J. Lea and C. C. Ashley, *Nature,* **275,** 236 (1978).
79. T. J. Lea and C. C. Ashley, *J. Membrane Biol.*, **61,** 115 (1981).
80. T. J. Lea, *J. Physiol. Lond.*, **330,** 52*P*. (1982).
81. U. Shoshan, D. H. MacLennan, and D. S. Wood, *Proc. Natl. Acad. Sci.*, **78,** 4828 (1981).
82. C. C. Ashley, D. G. Moisescu, and R. M. Rose, *J. Physiol.*, **241,** 104*P*. (1974).
83. C. C. Ashley, D. G. Moisescu, and R. M. Rose, "Kinetics of calcium during contraction: Myofibrillar and SR fluxes during a single response of a skeletal muscle fibre," in *Calcium Binding Proteins*, Drabikowski et al., Eds., Elsevier, Amsterdam, 1974, pp. 609–642.
84. L. B. Nanninga, *Biochim. Biophys. Acta*, **54,** 338 (1961).
85. S. M. Cohen and C. T. Burt, *Proc. Natl. Acad. Sci. USA*, **74,** 4271 (1977).
86. F. J. Brinley, A. Scarpa, and T. Tiffert, *J. Physiol. Lond.*, **266,** 545 (1977).
87. P. Hess, P. Metzger, and R. Weingart, *J. Physiol. Lond.*, **333,** 173 (1982).
88. J. D. Johnson, S. C. Charlton, and J. D. Potter, *J. Biol. Chem.*, **254,** 3497 (1979).
89. E. J. Krebs, *Cell Calcium*, **2,** 295 (1981).
90. S. Shenolikar, P. T. W. Cohen, P. Cohen, A. C. Nairn, and S. V. Perry, *Eur. J. Biochem.*, **100,** 329 (1979).
91. E. M. V. Pires and S. V. Perry, *Biochem. J.*, **167,** 137 (1977).
92. C. Picton, C. G. Klee, and P. Cohen, *Cell Calcium*, **2,** 281 (1981).
93. J. R. Dedman, J. D. Potter, R. L. Jackson, J. R. Johnson, and A. R. Means, *J. Biol. Chem.*, **252,** 8415 (1977).
94. O. Shimomura, F. H. Johnson, and Y. Saiga, *J. Cell. Comp. Physiol.*, **59,** 223 (1962).
95. E. B. Ridgway and C. C. Ashley, *Biochem. Biophys. Res. Comm.*, **29,** 229 (1967).
96. C. C. Ashley and E. B. Ridgway, *Nature (Lond).*, **219,** 1168 (1968).
97. D. G. Moisescu, *Nature (Lond).*, **262,** 610 (1976).
98. D. G. Moisescu and R. Thieleczek, *J. Physiol.*, **275,** 241 (1978).
99. D. G. Stephenson and D. A. Williams, *J. Physiol. Lond.*, **317,** 281 (1981).
100. C. C. Ashley and D. G. Moisescu, *J. Physiol.*, **233,** 8*P*. (1973).
101. C. C. Ashley and D. G. Moisescu, *J. Physiol.*, **231,** 23*P*. (1973).
102. J. Lignon and C. C. Ashley, *J. Muscle Res. Cell Motil.*, **1,** 457 (1980).
103. J-M. Gillis, "The biological significance of muscle parvalbumins," in *Calcium Binding Proteins*, Siegel et al., Eds., Elsevier/North Holland, Amsterdam, 1980, pp. 309–311.
104. M. B. Cannell, *J. Physiol.*, **326,** 70*P*. (1982).
105. C. C. Ashley, *Biochem. Soc. Trans.*, **10,** 212 (1982).
106. A. Weber and J. M. Murray, *Physiol. Rev.*, **53,** 612 (1973).
107. J. M. Regenstein and A. G. Szent-Györgyi, *Biochemistry*, **14,** 917 (1975).
108. R. D. Bremel and A. Weber, *Nature New Biol.*, **238,** 97 (1972).
109. S. M. Baylor, W. K. Chandler, and M. W. Marshall, *J. Physiol.*, **287,** 23 (1979).
110. W. Hastings, G. Mitchell, P. H. Mattingly, J. R. Blinks, and N. van Leeuwen, *Nature, Lond.*, **222,** 1047.
111. G. Loschen and B. Chance, *Nature New Biol.*, **233,** 273 (1971).

112. I. R. Neering and W. G. Weir, *Fed. Proc.*, **39,** 1806 (1980).
113. D. G. Stephenson and P. J. Sutherland, *Biochim. Biophys. Acta*, **678,** 65 (1981).
114. C. C. Ashley and P. J. Griffiths, *Quart. Rev. Biophys.*, in preparation.
115. B. R. Jewell and J. C. Rüegg, *Proc. Roy. Soc. B*, **164,** 428 (1967).
116. M. Endo, M. Tanaka, and S. Ebashi, *Proc. XXIV Int. Congr. Physiol. Sci.*, **7,** 126 (1967).
117. F. J. Julian, *J. Physiol. Lond.*, **218,** 117 (1971).
118. L. E. Ford and R. J. Podolsky, *J. Physiol. Lond.*, **223,** 1 (1972).
119. A. Fabiato and F. Fabiato, *J. Physiol. Lond.*, **276,** 233 (1976).
120. A. Fabiato, *Fed. Proc.*, **41,** 2223 (1982).
121. M. Endo, *Physiol. Rev.*, **57,** 71 (1977).
122. M. Endo, *Cold Spring Harb. Symp. Quant. Biol.*, **37,** 505 (1973).
123. M. Endo and M. Iino, *J. Muscle Res. Cell. Motil.*, **1,** 89 (1980).
124. M. Endo, T. Kitazawa, S. Yagi, M. Iino, and Y. Kakuta, "Some properties of chemically skinned smooth muscle fibres," in *Excitation-Contraction Coupling in Smooth Muscle*, R. Casteels, T. Godfraind, and J. C. Rüegg, Eds., Elsevier, Amsterdam, 1977, pp. 199–209.
125. A. Fabiato and F. Fabiato, *J. Physiol. Lond.*, **249,** 469 (1975).
126. L. L. Costantin and R. J. Podolsky, *Fed. Proc.*, **24,** 1141 (1965).
127. R. A. Podolin and L. E. Ford, *J. Muscle Res. Cell Motil.*, **4,** 263 (1983).
128. E. Marban, T. S. Rink, R. W. Tsien, and R. Y. Tsien, *Nature, Lond.*, **286,** 845 (1980).
129. G. Elliott and I. Matsubara, *J. Mol. Biol.*, **72,** 657 (1972).
130. D. W. Maughan and R. E. Godt, *J. Gen. Physiol.*, **77,** 49 (1981).
131. A. M. Gordon, R. E. Godt, S. K. B. Donaldson, and C. E. Harris, *J. Gen. Physiol.*, **62,** 550 (1973).
132. J. Crank, *The mathematics of diffusion*. Clarendon Press, Oxford, 2nd., Edit 1975.
133. P. D. Smith, R. L. Berger, and R. J. Podolsky, *Biophys. J.*, **17,** 159a (1977).
134. J. R. Blinks, F. G. Prendergast, and D. G. Allen, *Pharm. Rev.*, **28,** 1 (1976).
135. O. Shimomura and F. H. Johnson, *Biochemistry, NY*, **8,** 3991 (1969).
136. F. Prendergast and K. G. Mann, *Biochemistry*, **17,** 3448 (1978).
137. D. G. Stephenson, I. R. Wendt, and Q. G. Forrest, *Nature (Lond.)*, **289,** 690 (1981).
138. G. Cecchi, P. J. Griffiths, and S. R. Taylor, *Science NY*, **217,** 70 (1982).
139. Y. E. Goldman, M. G. Hibberd, J. A. McCray, and D. R. Trentham, *Nature*, **300,** 701 (1982).
140. T. J. Lea and C. C. Ashley, *Biochim. Biophys. Acta*, **681,** 130 (1982).
141. R. D. Keynes, E. Rojas, R. E. Taylor, and J. Vergara, *J. Physiol.*, **229,** 405 (1973).
142. C. C. Ashley and M. V. Thomas, *J. Physiol.*, **334,** 10*P*. (1983).
143. T. J. Lea, *J. Physiol.*, **334,** 39*P*. (1983).
144. C. C. Ashley, P. C. Caldwell, A. K. Campbell, T. J. Lea, and D. G. Moisescu, "Calcium movements in muscle," in *Calcium in Biological Systems, Soc. exper. Biol. Symp.*, **XXX,** 397 (1976).
145. C. C. Ashley, L. M. Castell, and D. Harvey, *J. Physiol.*, **344** (1983), in press.
146. H. Reuter, *Nature*, **301,** 574 (1983).
147. K. S. Lee and R. W. Tsien, *Nature*, **302,** 790 (1983).

CHAPTER 4

Exocytosis

RONALD P. RUBIN

Professor of Pharmacology
Medical College of Virginia
Richmond, Virginia

CONTENTS

1 INTRODUCTION

Studies over the past two decades have clearly established that calcium has pride of place among the factors that control the discharge of secretory material from the cell. The pivotal role of calcium as a mediator of the secretory process—which can be demonstrated throughout the phylogenetic scale—has been affirmed in regard to the release of such diverse substances as neurotransmitters, hormones, enzymes, and other proteins (1). However, there is variability from cell to cell with regard to the source of calcium that is utilized for secretion. In electrically excitable cells, the entry of calcium from the extracellular medium is the critical link in the main pathway leading to secretion (2). In exocrine glands, and perhaps in the blood platelet and neutrophil as well, the mobilization of a pool of membrane-bound calcium provides the trigger for initiating the release response (1). Finally, calcium is sequestered in cells in mitochondria and endoplasmic reticulum; and during the activation of secretion calcium may be released from these intracellular pools to trigger the secretory response. But whatever its source, it is clear that an increase in cellular ionic calcium provides the link between the events that occur following stimulation and the actual release response itself—a sequence commonly known as stimulus-secretion coupling (3).

2 HISTORICAL PERSPECTIVE

The characteristics of the secretory process must be understood if the role of calcium is to be defined, and the ubiquity of calcium's action denotes some common feature or property of the secretory process that is calcium sensitive. The "quantal" nature of neurotransmitter release, revealed by the elegant electrophysiological studies of Sir Bernard Katz and his colleagues (4), focused on the importance of the secretory organelle as the primary source of secretory product. Favoring this concept were the subsequent findings that following stimulation not only are nerve endings depleted of synaptic vesicles, as demonstrable by morphological analysis (5), but also the yield of vesicular neurotransmitter is decreased (6).

The observation that the catecholamines of the adrenal medulla are mainly localized in subcellular structures, which could be separated from other cellular organelles by differential centrifugation (7, 8), provided the impetus for morphological and biochemical investigations. These inquiries led to the conclusion that during activation, the secretory vesicle or granule moves to the periphery of the cell, the granule membrane fuses with the plasma membrane, and the soluble products of the granule are extruded into the extracellular medium (9). This phenomenon is known as exocytosis, and appears to be a common feature of most secretory cells (3). The vesicle membrane is retrieved in the form of coated vesicles and large vacuoles, which are reprocessed to form vesicles able to refill again and be reused (10, 11).

The importance of exocytosis as the mechanism for secretion was first established from studies on the adrenal medulla performed by Douglas and his associates, Kirshner and his collaborators, and a group working in England (3, 9). The catecholamine-containing secretory granules of the adrenal medulla contain not only the hormones epinephrine and norepinephrine, but also adenine nucleotides (ATP), calcium, various proteins (including dopamine beta hydroxylase—both membrane-bound and soluble form), and chromogranin. Douglas made a quantitative study of the simultaneous release of adenine nucleotides with catecholamines and determined that the molar ratio of catecholamine to ATP (4:1) found in the adrenal perfusate paralleled the ratio in isolated secretory granules. This was followed by the demonstration that secretion from the adrenal medulla was accompanied by the release of soluble proteins identified as chromogranin and soluble dopamine beta hydroxylase (3). Kirshner found that only the soluble dopamine beta hydroxylase is released into the medium, whereas the membranous form of the enzyme is retained within the cell, thus further establishing that the secretory granule itself is not discharged into the extracellular medium (9). The physiological relevance of exocytosis was affirmed illustrating chromogranin release from the adrenal gland *in vivo* after splanchnic nerve stimulation (12).

Electron microscopic evidence for exocytosis has been found in many secretory systems, including the adrenal medulla, the neurohypophysis, the endocrine pancreas, the mast cell, and the adenohypophysis (13–16). The infrequency with which exocytotic images are demonstrable morphologically might dictate caution in elaborating exocytosis as a basic mechanism for secretion. However, the biochemical evidence for its existence in the adrenal medulla, as well as in the neurohypophysis and the parathyroid gland (11, 17), is convincing. Additionally, freeze-fracture analysis facilitates the demonstration of exocytosis (16), and quick-freezing, which rapidly arrests structural changes, allows temporal and quantitative correlation of the incidence of exocytotic images with enhanced neurotransmitter release (18, 19). These results convey the indisputable conclusion that exocytosis is the mechanism by which secretory organelles discharge their product into the extracellular fluid (Fig. 1).

The importance of calcium in the secretory process, together with the evidence that secretion occurs by exocytosis, forges a link between calcium-dependent secretion and the exocytotic process. This conclusion is supported by pharmacologic evidence wherein tyramine, an indirectly acting sympathomimetic amine, releases catecholamines from sympathetic nerves by a mechanism that does not involve the release of adenine nucleotides and/or

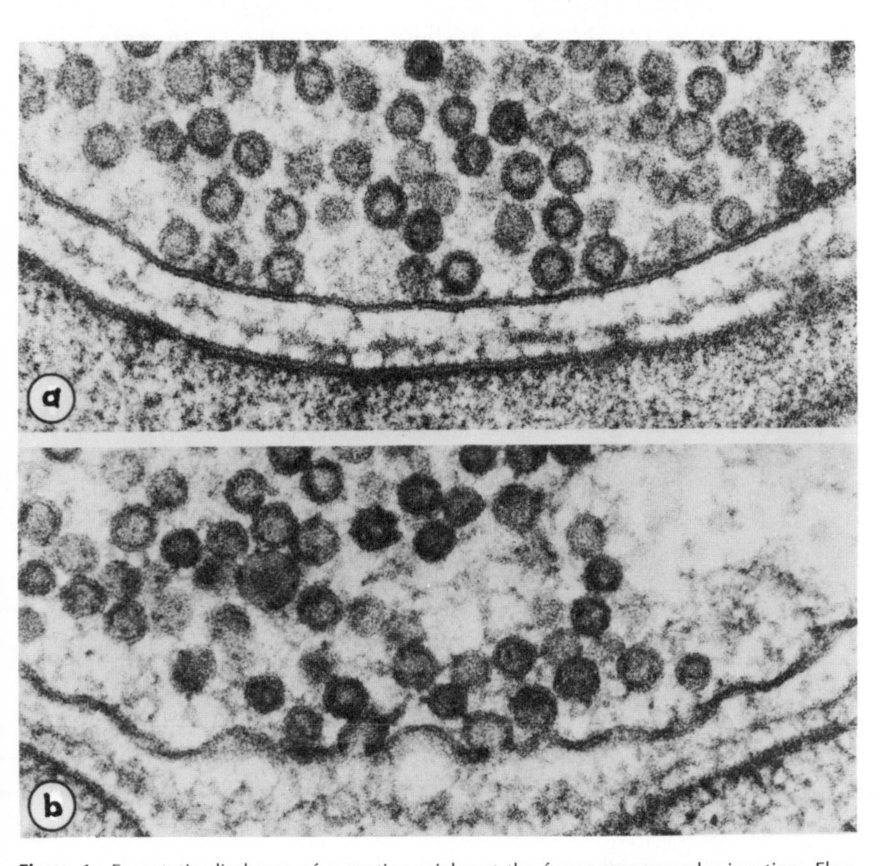

Figure 1 Exocytotic discharge of synaptic vesicles at the frog neuromuscular junction. Electronmicrographs of nerve terminals at rest (*a*) and stimulated in the presence of 4-aminopyridine (*b*) to increase quantal release of transmitter. The terminals were frozen 5 to 6 msec after the stimulus. Pockets of the same size and curvature as synaptic vesicles occur in the stimulated nerve terminal, indicating that they are synaptic vesicles caught in the act of exocytosis ($\times$ 180,000). (From ref. 18.)

soluble protein, that is, nonexocytotic (20). The tyramine-induced release of neurotransmitter, in contrast to that induced by nerve stimulation, is not dependent upon the presence of extracellular calcium (21). Moreover, pharmacological manipulation, using reserpine and monoamine oxidase inhibitors, can markedly enhance the extravesicular store of catecholamines in sympathetic nerves. Depolarization releases this extravesicular store by a calcium independent mechanism, in contrast to the release of vesicular catecholamine that is calcium dependent; this indicates that the nonvesicular store does not directly participate in the release process under normal physiological conditions (22, 23). Such findings discredit the arguments of those who envision the extragranular (or cytoplasmic) pool of secretory product as the readily releasable form (24, 25). We must be aware that a significant proportion of the secretory material may be found as an extra-granular pool in certain cells (25), and that the secretory granules of a given cell are not equally susceptible to being utilized (26, 27). However, it is clear that the secretory granule represents the primary source of the product discharged by exocytosis.

3 GENERAL FEATURES OF MEMBRANE FUSION

Since the action of calcium in the secretory process is linked to the fusion of granule membrane with cell membrane, elucidation of the fundamental action of calcium in secretion might lead to a better understanding of these phenomena. Although membrane fusion was first recognized by George Palade in 1959 as an essential event in the secretory process (28), it is not confined to the secretory cells, but occurs in a variety of cellular functions, such as cell fusion and the uptake and digestion of extracellular material (pinocytosis and phagocytosis) (29–31). Specific details of this fundamental process are not known. However, some general statements can be made with regard to membrane fusion as it pertains to exocytosis, a two-phase process involving the mobilization of secretory granules to the membrane site and the actual fusion-fission process itself.

Membrane components of the secretory organelle are distinct from those of the plasma membrane, and it is the distinctive components of each of these membrane systems that render selectivity to the fusion-fission process. In membranes the lipids are arranged as a bilayer existing in a fluid or gel phase; the proteins are either inserted into the lipid bilayer (integral proteins) or bound loosely (peripheral proteins) to the surfaces of the bilayer. Interaction between these macromolecular components to permit fusion requires the introduction of local "disorder" into the membranes. Some changes in the lipoidal portion of the membrane occur, allowing the membranes to fuse and form a continuous structure, so that their internal contents intermix. Perhaps equally important, membrane proteins may help to create domains of certain phospholipids that promote fusion. By interacting with these macromolecules in a relatively unusual way, calcium modifies the physical and chemical properties of membranes to promote the fusion of biological membranes (29, 32). Calcium may promote the apposition of adjacent membranes or induce destabilization of apposing membranes at specific sites where fusion can take place.

4 MODEL SYSTEMS

Elucidation of the molecular mechanism of membrane fusion has been hindered by the lack of suitable systems for studying this process. Investigations with intact cells are unable to yield data that could be reliably interpreted on the basis of molecular processes. Therefore, model systems have been developed for elucidating the mechanism of exocytosis. These model systems allow us to define: (a) the steps involved in the membrane fusion reaction, and (b) the mechanism by which calcium activates the process.

4.1 Secretory Granule

The first, and perhaps simplest, model system used for probing the exocytotic process is the secretory granule itself. Secretory stimuli that increase intracellular calcium levels not only promote fusion of the granule membrane with the plasma membrane but also induce secretory granules to fuse with one another, a phenomenon called compound exocytosis (14). Similarly, isolated secretory vesicles (or granules) are predisposed to fuse in the presence of calcium (33, 34) (Fig. 2), and this *in vitro* fusion process is envisioned both as the step prior to fission and a representation of the process occurring in the intact cell (35). The calcium-induced granule fusion is enhanced by a specific cytoplasmic protein termed synexin (36). The physiological relevance of these findings is supported not only by the fact that secretory granules isolated from several different tissues can fuse with one an-

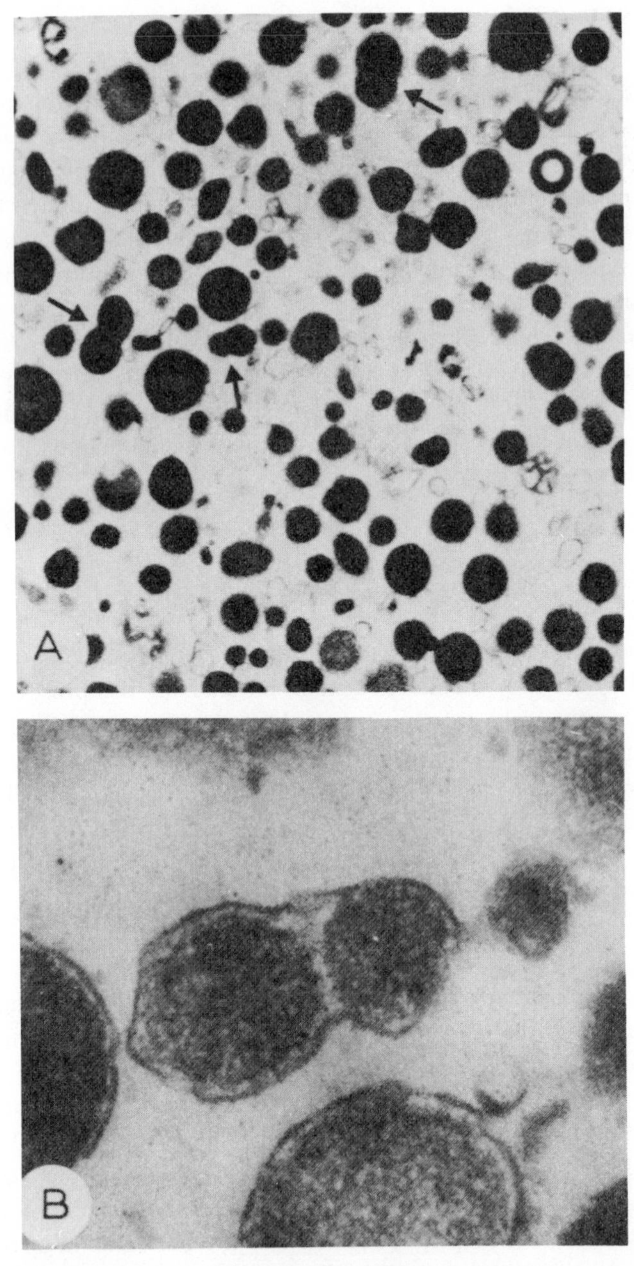

Figure 2 Membrane fusion of secretory vesicles. (*a*) Electronmicrograph of isolated bovine adrenal medullary secretory vesicles (granules) demonstrating fusion (arrows) following incubation with 100 μM calcium ($\times$ 14,000). (*b*) In spray droplets rapidly frozen in liquid propane after incubation with 100 μM calcium, intervesicular fusion is also indicated by the presence of membrane surrounding two electron-dense cores of a "twinned" vesicle ($\times$ 80,000). (From ref. 33.)

other in the presence of micromolar concentrations of calcium, but also that magnesium, which cannot replace calcium in promoting secretory activity in the intact cell, is a much weaker fusogen (33).

After this fusion process, the barrier separating the vesicle interior and the extracellular fluid eventually undergoes fission. This fission process is viewed as an ATP-dependent process involving osmotic lysis (35). In accord with this view are the findings that magnesium, ATP, plus chloride induce the release of catecholamines from chromaffin granules by osmotic lysis. This release is suppressed by raising the osmotic strength of the medium or by adding inhibitors of anion transport. According to this model, fusion induced by calcium and synexin results in anion transport, producing an increase in osmotic strength inside the granule; fission then occurs by hypotonic lysis. Although the studies of Pollard and his associates have focused on medullary catecholamine release, as well as secretion by the parathyroid gland and the blood platelet, this theory is not applicable to peripheral nerves, since at the neuromuscular junction an increase in osmotic strength increases transmitter release, and inhibitors of anion transport are ineffective (35–37). Nevertheless, recent studies, in which synexin recognition sites were found not only on the chromaffin granules but also on the internal surface of the plasma membrane (38), support the theory that specific proteins may permit calcium to form a bridge between the membrane at the surface of the cell and the membrane of the secretory granule. Synexin, or proteins with a similar function, may thus facilitate fusion by providing a connecting link between the two apposed membranes.

4.2 Reconstituted Systems

The absence of the second principal component of the exocytotic process, the plasma membrane, has fostered doubts about the validity of the isolated secretory granule as a model for exocytosis, particularly since calcium does not produce consistent or marked effects on the release of secretory product from various isolated granule preparations (39). On the other hand, ATP elicits release from granules by a lytic action (40). Both calcium and ATP play critical roles in the activation of secretion—and without these two components membrane fusion would not occur (1)—calcium-ATP interactions, however, do not focus solely on the secretory granule.

There have, therefore, been several attempts to reconstitute *in vitro* secretory systems comprised of secretory granules and plasmalemma. Cell-free models for exocytosis have been developed for the adrenal medulla as well as the endocrine and exocrine pancreas (41–43). Fusion between isolated granules and plasma membrane from secretory cells has been induced by the physiological secretagogue or by calcium (42, 44). Moreover, Konings and DePotter (43) found that the addition of bovine adrenal medullary plasma membrane stimulates catecholamine release from isolated chromaffin granules when the concentration of calcium is greater than $2 \times 10^{-7}M$, which corresponds to the ionic calcium concentration in cells during enhanced activity. A system of mouse islet granules and cod islet plasma membranes also exhibited a glucose-induced calcium-dependent release of insulin optimized by the addition of ATP (41). Such findings not only substantiate the concept that fusion of the granule membrane with the plasma membrane is the ultimate step in exocytotic secretion, but also reveal the existence of specific recognition sites on the interacting membranes that express their actions through calcium and ATP.

4.3 Artificial Membrane Systems

Cell-free systems, which include the major constituents of the exocytotic process (i.e., the secretory granule and plasma membrane), should help to elucidate the molecular mechanism of exocytosis. However, even simpler systems are needed to identify the specific biochemical components of each membrane system that are critical for the fusion-fission process itself. Hence, the use of artificial membrane systems to probe the mechanism of exocytosis has attained popular appeal, despite the fact that these simple membrane systems are incomplete, relative to biological membranes, and do not represent a totally reconstituted system.

Studies with liposomes have revealed that acidic phospholipids and divalent cations are important elements in membrane fusion, and that the role of phospholipids in membrane fusion is related to their ability to form intermembrane complexes with calcium (45). The affinity of calcium for membrane-binding sites appears to be influenced by membrane phospholipid composition; and calcium particularly influences the structure of membranes containing acidic phospholipids. Accordingly, calcium causes a phase transition of the lipid bilayer into a highly ordered crystalline structure; and the ability of calcium to induce phase changes correlates with its ability to produce fusion of acidic phospholipid vesicle membranes (45, 46). The particular importance of acidic phospholipids is underscored by the finding that phosphatidylcholine, a neutral phospholipid, inhibits calcium-induced fusion of phosphatidylserine vesicles (47). Thus, the lipid phase transitions induced by calcium involve a phase separation of mixed acidic and neutral phospholipids, which by altering the stability of the membranes may make them more susceptible to fusion.

The relationship between calcium-induced fusion of phospholipid vesicles and naturally occurring fusion phenomena remains problematic, since the calcium-induced fusion of these artificial membrane models requires millimolar concentrations of calcium (48). The relatively high concentrations of calcium may be necessitated by the absence of proteins, because phospholipids alone do not have the sensitivity or the specificity for calcium that is manifested in biological membranes. The calcium sensitivity and specificity of fusion apparently arises from protein components, since the sensitivity of the fusion process of lipid vesicles is heightened by the addition of calcium-binding proteins (49).

The asymmetric distribution of phospholipids in biological membranes is also an important element in providing the selectivity for fusion phenomena. The outer face of the plasma membrane is assumed to be inherently resistant to fusion, resulting from the heterogeneous composition of the membrane. Negatively charged phospholipids tend to be on the inside of the membrane and on the outside of some intracellular vesicles (50). This asymmetry provides the potential for cellular control of fusion events. Calcium influx initiated by the primary signal would bring calcium in contact with the acidic phospholipid head group on the cytoplasmic side of the plasma membrane and the abutting synaptic vesicle. Calcium could then trigger a phase change with the formation of crystalline domains, and thus induce fusion at the domain boundaries (45).

In addition to the interaction of calcium with acidic phospholipids of the cell membrane, the hydrolysis of membrane ATP may also be required during membrane fusion (29). Alan Poisner (51) proposed some years ago that during exocytosis plasmalemmal ATP is hydrolyzed by granular ATPase. Plattner et al. (52, 53), using the *Paramecium* as a model system for analysis of the mechanism of exocytosis and membrane fusion, have demonstrated calcium-activated ATPase activity at the specific site where specialized secretory vesicles—called trichocysts—attach to the plasma membrane. Capitalizing on the fact that in this particular system membrane fusion and exocytotic discharge can be dissociated, Plattner and his associates found that attachment of vesicles does not require ATPase, but discharge does (52, 53). Thus, local ATPase hydrolysis may promote fusion in a direct way by "destabilizing" membranes. Although the removal of calcium may also add to the instability of the apposing membranes (29), a more direct role for calcium presumably involves the bridging of anionic sites on the cytoplasmic side of the cell membrane and on the secretory vesicle.

4.4 Electrophoretic Model

Molecular constraints to granule-cell membrane interaction may depend upon the nature of the granule surface and its electrostatic charge. The electrophoretic model avers that exocytosis is initiated by an electrophoretic migration of the vesicles toward the plasma membrane (54). Secretory organelles possess a net negative charge that is decreased by calcium ions (55). Thus, calcium-induced screening of negative charges on the interior of the cell may reduce electrostatic repulsion between vesicles and membranes, and enhance the approach of the vesicle to the membrane so that fusion and secretion occur. The credibility of this model is compromised by the fact that it does not reflect the selective action of calcium; for magnesium, an inhibitor of calcium-dependent secretion, also neutralizes the negative charges of the secretory organelle (55).

5 CYTOPLASMIC COMPONENTS

5.1 Microtubule-Microfilament System

Investigations of simple model systems have greatly enhanced our understanding of membrane fusion phenomena and exocytosis in particular; however, additional insight into the nature of the mechanism of exocytosis in the intact cell can be gleaned by considering various cytoplasmic components of the cell that may participate in the secretory process. An involvement of the microtubular-microfilament (cytoskeletal) system in the migration of secretory organelles has received considerable attention, based primarily on its presence along the pathway traversed by the secretory granule and by the interference with secretion after its pharmacologic disruption. The microtubules, which are mainly composed of the protein tubulin, have been assigned a role in the secretory process, perhaps to convey the secretory granule to the cell membrane (56, 57). A shift in the equilibrium between the depolymerized and polymerized forms of microtubular protein during heightened secretory activity may be crucial for the migration of the granules to the cell surface.

Colchicine, which causes disaggregation of microtubules, has been widely used to study the role of microtubules in the secretory process; and the disorganization of the microtubule system by colchicine can lead to the arrest of secretion (58, 59). However, these inhibitory effects may be a consequence of action on the packaging and/or mobilization of the secretory granule rather than on the exocytotic process itself (60). Moreover, colchicine can bind to other protein components of the cell, so the disruptive effects of this agent may be exerted on a number of cellular structures.

Cell surface events may be controlled by cytoplasmic structures contiguous with the plasma membrane, particularly the microfilament system (61), and drugs that alter this system may intervene with phenomena taking place either intracellularly or at the cell surface. Cytochalasin B causes dissolution of microfilaments and depresses secretory activity in certain systems (62–64), but enhances the secretory response in others (65–67). The propensity of these pharmacological agents to affect other aspects of cell function—taken together with their variable effects on secretory cells (68)—raises doubts that the microtubular-microfilament system plays a direct role in conveying the secretory granule to the periphery of the cell. Alternatively, it might be argued that the primary function of these organelles is to maintain the normal architecture of the cell rather than directly participating in the exocytotic process.

5.2 Actin and Myosin

Besides just considering actin as a structural component of the cell by comprising the microfilaments, it may play a more active role as a contractile element. Actin and myosin have been isolated from many types of secretory cells and bear remarkably similar properties to their analogs found in muscle (69–71). Moreover, the close association between the actions of calcium on stimulus-secretion coupling and excitation-contraction coupling points to the possibility that a basically similar molecular mechanism underlies both processes (3, 39).

Analysis of the secretory process as a contractile event has received wide support, particularly since actin and myosin have been found in all eukaryotic cells, where they are involved in cell motility (cytoplasmic streaming, phagocytosis, etc.) (61, 72). The origin of motive force for directing secretory organelles to the plasma membrane may involve cytoplasmic streaming, although "streaming motion" within animal cells is not as dramatic as that within plant cells (73). Polymerization and depolymerization, assembly and disassembly of polymers, and gelation and solation may be associated with the regulation of motility. Calcium regulates cytoplasmic structure (sol-gel) by causing gels to solate and contract locally (72). Thus, calcium, by dissolving the gel of actin filaments, may break these filaments and permit interaction with myosin. The dissolution of actin-related gels by calcium raises the possibility that by affecting actin-actin interactions, calcium causes solation of the cytoplasm to promote the interaction of the secretory organelle with the plasma membrane (74). These attractive theories notwithstanding, the functional importance of actin in secretion is still in doubt, particularly since the subcellular localization of this protein has not been clearly defined (70, 75).

In extending the analogy of stimulus-secretion coupling to muscle contraction, one must take into account the important fact that muscle contraction is seemingly accomplished by at least two basic mechanisms. In skeletal and cardiac muscle, the interactions of actin and myosin involve the troponin-tropomyosin system, in which calcium binds troponin, thereby altering protein conformation and enabling actin and myosin to interact. In smooth muscle, another form of regulation involves the phosphorylation of a specific pair of myosin light chains catalyzed by protein kinases (76). So viewed from this perspective, the control of secretion may involve the actomyosin regulation by phosphorylation, mediated by a calcium-dependent activation of protein kinase.

There is growing evidence that secretory activity is associated with the calcium-dependent activation of protein kinase and the phosphorylation of cellular proteins (76–80). In platelets, the time course of protein phosphorylation is similar to that of 5-hydroxytryptamine release (81). The additional findings that stimulators of the blood platelet release reaction (including thrombin, collagen, and the calcium ionophore A23187) enhance protein phosphorylation, and inhibitors (such as the calcium antagonists tetracaine and verapamil) obtund protein phosphorylation, imply that calcium-activated phosphorylation of proteins may be an essential feature in the activation of the secretory process (82, 83). Interactions of actin and myosin similar to those in muscle have been established in blood platelets (84). In fact, the analogy between muscle contraction and secretory activity is no more clearly exemplified than in the blood platelet, where phosphorylation of myosin light chain kinase may control the ability of the platelet to produce tension (76, 84). This would enable the contractile proteins not only to act as a cytoskeleton, but also to be responsible for powering various cellular movements, including migration of secretory organelles. But these intriguing results must be regarded cautiously, because the basic question as to whether calcium exerts a direct action on the actomyosin system of secretory cells to generate the motive force required in the various expressions of the exocytotic process, involving either vesicular movement or membrane fusion, still remains unresolved.

5.3 Calmodulin

During the last decade a ubiquitous cellular protein termed calmodulin (calcium activator protein) has been proposed as an intracellular calcium "receptor" (see Chapter 2). Calmodulin was first identified about 1970 as a calcium-dependent regulator of phosphodiesterase activity in brain (85). Subsequently, it has been shown to be distributed ubiquitously in plant and animal tissues and, because of its lack of tissue specificity, may be a receptor for transmitting signals initiated by calcium. During activation the increase in ionic calcium in the cell promotes the formation of a calcium-calmodulin complex, which is a necessary cofactor for enzyme activation.

There is a growing awareness of the potential importance of calmodulin in the control mechanisms involved in secretion. Calmodulin is being identified in a growing number of secretory organs, including the endocrine and exocrine pancreas, the parathyroid gland, and the neurohypophysis (34, 86–88). Phenothiazine antipsychotic drugs, such as trifluoperazine, have been employed to study the mechanism of action of calmodulin, since they bind to calmodulin with high affinity in a calcium-dependent fashion, and also inhibit secretory activity (89).

The most detailed studies in this area have been conducted by DeLorenzo (90), who has provided evidence implicating calmodulin in the mechanism regulating neurotransmitter release. Calmodulin can be isolated not only from synaptic vesicles of the brain, but also from the cytoplasm of synaptosomes (isolated nerve endings). In addition, calcium-calmodulin stimulates ATP-dependent neurotransmitter release from isolated vesicles, and trifluoperazine, while inhibiting neurotransmitter release, antagonizes the stimulatory effects of calmodulin on protein phosphorylation. From these findings emerged the proposal that calcium triggers neurotransmitter release by interacting with calmodulin to activate protein phosphorylating mechanisms that promote the interaction of the synaptic vesicle with the synaptic membrane (see Fig. 3). This intriguing concept gains favor by the presence of calmodulin and phosphorylating mechanisms in the exocrine pancreas (87, 91).

The additional finding that phenothiazines block enzyme secretion and ^{45}Ca release from rat pancreatic acinar cells (92) confirms the belief that calmodulin has a functional role in pancreatic enzyme secretion. Additionally, in the endocrine pancreas, a correlation between the activation and inhibition of calmodulin-induced phosphorylation of protein and insulin secretion suggests a similar mechanism is operative (93). In this latter study, evidence was provided that both calcium and cyclic AMP regulate secretion through the activation of distinct protein kinases, and that calmodulin is involved in calcium-mediated insulin secretion through a calmodulin-dependent protein kinase. Although little is known about the subcellular localization of the proteins that are phosphorylated during enhanced secretory activity, some insight into our understanding of this problem may be gleaned from studies on the adrenal medulla, the prototype model of

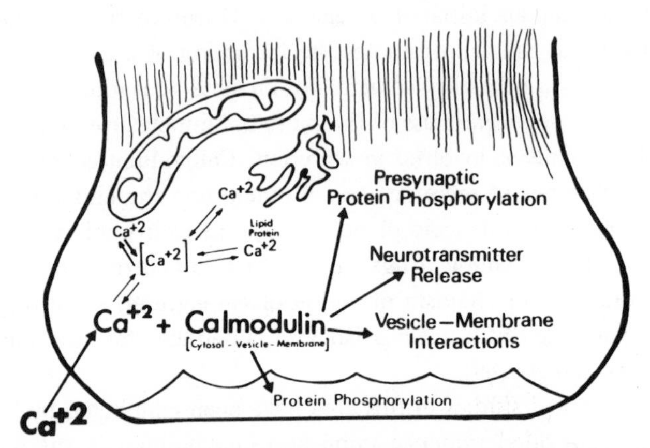

Figure 3 Schematic model of the calmodulin hypothesis for neurosecretion. Following depolarization dependent entry of calcium into the nerve terminal, calcium binds to calmodulin associated with the secretory vesicle or contained within the cytoplasm. The interaction of calcium and calmodulin could then trigger one or more processes, including protein phosphorylation, vesicle-membrane interactions, and/or neurotransmitter release. The regulation of assembly and disassembly of microtubules, by calcium–calmodulin mediated activation of tubulin kinase, may provide the basis for the role of protein phosphorylation in promoting vesicle-membrane interactions (Modified from ref. 90).

stimulus-secretion coupling. Cholinergic stimulation of isolated adrenal chromaffin cells induces a calcium dependent phosphorylation of proteins that precedes the secretory response (94); and the interaction of calmodulin with specific binding sites on the chromaffin granule membrane stimulates protein phosphorylation (95). These results provide evidence for a model of exocytosis involving a calcium-calmodulin complexation leading to phosphorylation of the secretory granule.

Calmodulin may also play an important role in the calcium-controlled assembly and disassembly of microtubules by affecting the activity of tubulin kinase, and in this way exerting an influence upon a wide variety of secretory cells (96). Although phosphorylation of the microtubules may be a critical control mechanism in promoting granule-microtubule interactions, calmodulin may regulate contractile proteins responsible for cell motility (actin and myosin) through the phosphorylation of myosin kinase (76, 97). We have already considered the possibility that in secretory cells myosin may play an important role in the calcium-dependent regulation of the hypothetical contractile apparatus important in exocytosis. Perceived from this aspect, the importance of calmodulin is heightened by the fact that this heat-stable, acidic protein closely resembles muscle troponin-C (98). These effects of calcium on protein phosphorylation—mediated through protein kinases that require calmodulin as a cofactor—may account for at least part of calcium's action on exocytosis. However, one must also be cognizant of the fact that there are protein kinase systems that are activated by cyclic AMP, cyclic GMP, as well as phospholipids (99, 100), and so one must not consider the protein-phosphorylating mechanisms that proceed through the calcium-calmodulin pathway to the exclusion of alternate pathways.

In summary, the recent studies concerned with calmodulin have provided what appears to be valuable information regarding the machinery that controls the secretory process. So even though most of the presently available data concerning the actions of calmodulin can be categorized as phenomenological, further developments in our understanding of how calmodulin is involved in the regulation of calcium-mediated secretion should provide keen insight into the nature of the mechanism of the secretory process.

5.4 Phospholipase-Mediated Phospholipid Turnover

The action of calcium on exocytosis may also be expressed through the enzymes that regulate phospholipid turnover. Membrane-bound phospholipases are likely to perform this function, although their physiological role has not yet been clearly defined. Changes in phospholipid turnover accompanying physiological (secretory) responses can be arbitrarily (and perhaps simplistically) classified into two groups: (a) those that are calcium-mediated, accompany the turnover of arachidonic acid in position 2 of phospholipids, and are presumably catalyzed by phospholipase A_2 (101–104); and (b) those that appear not to be calcium-mediated, primarily involve the turnover of the phospholipid headgroup, and are catalyzed by phospholipase C (105–108). The relevance of the calcium-dependent phospholipase activity to calcium-mediated exocytosis is apparent, not only because the enzyme(s) requires calcium for activity, but also because of the recent awareness of the possible role of arachidonic acid metabolites in cellular function, including secretion (109, 110). Arachidonic acid is metabolized through the cyclooxygenase and lipoxygenase pathways to prostaglandins and the hydroxy fatty acids which, like the cyclic nucleotides, seem to serve an important role in regulating calcium metabolism in the secretory cell (111, 112). In the platelet release reaction and the release of lysosomal enzymes from neutrophils, arachidonic acid mimics the effects of the primary stimulus for secretion (84, 112). Early studies focused on the products of the cyclooxygenase pathway, such as the classical prostaglandins (PGE or PGF), whereas more recent investigations implicate the lipoxygenase products (e.g., leukotrienes) as important intermediates in the molecular events associated with secretion (110, 112).

So, the pivotal role of calcium in the secretory process may be mediated through the activation of phospholipase A_2, triggered by the stimulus-induced redistribution of cellular calcium; this sets into motion the cascade of events resulting in arachidonic acid metabolism. There is also evidence that metabolites of arachidonic acid are important in the regulation of cyclic GMP levels in cells (113, 114). The calcium-dependent activation of arachidonic acid metabolism may, therefore, express its actions through the calcium regulation of cyclic GMP levels in a secretory cell. However, the role of cyclic GMP in secretory cells, or in cell function in general, it still obscure. In fact, rather than promoting the activity of secretory cells, cyclic GMP may somehow feed back as a negative modulator of cellular function (114).

In addition to arachidonic acid, the lysophospholipids are a second product of the phospholipase A_2 reaction. Their potential importance in membrane fusion reactions rests with their well-known ability to destabilize membranes and to promote membrane fusion (32). In addition, lysophosphatidylcholine (and phospholipase A_2) stimulates guanylate cyclase (115, 116), which may be an important component of the events that occur during or following secretion. Enzymes producing lysophospholipids via phospholipase A_2 activity are associated with cellular membranes, including the plasma membrane (117, 118). However, the physiological relevance of these effects of lysophosphatidylcholine to promote membrane fusion is somewhat dimmed by the finding that stimulation of secretory cells is not associated with increased turnover of lysophosphatidylcholine (119–121). On the other hand, the facilitatory action of phosphatidylserine on the release of histamine from mast cells induced by a variety of secretagogues (122) appears to be a consequence of its conversion of lysophosphatidylserine (123). Additionally, our own work on the adrenal cortex and neutrophils, which has demonstrated the importance of the phospholipase A_2 mediated turnover of phosphatidylinositol, favors the notion that the deacylation of phosphatidylinositol, and its resulting conversion to lysophosphatidylinositol, may be an important event in the activation of the secretory process (102–103).

Since lysophospholipids destabilize membranes and promote lysis, their concentration in the cell would have to be highly controlled; otherwise, levels of the lysophospholipids detrimental to the cell would be reached. Thus, a rapid mechanism for the reacylation of the lysophospholipids is necessary to control the concentration of cellular lysophospholipids. Also, as Lands and his associates proposed (125), the turnover of fatty acyl side chains of membrane phospholipids may alter membrane properties, and in this way influence the exocytotic process. The deacylation-reacylation reaction may increase the amount of unsaturated fatty acids in membrane phospholipids, and thereby alter the structure and function of the membranes. This would promote the exocytotic process, either by altering membrane permeability, or by affecting the activity of membrane-bound enzymes (126).

It is apparent, therefore, that an important aspect of calcium's action in the exocytotic process is related to an ability to alter phospholipid turnover of the secretory cell, perhaps through the activation of a calcium-dependent phospholipase A_2. A calcium-dependent phosphatidylinositol-specific phospholipase C may also be involved in certain secretory systems, such as the platelet (127). It has even been proposed that phospholipase C-mediated turnover of phosphatidylinositol is coupled to a phospholipase A_2-mediated deacylation of phospholipid to supply free arachidonic acid to the cyclooxygenase and lipoxygenase systems (108). Hence, the multiple ways in which the activation of phospholipases may alter phospholipid metabolism make it difficult to identify the specific component of this proposed series of reactions directly responsible for triggering the exocytotic process.

Signals initiated by calcium during activation of the secretory cell may be expressed through several different intracellular receptors. In this context, calcium-dependent protein phosphorylation stimulated by phospholipids has also been demonstrated in a number of different tissues (128). These findings, taken together with the phospholipid-sensitive, calcium-dependent protein kinase that has recently been identified (100), suggest an involvement of phospholipids in regulating calcium-dependent phosphorylation of endogenous substrate proteins. This system probably functions independently of the calmodulin-sensitive protein phosphorylation system.

Yet, there appear to be grounds for suspecting that calmodulin does not operate independently of phospholipids. Calmodulin reportedly affects the synthesis of arachidonic acid metabolites by promoting the action of calcium-dependent phospholipase A_2 as well as cyclooxygenase (129). Calmodulin may also be important for controlling levels of arachidonic acid metabolites by modifying the activity of enzymes that are involved in the degradation of prostaglandins (129). Thus, calcium-sensitive activator proteins may provide another important link between calcium and arachidonic acid metabolism in their interdependent actions to promote secretion.

6 PERSPECTIVE

In this overview we have considered the phenomenon of calcium-dependent exocytosis from a number of different aspects while attempting to avoid the pitfall of oversimplifying what is an apparently complex series of molecular events. The mechanism is sufficiently complex that no single approach is likely to reveal complete answers. The protean actions calcium exerts in the cell aver that we must realistically consider that the critical role of calcium in the exocytotic process may focus on more than one specific site. This assertion is supported by the fact that calcium not only directly affects the machinery that controls the secretory apparatus, but also critically modulates cyclic nucleotide and arachidonic acid metabolism, which in turn may be important for regulating the secretory process (1). The actions of cyclic nucleotides and arachidonic acid metabolites on secretory cells may, in the main, be exerted through the regulation of calcium metabolism in the cell. A feedback loop may therefore exist, wherein calcium controls the metabolism of various putative cellular mediators, which in turn regulate calcium metabolism in the cell. This concept then implies the existence of multiple calcium pools which are critical for the regulation of secretory activity (1, 111).

An alternative view of this multifaceted aspect of calcium's action envisions the activation of stimulus-secretion coupling as first involving the interaction of calcium with the acidic phospholipids of the plasma membrane, or through the activation of calcium-dependent phospholipases. Since exocytosis is clearly related to the increase in the ionic calcium of the cell, the interaction of calcium ion with various cytoplasmic components of the cell—perhaps calmodulin—may further serve to promote the important calcium-mediated actions necessary for the delivery of the secretory organelle to the plasma membrane and the subsequent discharge of its contents from the cell. Kretsinger (130) has postulated that processes in which calcium functions as a second messenger are mediated by calcium-binding proteins, implying that calcium-protein interactions explain, in molecular terms, the cellular actions of calcium. This statement may have validity with regard to the secretory process; it must not, however, be so narrowly interpreted as to exclude other critical factors, particularly phospholipids. Certainly, the final stage in fusion, the establishment of cytoplasmic continuity, presumably involves the rearrangement of both membrane lipids and proteins between contiguous membranes.

We have thus described what appears to be the myriad events associated with calcium-activated exocytotic secretion. Although our primary function here has been to clarify, it seems fair to state that an elucidation of the biochemical mechanism of exocytosis, and the role of calcium in this process, can only be ultimately achieved by a variety of experimental approaches to the problem, and the formulation of a unitary theory based on these diverse studies.

REFERENCES

1. R. P. Rubin, *Calcium and Cellular Secretion*, Plenum Press, New York, 1982.
2. J. A. Williams, *Fed. Proc.*, **40,** 128 (1981).
3. W. W. Douglas, *Brit. J. Pharmacol.*, **34,** 451 (1968).
4. B. Katz, *The Release of Neural Transmitter Substances*, Charles C Thomas, Springfield, Illinois, 1969.

5. B. Ceccarelli, W. P. Hurlbut, and A. Mauro, *J. Cell Biol.*, **54,** 30 (1972).
6. V. P. Whittaker in *Advances in Cytopharmacology*, Vol. 2, B. Ceccarelli et al., Eds., Raven Press, New York, 1973, p. 311.
7. H. Blaschko, P. Hagen, and A. D. Welch, *J. Physiol. (Lond.)*, **129,** 27 (1955).
8. N. A. Hillarp, S. Langerstedt, and B. Nilson, *Acta Physiol. Scand.*, **29,** 251 (1953).
9. O. Viveros in *Handbook of Physiology*, Vol. 6, H. Blaschko et al., Eds., American Physiological Society, Washington, D.C., 1975, p. 389.
10. J. E. Heuser and T. S. Reese, *J. Cell Biol.*, **57,** 315 (1973).
11. J. F. Morris, J. J. Nordmann, and R. E. J. Dyball, *Int. Rev. Exp. Path.*, **18,** 1 (1978).
12. H. Blaschko, R. S. Comline, F. H. Schneider, M. Silver, and A. D. Smith, *Nature*, **215,** 58 (1967).
13. M. G. Farquhar, *Mem. Soc. Endocrinol.*, **19,** 79 (1971).
14. W. W. Douglas, *Biochem. Soc. Symp.*, **39,** 1 (1974).
15. O. Grynszpan-Winograd in *Handbook of Physiology*, Vol. 6, H. Blaschko et al., Eds., American Physiological Society, Washington, D.C., 1975, p. 295.
16. L. Orci and A. Perrelet in *The Diabetic Pancreas*, B. W. Volk and K. F. Wellman, Eds., Plenum Press, New York, 1977, p. 171.
17. J. J. Morrissey, R. E. Shofstall, J. W. Hamilton, and D. V. Cohn, *Proc. Nat. Acad. Sci.*, **77,** 6406 (1980).
18. J. E. Heuser and T. S. Reese. *J. Cell Biol.*, **88,** 564 (1981).
19. J. E. Heuser, T. S. Reese, M. J. Dennis, Y. Jan, L. Jan, and L. Evans, *J. Cell Biol.*, **81,** 275 (1979).
20. I. W. Chubb, W. P. DePotter, and A. F. DeSchaepdryver, Naunyn Schmiedeberg's *Arch. Pharmacol.*, **274,** 281 (1972).
21. H. Thoenen, A. Huerlimann, and W. Haefely, *Eur. J. Pharmacol.*, **6,** 29 (1969).
22. A. R. Wakade and S. M. Kirpekar, *J. Pharmacol. Exp. Ther.*, **190,** 451 (1974).
23. D. A. Redburn, J. Stramler, and L. T. Potter, *Biochem. Pharmacol.*, **28,** 2091 (1979).
24. S. S. Rothman, *Amer. J. Physiol.*, **228,** 1828 (1975).
25. M. Israel and Y. Dunant, *Prog. Brain Res.*, **49,** 125 (1979).
26. H. Zimmermann, *Prog. Brain Res.*, **49,** 141 (1979).
27. M. W. Walker and M. G. Farquhar, *Endocrinology*, **107,** 1095 (1980).
28. G. E. Palade in *Subcellular Particles*, T. Hayashi, Ed., Ronald Press, New York, 1959, p. 64.
29. G. Poste and A. C. Allison, *Biochim. Biophys. Acta*, **300,** 421 (1973).
30. Z. Toister and A. Loyter, *J. Biol. Chem.*, **248,** 422 (1973).
31. R. D. Prusch and J. A. Hannafin, *J. Gen. Physiol.*, **74,** 523 (1979).
32. J. A. Lucy in *Membrane Fusion*, G. Poste and G. L. Nicolson, Eds., North Holland Publishing Company, Amsterdam, 1978, p. 268.
33. R. Ekerdt, C. Dahl, and M. Gratzl, *Biochim. Biophys. Acta*, **646,** 10 (1981).
34. N. A. Thorn, J. T. Russell, C. Torp-Pedersen, and M. Treiman, *Ann. NY Acad. Sci.*, **307,** 618 (1978).
35. H. B. Pollard, C. J. Pazoles, C. E. Creutz, and O. Zinder, *Int. Rev. Cytol.*, **58,** 159 (1979).

36. H. B. Pollard, C. E. Creutz, and C. E. Pazoles, *Rec. Prog. Horm. Res.*, **37,** 299 (1981).
37. S. Muchnik and R. A. Venosa, *Nature*, **222,** 169 (1969).
38. H. B. Pollard, C. E. Creutz, V. Fowler, J. H. Scott, and C. J. Pazoles, *Cold Spring Harbor Symp. Quant. Biol.*, **46,** 819 (1982).
39. R. P. Rubin, *Pharmacol. Rev.*, **22,** 389 (1970).
40. S. G. Oberg and M. R. Robinovitch, *J. Supramol. Str.*, **13,** 295 (1980).
41. B. Davis and N. R. Lazarus, *J. Physiol. (Lond.)*, **256,** 709 (1976).
42. S. Milutinovic, B. E. Argent, I. Schulz, and G. Sachs, *J. Membrane Biol.*, **36,** 281 (1977).
43. F. Konings and W. DePotter, Naunyn-Schmiedeberg's *Arch. Pharmacol.*, **317,** 97 (1981).
44. P. I. Lelkes, E. Lavie, D. Naquira, F. Schneeweiss, A. S. Schneider, and K. Rosenheck, *FEBS Lett.*, **115,** 129 (1980).
45. D. Papahadjopoulos, A. Portis, and W. Pangborn, *Ann. NY Acad. Sci.*, **308,** 50 (1978).
46. D. Papahadjopoulos in *Membrane Fusion*, G. Poste and G. L. Nicolson, Eds., North Holland Publishing Company, Amsterdam, 1978, p. 766.
47. N. Duzgunes, J. Wilschut, R. Fraley, and D. Papahadjopoulos, *Biochim. Biophys. Acta*, **642,** 182 (1981).
48. D. Papahadjopoulos, G. Poste, B. E. Schaeffer, and W. J. Vail, *Biochim. Biophys. Acta*, **352,** 10 (1974).
49. J. Zimmerberg, F. S. Cohen, and A. Finkelstein, *Science*, **210,** 906 (1980).
50. J. E. Rothman and J. Lenard, *Science*, **195,** 743 (1977).
51. A. M. Poisner in *Frontiers in Neuroendocrinology*, W. F. Ganong and L. Martini, Eds., Oxford University Press. 1973, p. 33.
52. H. Matt, M. Bilinski, and H. Plattner, *J. Cell Sci.*, **32,** 67 (1978).
53. H. Plattner, K. Reichel, H. Matt, J. Beisson, M. Lefort-Tran, and M. Pouphile, *J. Cell. Sci.*, **46,** 17 (1980).
54. G. J. Brewer, *J. Theor. Biol.*, **85,** 75 (1980).
55. E. K. Matthews, R. J. Evans, and P. M. Dean, *Biochem. J.*, **130,** 825 (1972).
56. P. E. Lacy and W. J. Malaisse, *Rec. Prog. Horm. Res.*, **29,** 199 (1973).
57. J. Wolff and J. A. Williams, *Rec. Prog. Horm. Res.*, **29,** 229 (1973).
58. A. C. Allison, *Ciba Found. Symp.*, **14,** 109 (1973).
59. E. Gillespie, *Ann. NY Acad. Sci.*, **253,** 771 (1975).
60. C. Patzelt, D. Brown, and B. Jeanrenaud, *J. Cell Biol.*, **73,** 578 (1977).
61. M. Clarke and J. A. Spudich, *Ann. Rev. Biochem.*, **46,** 797 (1977).
62. N. B. Thoa, G. F. Wooten, J. Axelrod and I. J. Kopin, *Proc. Natl. Acad. Sci.*, **69,** 520 (1972).
63. F. R. Butcher and R. H. Goldman, *J. Cell Biol.*, **60,** 519 (1974).
64. A. Khar, J. Kunert-Radek, and M. Jutisz, *FEBS Lett.*, **104,** 410 (1979).
65. E. VanObberghen, G. Somers, G. Devis, G. D. Vaughan, F. Malaisse-Lagae, L. Orci, and W. J. Malaisse, *J. Clin. Invest.*, **52,** 1041 (1973).
66. R. J. Haslam, M. M. L. Davidson, and M. D. McClenaghan, *Nature*, **253,** 455 (1975).

67. P. H. Naccache, H. J. Showell, E. L. Becker, and R. I. Sha'afi, *J. Cell Biol.*, **75,** 635 (1977).
68. E. F. Nemeth, and W. W. Douglas, Naunyn-Schmiedeberg's *Arch. Pharmacol.*, **302,** 153 (1978).
69. J. M. Trifaro, *Ann. Rev. Pharmacol. Toxicol.*, **17,** 27 (1977).
70. J. M. Trifaro, *Neuroscience*, **3,** 1 (1978).
71. R. E. Ostlund, J. T. Leung, and D. M. Kipnis, *J. Cell Biol.*, **77,** 827 (1978).
72. S. F. Hitchcock, *J. Cell Biol.*, **74,** 1 (1977).
73. K. R. Porter, in *Cell Motility* (Book A), R. Goldman et al., Eds., Cold Spring Harbor Laboratory, 1976, p. 1.
74. D. L. Taylor, S. B. Hellelwell, H. W. Virgin, and J. Heiple in *Cell Motility: Molecules and Organization*, S. Hatano et al, University Park Press, Baltimore, 1976, p. 363.
75. D. I. Meyer and M. M. Burger, *FEBS Lett.*, **101,** 129 (1979).
76. R. S. Adelstein, M. A. Conti, and M. D. Pato, *Ann. NY Acad. Sci.*, **356,** 142 (1980).
77. M. Lambert, J. Camus, and J. Christophe, *Biochem. Pharmacol.*, **24,** 1755 (1975).
78. W. Sieghart, T. C. Theoharides, S. L. Alper, W. W. Douglas, and P. Greengard, *Nature*, **275,** 329 (1978).
79. U. K. Schubart, N. Fleischer, and J. Erlichman, *J. Biol. Chem.*, **255,** 11063 (1980).
80. B. J. Baum, J. M. Freiburg, H. Ito, G. S. Roth, and C. R. Filburn, *J. Biol. Chem.*, **256,** 9731 (1981).
81. R. M. Lyons and J. O. Shaw, *J. Clin. Invest.*, **65,** 242 (1980).
82. R. J. Haslam and J. A. Lynham, *Biochem. Soc. Trans.*, **4,** 694 (1976).
83. R. J. Haslam and J. A. Lynham, *Biochem. Biophys. Res. Commun.*, **77,** 714 (1977).
84. M. B. Feinstein, G. A. Rodan, and L. S. Cutler in *Platelets in Biology and Pathology*, J. L. Gordon, Ed., Elsevier, Amsterdam, 1981, p. 437.
85. W. Y. Cheung, *Science*, **207,** 19 (1979).
86. M. C. Sugden, M. R. Christie, and S. J. H. Ashcroft, *FEBS Lett.*, **105,** 95 (1979).
87. D. C. Bartelt and G. A. Scheele, *Ann. NY Acad. Sci.*, **356,** 356 (1980).
88. E. M. Brown, B. F. Dawsonhughes, R. E. Wilson, and N. Adragna, *J. Clin. Endocrinol. Met.*, **53,** 1064 (1981).
89. F. Vincenzi, *Cell Calcium*, **2,** 387 (1981).
90. R. J. DeLorenzo, *Cell Calcium*, **2,** 365 (1981).
91. R. Jahn, C. Unger, and H. D. Söling, *Eur. J. Biochem.*, **112,** 345 (1980).
92. S. Heisler, L. Chauvelot, D. Desjardins, C. Noel, H. Lambert, and L. Desy-Audet, *Can. J. Physiol. Pharmacol.*, **59,** 994 (1981).
93. U. K. Schubart, J. Erlichman, and N. Fleischer, *J. Biol. Chem.*, **255,** 4120 (1980).
94. C. M. Amy and N. Kirshner, *J. Neurochem.*, **36,** 847 (1981).
95. R. D. Burgoyne and M. J. Geisow, *FEBS Lett.*, **131,** 127 (1981).
96. J. R. Dedman, B. R. Brinkley, and A. R. Means, *Adv. Cyclic Nucl. Res.*, **11,** 131 (1979).
97. C. B. Klee, T. H. Crouch, and P. G. Richman, *Ann. Rev. Biochem.*, **49,** 489 (1980).
98. A. R. Means and J. R. Dedman, *Nature*, **285,** 73 (1980).
99. Y. Nishizuka, Y. Takai, E. Hashimoto, A. Kishimoto, Y. Kuroda, K. Sakai, and H. Yamamura, *Mol. Cell Biochem.*, **23,** 153 (1979).

100. J. F. Kuo, R. G. G. Andersson, B. C. Wise, L. Mackerlova, I. Salomonsson, N. L. Brackett, N. Katoh, M. Shoji, and R. W. Wrenn, *Proc. Natl. Acad. Sci.*, **77,** 7039 (1980).
101. B. Haye and C. Jacquemin, *Biochim. Biophys. Acta*, **487,** 231 (1977).
102. M. P. Schrey and R. P. Rubin, *J. Biol. Chem.*, **254,** 11234 (1979).
103. R. P. Rubin, L. E. Sink, and R. J. Freer, *Biochem. J.*, **194,** 497 (1981).
104. Z. Naor and K. J. Catt, *J. Biol. Chem.*, **256,** 2226 (1981).
105. J. M. Trifaro, *Mol. Pharmacol.*, **5,** 420 (1969).
106. Y. Oron, M. Lowe, and Z. Selinger, *Mol. Pharmacol.*, **11,** 79 (1975).
107. L. M. Jones, S. Cockcroft, and R. H. Michell, *Biochem. J.*, **182,** 669 (1979).
108. M. M. Billah, E. G. Lapetina, and P. Cuatrecasas, *J. Biol. Chem.*, **255,** 10227 (1980).
109. R. P. Rubin and S. G. Laychock, *Ann. NY Acad. Sci.*, **307,** 377 (1978).
110. R. I. Sha'afi and P. H. Naccache in *Advances in Inflammation Research*, Vol. 3, G. Weissmann, Ed., Raven Press, New York, 1981, p. 115.
111. R. P. Rubin in *New Perspectives on Calcium Antagonists*, G. B. Weiss, Ed., Waverly Press, Baltimore, 1981, p. 147.
112. P. H. Naccache, R. I. Sha'afi, P. Borgeat, and E. J. Goetzl, *J. Clin. Invest.*, **67,** 1584 (1981).
113. N. D. Goldberg, G. Graff, M. K. Haddox, J. H. Stephenson, D. B. Glass, and M. E. Moser, *Adv. Cyclic Nucl. Res.*, **9,** 101 (1978).
114. F. Murad, W. P. Arnold, C. K. Mittal, and J. M. Braughler, *Adv. Cyclic Nucl. Res.*, **11,** 175 (1979).
115. W. Y. Shier, J. H. Baldwin, M. N. Hamilton, R. T. Hamilton, and N. M. Thanass, *Proc. Natl. Acad. Sci.*, **73,** 1584 (1976).
116. D. Aunis, M. Pescheloche, and J. Zwiller, *Neuroscience*, **3,** 83 (1978).
117. J. D. Newkirk and M. Waite, *Biochim. Biophys. Acta*, **298,** 562 (1973).
118. M. P. Schrey, R. C. Franson, and R. P. Rubin, *Cell Calcium*, **1,** 91 (1980).
119. R. R. Baker, M. J. Dowdall, and V. P. Whittaker, *Brain Res.*, **100,** 629 (1975).
120. J. Meldolesi, N. Borgese, P. DeCamilli, and B. Ceccarelli in *Membrane Fusion*, G. Poste and G. L. Nicolson, Eds., North Holland Publishing Company, Amsterdam, 1978, p. 510.
121. G. Arthur and A. Sheltawy, *Biochem. J.*, **191,** 523 (1980).
122. J. C. Foreman, L. G. Garland, and J. L. Mongar, *Symp. Soc. Exp. Biol.*, **30,** 193 (1976).
123. T. W. Martin and D. Lagunoff, *Nature*, **279,** 250 (1979).
124. R. P. Rubin, *Fed. Proc.*, **41,** 2181 (1982).
125. W. E. M. Lands and C. G. Crawford in *The Enzymes of Biological Membranes*, A. Martonosi, Ed., Plenum Press, New York, 1976, p. 3.
126. H. Sandermann, *Biochim. Biophys. Acta*, **515,** 209 (1978).
127. S. Rittenhouse-Simmons, *J. Clin. Invest.*, **63,** 580 (1979).
128. R. W. Wrenn, N. Katoh, and J. F. Kuo, *Biochim. Biophys. Acta*, **676,** 266 (1981).
129. P. Y. K. Wong, W. H. Lee, and P. H-W. Chao, *Ann. NY Acad. Sci.*, **356,** 179 (1980).
130. R. H. Kretsinger in *Calcium Transport in Contraction and Secretion*, E. Carafoli et al., Eds., North Holland Publishing Company, Amsterdam, 1975, p. 469.

CHAPTER 5

Calcium in Mineralized Tissues

E. D. EANES
J. D. TERMINE

Mineralized Tissue Research Branch
National Institute of Dental Research
National Institutes of Health
Bethesda, Maryland

CONTENTS

1 INTRODUCTION

Calcium is a major chemical constituent of the skeletal tissues of vertebrate organisms. Well over 20% by weight of most hard tissue is comprised of this ion alone. Calcium is found predominantly in the mineral fraction of these tissues, combined with phosphate in the form of inorganic salts. These salts provide the hardness and rigidity that uniquely characterize normal, healthy skeletal tissue. In addition, they are the principal storehouse for conserving body calcium. In this latter capacity, skeletal mineral plays an important physiological, as well as mechanical, role as an integral part of the body's calcium regulatory machinery. In this chapter, the mechanisms of skeletal calcification are discussed. Because of the complexity and heterogeneity of mineral/tissue interactions, only normal mineralization processes are covered. This is followed by a brief discussion of the role of skeletal mineral in calcium homeostasis. Before presenting the biology of mineralization, however, many of the physical and chemical aspects of the mineral salts themselves are reviewed.

2 STRUCTURE AND CHEMISTRY OF SKELETAL MINERAL

2.1 Mineral Structure

DeJong (1) in 1926 was the first to show, with the then relatively new technique of x-ray diffraction, that the mineral in bone bore a close structural resemblance to the apatitic calcium phosphates. In this pioneering diffraction study, deJong also observed, however, that the apatite crystals in bone were extremely minute and ill-defined, with maximum dimensions considerably less than 0.1 μm. Because crystals of such small size do not readily lend themselves to structural analysis, progress in refining and extending the initial observations of deJong, by direct x-ray examination of biological apatites, has been slow and difficult. Even today, after nearly 60 years of dramatic technical advances in x-ray crystallography, researchers (2) are only beginning to be successful in applying the powerful structure solving techniques of this field to enamel crystallites, the largest of the biological apatites. The smaller bone and dentin apatites are still not amenable to detailed structure analysis. For this reason, much of what we currently know about the main structural features of the biological apatites has been inferred by comparison with well-crystallized, nonbiological analogs. Several excellent reviews (3–8) have appeared during the past decade or so, in which the atomic scale features of these apatites and their applicability to the biological apatites are thoroughly discussed. In this section, mainly those features that relate to the distribution of Ca^{2+} ions in the apatite structure are emphasized.

The basic structural features of the apatite lattice were first worked out for geological fluorapatite ($Ca_5(PO_4)_3F$) by Mehmel (9) and, independently, Náray-Szabó (10) in 1930. However, the detailed spatial arrangement of the constituent ions in the apatite structure was not firmly established until about 25 years later by Posner et al. (11) from x-ray diffraction studies on synthetically prepared single crystals of hydroxyapatite ($Ca_5(PO_4)_3OH$), and by Kay et al. (12) from neutron diffraction studies on geological material. As these studies revealed, the most striking feature of the hydroxyapatite structure is the hexagonal arrangement of Ca^{2+} and PO_4^{3-} ions about columns of monovalent OH^- ions. Figure 1 shows a one unit thick layer of this arrangement projected onto a plane perpendicular to the column axis. Figure 2 shows the hexagonal framework of the Ca^{2+} ions only. As seen from these figures, the Ca^{2+} ions are not all equivalent, but instead are distributed among two distinctly different lattice sites. Three out of every five calcium ions (Ca_{II}) form equilateral triangles centered on, and perpendicular to, the hexad OH^- axis. Successive triangles are spaced 3.44 Å apart along this axis [at 0.25 (dotted line) and 0.75 (dashed line), Fig. 1] and are rotated 60° about it. The remaining calcium ions (Ca_I) occupy sites at 0 and 0.5 units along columns paralleling the hexad OH^- axis, but separated laterally from it by intervening phosphate groups. Each triangular Ca_{II} ion is irregularly surrounded by 6 oxygens from 5 different PO_4^{3-} groups and by an OH^- ion. Each columnar Ca_I ion is coordinated with 9 oxygens contributed by 6 neighboring PO_4^{3-}'s. Average Ca_I-O and Ca_{II}-O distances are 2.56 and 2.45 Å, respectively. The sevenfold coordinated triangular Ca^{2+} site, therefore, is the smaller of the two Ca^{2+} positions in the apatite lattice. However, the available space at both Ca sites is greater than that found in some other inorganic Ca compounds, such as $Ca(OH)_2$, where the Ca-O distance is only 2.37 Å (13). Distances to the P centers of neighboring PO_4^{3-} groups and to other Ca sites are more variable than the closer Ca-O distances. They range from 3.09 to 3.67 Å for Ca-P, and from 3.44 to 4.17 Å for Ca-Ca distances.

2.2 Mineral Chemistry

Hydroxyapatite is generally considered the chemical prototype for the biological apatites. Hard tissue mineral, however, contains a variety of chemical species

Figure 1 The hydroxyapatite structure projected onto the *a*, *b* plane perpendicular to the hexad axis. The single circles represent the Ca^{2+} ions, the double circles OH^- ions, and the triangles the tetrahedral PO_4^{3-} groups. Two of the oxygens in each PO_4^{3-} group superimpose in this projection. Their centers are directly above and below the triangular apex closest to the CA^{2+} positions.

The numbers within the symbols indicate, as fractions of one unit layer thickness (6.88 Å), the elevation of the ions from the plane of the figure. The distance between adjacent OH^- columns is 9.43 Å. The dashed and solid lines outline the calcium triangles at 0.25 and 0.75, respectively.

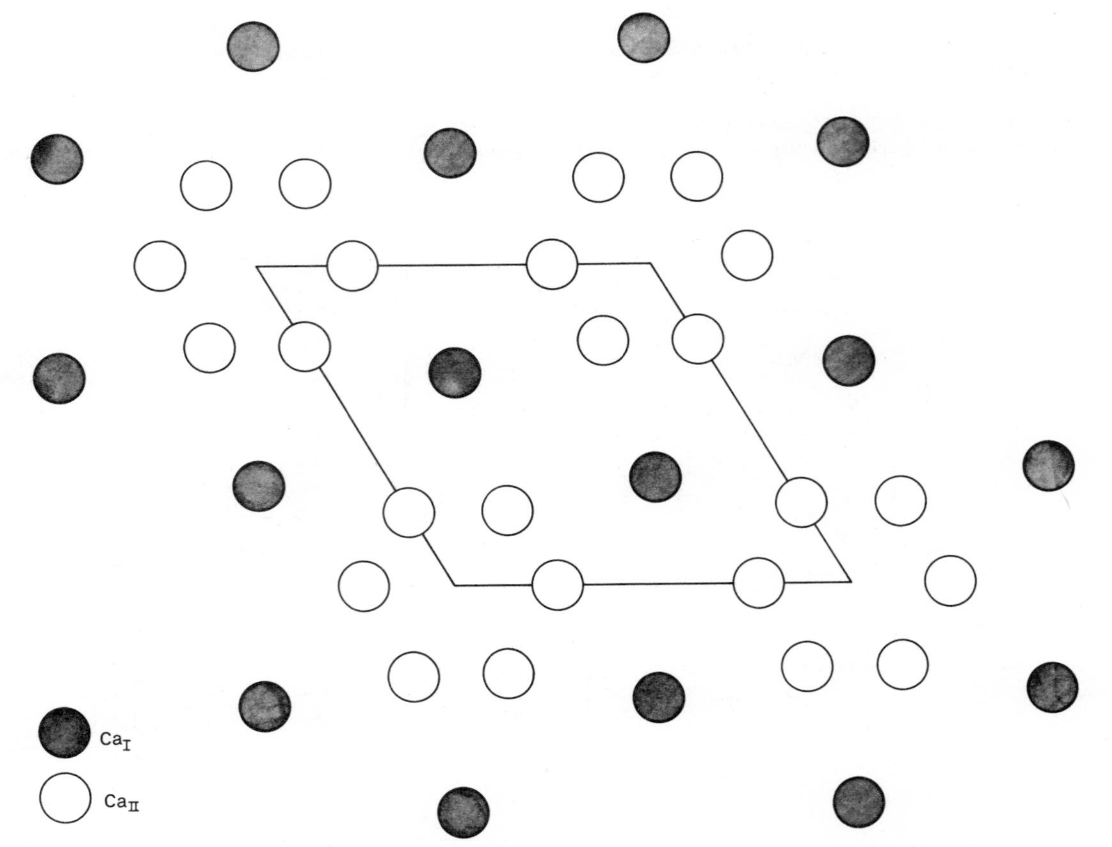

Figure 2 The calcium arrangement in the hydroxyapatite structure, shown in the same perspective as in Figure 1. The closed circles represent the columnar Ca_I ions and the open circles the triangular Ca_{II} ions. The solid lines drawn between the centers of the triangular Ca_{II} clusters outline the unit cell edges of the apatite lattice.

not normally found in pure hydroxyapatite. The most common inorganic impurities include CO_3^{2-}, HCO_3^-, F^-, Cl^-, Mg^{2+}, Na^+, K^+, H^+, and H_2O (14). Some 40 other species (mostly elements) are present in small and widely varying amounts. It is not always known, however, to what extent biomineral composition data can be correlated with lattice substitutions that result in departures from apatite stoichiometry. Some of the impurities may be adsorbed on crystal surfaces, trapped within gross lattice defects such as dislocations and inclusions, or concentrated in nonapatite phases. Nevertheless, many structural-chemical studies (6, 7) have shown that much of the observed compositional variations in hard tissue mineral can be interpreted in terms of crystalline apatite models.

Divalent cations, such as Sr^{2+} and Ba^{2+}, can easily replace Ca^{2+} in the apatite lattice. On the other hand, Mg^{2+}, the most abundant divalent cation impurity in hard tissue mineral, appears to be mostly outside the body of the apatite crystal (14). This ion is too small to fit easily into either of the Ca^{2+} positions. Any Mg^{2+} that does substitute probably locates at the smaller triangular Ca^{2+} site. Another factor that excludes Mg^{2+} from the apatite lattice is its inability, compared to other divalent cations, to readily desolvate at surface growth sites during crystal formation. Unless completely stripped of its hydration layer, Mg^{2+} cannot subsequently be incorporated into the lattice structure. The ions Na^+ and K^+ also can locate at Ca^{2+} sites, although the extent to which they substitute is small, compared to their relative abundance in surrounding skeletal fluids. The accompanying loss of positive charge by the exchange of Na^+ or K^+ for Ca^{2+} requires, however, a concomitant reduction in negative charge to maintain electrical balance in the lattice. As discussed below, one mechanism by which this can be accomplished is through the protonation of an adjacent PO_4^{3-} ion. There has been some interest in the possibility that hydronium ions, H_3O^+, may replace Ca^{2+} ions. Its size does not preclude such a substitution. It is unlikely, however, that H_3O^+ would be stable in a basic salt like apatite; since H_3O^+ is usually found only in very proton-rich substances, such as crystalline hydrates of strong acids (3).

Although there is no evidence for direct substitution, anion impurities such as CO_3^{2-}, Cl^-, and F^- can have a measurable effect on the position and/or occupancy of the Ca^{2+} lattice sites. As an example, carbonate, the most abundant anion impurity in bioapatites, locates in sites adjacent to the calcium positions. Most of the carbonate appears to fill sites normally occupied by PO_4^{3-} (2, 16). The planar shape of the smaller CO_3^{2-} allows it to occupy a position corresponding to one of the faces of the tetrahedral PO_4^{3-} space, leaving a void at the opposing apex. Electroneutrality could be preserved by filling this void with a monovalent anion, such as OH^- or F^-. But the concomitant exchange of a monovalent cation, such as Na^+ or K^+, for a nearby Ca^{2+} would also maintain electrical balance, as would a Ca^{2+} vacancy coupled with the protonation of the CO_3^{2-}. A minor amount of carbonate (from 5 to 10% in enamel) substitutes for two OH^- ions, and occupies a site along the hexad axis about midway between the two empty OH^- sites (2, 17, 18). Although none of the carbonate atoms actually occupy the vacated OH^- positions, the carbonate oxygens are sufficiently close to strain the triangular Ca_{II} ions adjoining these positions.

The exact location of the Cl^- anion in bioapatites is not established, but, from comparison with the chlorapatite structure (19), placement along the hexad column approximately halfway between two adjacent OH^- sites is most probable. However, only one of these two OH^- sites is required to be vacated by the substituting Cl^- to maintain proper charge balance. The very close proximity of this rather large anion to the other OH^- would subject this latter anion to some steric stress. Similar stresses on adjacent Ca^{2+} ions, displacing them slightly from their normal positions, would also occur. Such disruptive perturbations may explain why bioapatites discriminate against Cl^- (20), despite it being the most abundant anion in biological fluids.

In contrast to the Cl^- anion, F^- readily substitutes in the hydroxyapatite structure. Fluoride's strong affinity for the OH^- sites in the apatite lattice (21) accounts for why nearly all the dietary fluoride retained by the body is found in this component of hard tissue mineral. However, the F^- for OH^- substitution is not strictly isomorphous. These two ions occupy slightly different positions along the hexad axis. Fluoride prefers to locate at the center of the Ca^{2+} triangle, whereas the oxygen of the polar OH^- group is about 0.3 Å away. Of these two positions, the F^- location is more stable because the occupying anion cannot form an electric moment with the encircling Ca-triangle (21). Thus, the more favorable geometry of the F^- lattice site may be an important contributing factor to the well-known stability this ion imparts to the apatite structure.

Although certain substitutions, most notably F^-, seem to protect bioapatites against chemical change, structural impurities generally appear to have a deleterious impact on the chemical and physical characteristics of these minerals. One of the more measurable effects of this overall weakening of the structural integrity and stability of bioapatites is increased solubility (8). However, as we discuss in a later section, local and systemic biological factors may mask these deleterious effects *in vivo*.

Chemical studies on synthetic apatites prepared at physiological pH's suggest that the biological apatites may possibly deviate from the model hydroxyapatite structure in another important respect; namely, the amount of Ca^{2+} in these apatites may be less than can be accounted for solely by impurity substitutions such as those just described. Several theories have been proposed to account for this so-called Ca-deficiency. Posner and coworkers (22–24) originally suggested that the calcium loss is balanced by the protonation of neighboring PO_4^{3-} groups. More recent schemes maintain electroneutrality in the presence of Ca^{2+} vacancies by a combination of proton additions and OH^- deficiencies (25–29). In addition, Brown and associates (30–32) have advanced the alternative proposal that the Ca-deficiency in bioapatites may be due to interlayered mixtures of stoichiometric hydroxyapatite and octacalcium phosphate, an acid calcium phosphate salt with a molar Ca/P ratio of 1.33.

Unfortunately, technical and interpretative difficulties associated with the analytical methods used for assaying H^+ and OH^- in apatites have, in the past, frustrated attempts to experimentally test the correctness of these proposals. Recently, however, Meyer (33) developed a titration method for determining the OH^- content, which he used to demonstrate that the chemical formula that best represented the observed compositional data for a number of synthetically prepared low carbonate apatites was $Ca_{10-x-y}(HPO_4)_x(PO_4)_{6-x}(OH)_{2-x-2y}$ (34). This formula was first proposed by Kuhl and Nebergall (27). Implicit in this formula are two types of calcium vacancies. In the first, or x-type, the Ca^{2+} loss is compensated for by the loss of an OH^- ion and the addition of a H^+ ion to an adjoining PO_4^{3-}. The second, or y-type, is electrically balanced solely by OH^- deficiencies. Both types of Ca^{2+} vacancies, therefore, involve compensating OH^- losses. This would suggest that the vacancies are concentrated among the triangular Ca_{II} sites that border the OH^- channels. Columnar Ca_I vacancies would be unlikely, on the basis of this model, because intervening PO_4^{3-} groups between empty OH^- and Ca_I^{2+} sites would create disruptive local charge imbalances, even if overall electroneutrality is maintained. In contrast to the Ca-deficient apatites described above, Posner and Perloff (22) reported that in an x-ray diffraction study of a cation-deficient lead apatite, the empty lead sites were found only along the columnar positions, and not among the triangular cation sites. Whether this contrast in location of vacant cation sites represents a fundamental difference in the structural properties of these two apatites, or whether it points up the inherent difficulty in deducing structural information from compositional data alone, remains to be established. In any case, despite impressive advances in this area, the situation described above illustrates that problems remain in establishing the exact structural locations of compositional deficiencies in chemically well-defined apatites.

The situation regarding the nature of the Ca^{2+} deficiency in the biological apatites is even more murky. First, the exact extent to which biological apatites may be deficient in Ca^{2+} is difficult to determine directly. Second, even though Meyer's OH^- assay (33) has proven to be extremely useful in testing nonstoichiometry theories in selected low carbonate apatites, the uncertainty of the ionic form of carbonate in bioapatites (CO_3^{2-} or HCO_3^-), together with the interfering effect this ion has on H^+ assays (35, 36), currently makes his method unsuitable for the analysis of OH^- in these high carbonate-containing structures. Using a different approach (neutron diffraction), Young and Spooner (37) were able to demonstrate a 20% or greater hydrogen deficiency at normal OH^- positions in tooth enamel apatite. More recent infrared analysis (2) indicated that this deficiency was most probably the result of a loss in structural OH^-. However, it was not possible to determine the extent to which this loss could be accounted for by vacancy formation as compared to substitution by other ions. Neither was it possible to correlate the OH^- loss with the presence of Ca^{2+} vacancies in this material.

3 NONAPATITIC MINERAL PHASES IN SKELETAL TISSUE

Another complication in interpreting the crystal chemistry of hard tissue mineral is the possible presence of nonapatitic mineral components. Although apatite is the dominant, thermodynamically stable phase, there have been numerous reports, in the past several years, on the existence of other calcium phosphate salts in skeletal tissue. Most prominently mentioned in these accounts are amorphous calcium phosphate (ACP), octacalcium phosphate (OCP), and dicalcium phosphate dihydrate (DCPD), also known as brushite. Of these three phases, ACP has received by far the greatest attention.

Robinson and Watson (38) were the first to present evidence for an amorphous mineral phase in bone. In their electron microscopic studies on human rib bone, they described an amorphous haze in areas adjacent to calcification fronts, which they concluded was a noncrystalline inorganic salt. A number of similar observations (39–41) have been reported since then on a variety of bone types. In all these studies, the amorphous bone mineral was seen concentrated primarily in preosseous regions, and occurred either as small spheroidal particles or as a homogeneous haze. The first physical evidence that an amorphous phase may be a separate mineral component in bone came from the x-ray diffraction studies of Harper and Posner (42) and Termine and Posner (43). These investigators observed that the bone mineral of several animal species uniformly exhibited markedly less intense integrated x-ray diffraction patterns than fully crystalline synthetic apatites comparable in chemistry, ultrastructure, and other x-ray features. From these findings, they established that the amount of fully crystalline apatite in mature bone was only 60 to 70% of the total mineral content. This figure was found to be even lower in young bone tissue (43). Since no other crystalline phases were observed, the remaining percentage was attributed to the presence of amorphous mineral components in the bone specimens. Similar estimates of crystalline and amorphous material in bone were obtained by infrared (44) and electron spin resonance spectroscopy (45).

However, some more recent studies (46, 47) questioned whether all of the x-ray amorphous fraction of bone mineral is truly noncrystalline. The possibility was raised that some of the x-ray amorphous pool may include crystalline material too poorly developed structurally to be discernible by the physical methods used. Nevertheless, the electron microscopic evidence clearly demonstrates that at least part of the x-ray amorphous fraction is comprised of mineral particles sufficiently distinct from the dominant apatite phase to be classified as a separate component.

Brown (31, 32) was the first to bring attention to the possibility of octacalcium phosphate ($Ca_8H_2(PO_4)_6 \cdot 5H_2O$) as a mineral component in hard tissue. Although there is no direct physical evidence for this phase *in vivo*, circumstantial evidence argues for its possible importance as a factor in the mineralization process. First, the fact that a lower crystal surface energy kinetically favors OCP formation over that of the more stable apatite, and the observation that the distinctive platy shape of bone crystals is atypical for geoapatites but characteristic of OCP, suggest that this salt may be a template phase for apatite growth *in vivo*. Second, in a recent chemical study of rat incisor enamel, Gruninger et al. (48) found that the effect of endogenous Mg^{2+} and F^- on the amount of acid phosphate in rapidly mineralizing areas of this tissue was consistent with this chemical species being concentrated in a transient OCP precursor phase. Finally, as detailed below, probably the most compelling evidence for the involvement of OCP in early mineral formation comes from chemical modeling studies of apatite growth in physiological-like synthetic solutions.

Several investigators (49–52) have proposed that brushite ($CaHPO_4 \cdot 2H_2O$) should be considered a possible mineral component in bone. Experimental evidence for its occurrence in bone, however, is at best tentative. Roufosse et al. (53) recently identified this phase by x-ray diffraction in embryonic chick bone. Its presence, however, could be firmly established only in less mineralized fractions of powdered bone that were separated from higher density fractions by differential centrifugation in bromoform-toluene. They were unable to detect brushite in untreated bone powder. Although the x-ray evidence is unequivocal, it cannot be ruled out with certainty that this phase was not formed artifactually as a result of the fractionation procedure.

4 SYNTHETIC MODEL STUDIES

The apparent concentration of nonapatitic calcium phosphates in less mature, incompletely mineralized tissue suggests that these phases may be precursors to the apatite phase. However, more definitive data on the possible roles these phases play as intermediates in apatite formation are extremely difficult to obtain in the biological setting, primarily because the sequence of precipitation events leading to apatite is obscured by the metabolic complexity of the overall bone forming process. Recognizing this difficulty, several investigators have resorted to synthetic model systems in attempts to separate the physicochemical and physiological factors underlying the mineralization process.

Watson and Robinson (54) were among the first to use synthetic calcium phosphate preparations to better understand the crystal growth of biological apatites. They found that the first precipitate to form, upon mixing sufficiently concentrated solutions of calcium and phosphate at neutral pH, was an extremely fine solid with no recognizably crystalline features when examined by electron microscopy and no resolvable electron diffraction pattern. This phase was also very short-lived, and rapidly changed into an apatite that very closely resembled bone apatite in crystal size and habit. In a later study, Bachra et al. (55) observed that coprecipitation of carbonate interfered with the crystallization of this amorphous phase to apatite. However, it was not until 1965, when lyophilization techniques were introduced to minimize post-reaction changes in solids sampled for analysis (56, 57), that further details on the nature of the amorphous phase and on its transformation to apatite were established.

Several lines of evidence suggest that synthetic amorphous calcium phosphate is a distinct chemical and structural entity with no known crystalline counterpart. When initially precipitated, ACP is a highly hydrated flocculent mass of loosely associated Ca and PO_4 ions (58) (Fig. 3*a*). It rapidly and spontaneously desolvates in solution to form irregularly shaped clusters of fused spher-

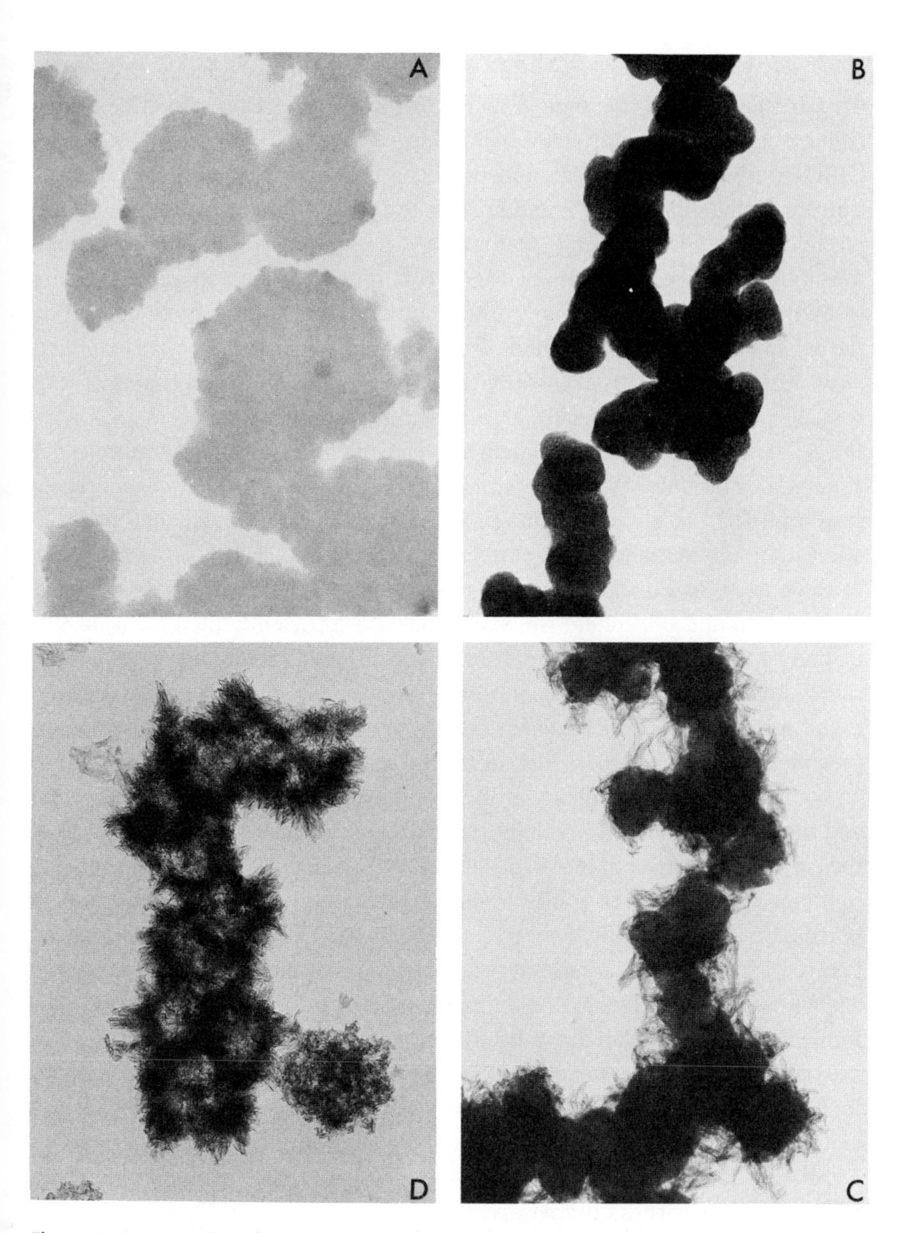

Figure 3 Transmission electron micrographs of calcium phosphate salts at different stages of ACP development and ACP-OCP conversion. The irregularly shaped low (*a*) and high (*b*) contrast material represent ACP during early and late stages of maturation, respectively. (*c*) First crystalline material shown in association with ACP clusters. (*d*) Appearance of crystalline clusters at completion of conversion.

oidal particles (59) (Fig. 3*b*). At no time during this consolidation process do the particles develop the angular shape or flat surfaces characteristic of crystalline calcium phosphates. The chemical formula for ACP(60), $Ca_3(PO_4)_{1.87} - (HPO_4)_{0.2}nH_2O$, is compositionally between that of apatite and the acid calcium phosphate phases OCP and DCPD; the apparent invariance in this composition over a wide pH range precludes ACP as a mixture of these crystalline salts (60). Also, recent studies by Meyer (61) show that the apparent solubility of ACP can be described by an isotherm quite different from those obeyed by the crystalline calcium phosphates. Finally, a recent calcium K-edge extended x-ray absorption fine structure (EXAFS) analysis (62) indicated that ACP can best be described structurally as a continuous, random network of Ca^{2+}, PO_4^{3-}, and H_2O interactions. There was no evidence for local structural order about Ca extending beyond the immediate Ca–O shell. If ACP was a cryptostructural form of any crystalline phase, a well-defined second shell of Ca–P and Ca–Ca interactions would have been observed.

In general, ACP is observed only in synthetic preparations where the solution calcium and phosphate concentrations are sufficiently high for solid formation to occur spontaneously. Studies with physiologically balanced preparations show that calcium and phosphate concentrations comparable to those found in serum and skeletal fluids are at the lower limit for such spontaneous precipitation reactions (64). Moreover, at these concentrations delays of up to 24 hours can follow the mixing of the reactants before the ACP first appears; and the actual amount of precipitate formed can be quite low, from 3 to 5% of the calcium and phosphate initially available for reaction. In addition, seeding solutions at similar, or even higher, supersaturations with apatite can bring on immediate and substantial new crystal development in lieu of free ACP formation (65). These observations suggest that the amorphous material present in actively mineralizing areas of skeletal development probably could not have formed unless local skeletal calcium and phosphate concentrations were at least minimally in excess of normal systemic fluid levels. Possible mechanisms for effecting such localized solution changes *in vivo* are discussed in the section on early endochondral bone formation.

As already indicated, spontaneously precipitated ACP is not stable when kept in contact with its preparative solution, and in time will transform into crystalline phases. Originally, it was thought that the first crystals were apatitic. At pH's above 9 this seems to be the case (60), but at physiological pH's an OCP-like crystalline phase appears to form preferentially (66).

Electron micrographs (Fig. 3*c*) of solids sampled during the early stages of the ACP to OCP conversion suggest that the surface of the amorphous spheroids provided the majority of initial sites for OCP nucleation (58). However, subsequent crystal growth and proliferation, in addition to being quite rapid, was generally directed away from the amorphous surface, so that at the completion of the conversion very few, if any, of the enveloping crystals actually penetrated the space previously occupied by the amorphous particles (Fig. 3*d*). Such a pattern of crystal genesis suggests that the conversion is not an *in situ* solid state process. Instead, the results support a solution-mediated mechanism, that is, the OCP crystals, once nucleated, grow primarily by direct addition of Ca and PO_4 ions from the solution phase. Solubility data show that during this period the solution is maintained at a level of supersaturation sufficient to sustain OCP growth by continuous dissolution of the ACP phase (66). Once this phase is consumed, however, solution Ca and PO_4 concentrations drop to the solubility level for OCP and further growth ceases.

Several studies (67–69) have shown that OCP is also an important transitional phase in induced precipitations. It is the crystalline phase that forms in lieu of free ACP when physiologically balanced synthetic calcifying solutions are seeded with apatite. In these solutions, the seed apatite plays a role similar to ACP in spontaneous precipitation reactions in that the apatite surface influences only the primary nucleation of OCP and not its subsequent growth. Electron micrographs reveal that the OCP crystals loosely cluster about the seed crystals, rather than tightly adhere to the surface as contiguous epitaxial overgrowths (65).

In spontaneous precipitation reactions, and in the seeding of metastable solutions comparable to physiological fluids, the OCP intermediate is apparently the key factor in understanding the emergence of the final apatite phase. Although the details are complex and not fully elucidated, the transition from OCP to apatite appears to be in part an *in situ* process. X-ray evidence indicates that the formation of interlayered crystals of these two phases is involved in this transition (66, 70). Brown (32, 70) postulates that such hybrid crystals result from the hydrolytic rearrangement of OCP surface layers directly into apatite. Apparently, the close structural similarities and low interfacial energies between these two phases readily allow this unusual solid-state-like change to take place. The transition from OCP to apatite by way of interlayered structures also may account for the observation that the crystals of the final apatite phase retain the plate-like habit of their OCP predecessors (66).

One other aspect of OCP's role in apatite formation is worth noting. Whether initiated by ACP or by seed apatite, OCP nuclei proliferate into masses of many small submicron crystals rather than develop into a few large crystals (58, 65). This proliferation appears to occur through an autocatalytic process in which crystals, upon reaching a certain size, generate new crystals rather than continue to grow (57, 71). The mechanism for this secondary nucleation process is not known, but it may be tied in with the tendency for OCP surfaces to hydrolyze *in situ* into apatite. Although the lattice match between these two structures is very close, it is not perfect. As a consequence, the hydrolysis could lead to surface roughness and eventual particle fragmentation. Another factor is the apatitic nature of the transformed surface, which could promote new OCP crystal formation. As already indicated, apatite is an excellent seed material for this phase.

The time required to complete the various phase changes in synthetic precipitation reactions is generally quite short (63). In spontaneous precipitations at 37°C and pH 7.4, for example, the entire mass of ACP can transform into crystalline OCP within 6 min, and the OCP, in turn, can become predominantly apatite by 25 min. In a like manner, the OCP stage in seeded preparations can be equally transient. However, a number of substances commonly present in skeletal tissue can affect the rate of these phase changes in synthetic systems. Most of these substances, such as magnesium (72, 73), carbonate (55, 72), pyrophosphate (74), citrate (75), adenine nucleotides (76), phospholipids (77, 78), proteoglycans (79), and phosphoproteins (76) delay apatite formation by extending the lifetime of the precursor phases, often by several hours. Fluoride appears to be an exception in that it can greatly curtail the lifetime of OCP by promoting its hydrolysis to apatite (80). The amount of these substances needed to measurably affect precursor lifetimes is usually quite small ($<0.1\mathrm{m}M$), which suggests that they interfere primarily with nucleation and growth processes occurring at precursor surfaces, rather than stabilizing the precursors by being structurally incorporated. The initial nucleation of OCP on ACP, or the hydrolysis of OCP to apatite, are two surface-oriented reactions that may be affected by transient adsorption of these substances at active surface sites. Although their stabilizing effect on ACP may partly explain the possible presence of this phase in mature skeletal tissue, in general, the regulatory action of these substances is difficult to demonstrate *in vivo*. As stated at the beginning of this section, the principal reason for such difficulty is that biomineral formation is a physiologically controlled process closely tied to overall skeletal tissue formation and maturation.

5 MECHANISMS OF MINERALIZATION

The mineralization of skeletal tissue in vertebrate organisms is a complex affair. The once commonly held view that biological calcification could be described simply as the nucleation and growth of apatite within an extracellular, collagenous framework is now seen as inadequate. Although the apatite/collagen association is well-defined in skeletal tissue, the discovery of nonapatitic biominerals, together with the results of the synthetic model studies, indicates that a number of mineralization events may precede the appearance of the apatite phase. In addition, these preapatitic events may not always occur within a collagenous setting. The primary involvement of noncollagenous tissue structures is particularly evident in enamel and in chondroid calcifications. Because the pattern of calcification differs in cartilage, bone, teeth, and other mineralized tissues, both in the manner in which the mineral is deposited and in the nature and involvement of nonmineral tissue components, it is not possible to present a single theory of calcification. Instead, a number of representative hard tissue calcifications are described and discussed.

5.1 Matrix Vesicle Calcifications

The initial loci for mineral formation in growth plate cartilages appear to be globular membrane-bound bodies of cellular origin known as matrix vesicles. These submicron-size structures were first observed, independently, by Anderson (81, 82) and Bonucci (83, 84) in the extracellular space of epiphyseal cartilage tissue. Matrix vesicles apparently originate in the proliferative zone of this tissue as bud-like outgrowths of the plasma membrane of chondrocytes (Fig. 4*a, b*). Mineral deposits first appear associated with these structures in the hypertrophic zone. Mineral-containing matrix vesicles have been observed, subsequently, in a number of other calcifying tissues. They occur most frequently in tissues undergoing rapid mineral deposition, as, for example, early (mantle) dentin (85, 86) and embryonic woven bone (87, 88). In lamellar bone, where mineralization proceeds differently, matrix vesicles are seen much less frequently, if at all.

Matrix vesicles appear to promote mineralization by concentrating calcium and possibly phosphate in their membrane and interior spaces. Most of this uptake, especially that of Ca^{2+}, occurs after the vesicles are released into the extracellular matrix space (89). However, the exact manner by which matrix vesicles can accumulate Ca^{2+} to precipitable levels is not clear. One possibility is that

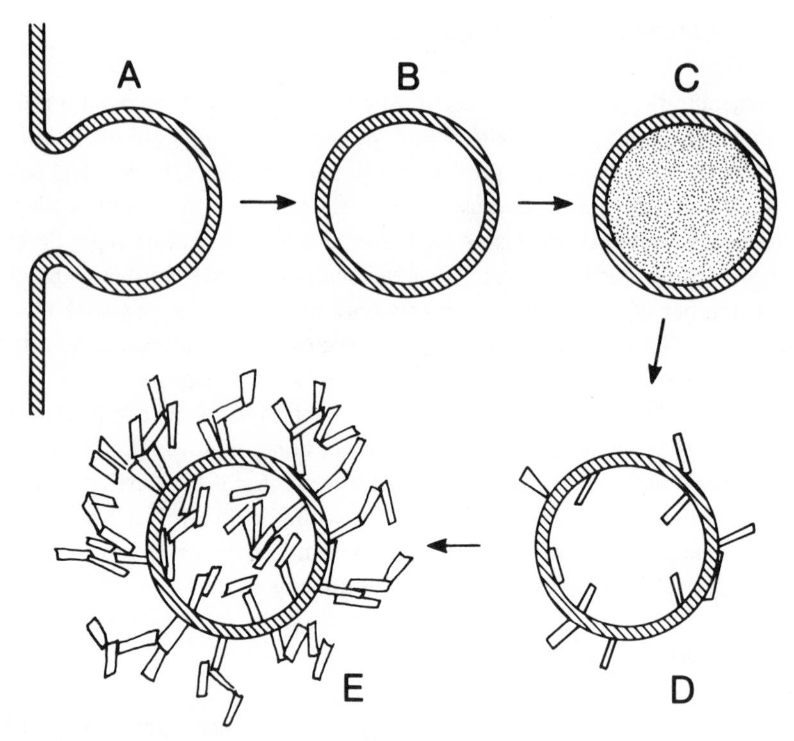

Figure 4 Schematic illustrating (*a*) the formation of a matrix vesicle from the plasma membrane of a proliferating chondrocyte, (*b*) its separation into an independent extracellular organelle, (*c*) the intravesicular accumulation of calcium and phosphate to form an amorphous mass (stippled area), (*d*) the appearance, and (*e*) the proliferation of crystalline particles in and around the vesicle.

Ca^{2+} may be transported across the vesicle membrane by energy-dependent enzymatic pumps (90, 91). The data are equivocal, however, on whether such enzymes are actually involved directly in vesicle Ca^{2+} uptake (89, 92–94). One factor which suggests that ATP-dependent transport cannot contribute to net Ca^{2+} accumulation in matrix vesicles is that the enclosing membrane is derived from the plasma membrane without inversion. Pumps driven by ATP-ase would thus be expected to remove Ca^{2+} from the vesicle interior rather than add Ca^{2+} to it.

Wuthier (89) has suggested that Ca^{2+} enrichment may occur through the formation of stoichiometric mineral complexes with components of the vesicle membrane. One such complex, Ca-phospholipid-PO_4, has been isolated from epiphyseal cartilage (95). According to Boskey and Posner (96), the particular association between Ca, PO_4, and membrane phospholipids may provide a seeding site for direct mineral formation. Wuthier (89), however, observed that Ca^{2+} accumulation during vesicle maturation is accompanied by intravesicular K^+ losses. This finding suggests that membrane components may transport and release Ca^{2+} into the vesicle interior by ionophore-like, cation-exchange mechanisms. Two phospholipids found in matrix vesicle membranes, phosphatitic acid and diphosphatidylglycerol, possess relatively strong ionophore properties *in vitro* (97, 98). The ionophoric activity of these two phospholipids *in vivo* would depend, however, on how mobile they are within the membrane bilayer. Perhaps when complexed with phosphate, these two phospholipids could have sufficient membrane fluidity to facilitate cation transport. Along these lines Yaari and Shapiro (99) recently demonstrated that inorganic phosphate can enhance phospholipid-mediated Ca^{2+} transport in lipid-aqueous model systems.

Matrix vesicles apparently can accumulate appreciable quantities of Ca^{2+} and inorganic phosphate before crystalline deposits are observed (Fig. 4*c*). Electrolyte analysis of crystal-free vesicles isolated from the proliferating zone of epiphyseal cartilage reveals interior Ca^{2+} and phosphate concentrations of 50 and 30m*M*, respectively (89). Such ion levels imply the presence of amorphous solid material, as they exceed by severalfold the metastable limit for spontaneous precipitation in calcium phosphate solutions. Ultrastructural studies (40, 100) of newly mineralizing tissue also frequently reveal amorphous-like granular material within, as well as around, the vesicles.

The later appearance of crystalline particles within and around matrix vesicles (Fig. 4*d*) suggests a precursor mineral conversion process similar to that described for the synthetic systems. However, the possibility cannot be ruled out that other sites favorable for crystal nucleation, perhaps the inner surfaces of the vesicle membranes, may exist within the organelle. In this situation, the amorphous phase may simply be a reservoir of Ca and PO_4 ions for the growing crystals.

Regardless of origin, once established the crystals rapidly proliferate to form clusters throughout and around the vesicles (Fig. 4*e*). As they expand, the mineral clusters eventually coalesce into seam-like structures (41, 88). Closer to more heavily mineralized areas of tissue, the vesicular remnants of the fully crystalline mass are no longer evident. Although type II collagen is present in the regions where the crystal clustering occurs, it is unorganized and shows no clear association with the crystals (41, 88, 100).

The first crystals in vesicular calcifications often appear to have a plate-like shape, suggestive of OCP or an OCP precursor. This intermediate could, therefore, have an important role in the crystal proliferation process. As discussed in an earlier section, OCP-apatite interactions favor secondary crystallization over continued crystal growth. Therefore, with a steady influx of reactant ions maintaining the surrounding fluid milieu in a fairly constant state of supersaturation, the conversion of the first crystals in the cluster to apatite could trigger the spherulitic expansion of the cluster through subsequent repetitive cycles of epitaxial OCP nucleation, growth, and hydrolytic transformation to apatite.

In dentin, the mineral clusters resulting from matrix vesicle calcification remain as a thin mineral line at the dentin-enamel junction and persist throughout the life of the tooth. In growth plate cartilage and embryonic bone, on the other hand, vesicle-derived mineral clusters are quickly resorbed by osteoclastic action and are subsequently replaced by nonvesicular lamellar bone, as described later.

5.2 Tooth Enamel Mineralization

The developing tooth bud contains two adjacent mineralizing tissue systems that form in concert throughout tooth development. These are an inner layer of mesenchymal origin, dentin, and an outer layer of epithelial origin, enamel. Tooth enamel represents the only naturally occurring epithelial mineralization system in mammals. Several features of enamel mineralization are unique to that tissue. In its final form, human tooth enamel is over 98% mineral and retains less than 0.1% protein in the adult tissue (101). During enamel formation and maturation, there is both a decrease in total matrix protein content and a change in the composition of the mineralizing enamel matrix (102–104). The enamel mineral phase forms adjacent to cell processes in early secretory enamel, which consists of 20 to 30% protein (102, 103). Matrix vesicles or other extracellular organelle material are never seen in developing enamel, and the enamel apatite crystals form directly in the extracellular organic matrix space. As mineralization in the matrix space increases and total protein content diminishes, there is a concomitant shift in matrix protein population from larger molecules (which dominate the early secretory matrix) to progressively smaller and smaller ones (104–108). Thus, the matrix proteins are gradually degraded as more enamel apatite is formed in the extracellular matrix space. In the hamster, this entire process takes place in a period of less than 10 days (109).

Enamel hydroxyapatite mineral is unique among mammalian calcifications. Nowhere else in mammalian biology are bioapatite crystallites so large (both in length and in cross section) or so highly cooriented. Only in invertebrate systems are similar mineral structures observed, and in those species such structures are composed of calcium carbonate, not calcium phosphate. In fact, until recently it was impossible to produce apatite crystals resembling those of enamel under synthetic conditions even remotely approximating those found in physiological systems. Early enamel crystals are extremely thin and ribbon-like in shape (110). These grow into very long (1 to 10 μ) crystallites with hexagonal cross sections (300×600 Å) (110–112). The crystals are packed in prism groupings, in which the long axes of all crystallites adjacent to one another within a given prism are coaligned. The prisms themselves are arranged in patterns characteristic for each mammalian species (113).

As mentioned above, enamel mineralization is more reminiscent of invertebrate mineralizations, at least with respect to the arrangement of the final mineral phase, than of collagen-centered mammalian mineralizations. Since the mineralization pattern of the invertebrates generally involves the deposition of discretely ordered inorganic mineral crystals in a protein matrix, it is not unreasonable to assume that a similar process may also occur in dental enamel. Recent work has shown that in at least five different species (cow, pig, hamster, sheep, and human) the developing enamel matrix consists of two different protein classes (107, 114). These are the hydrophobic amelogenins, which account for 80 to 90% of the total secretory enamel matrix protein, and the enamelins, acidic glycoproteins closely associated with the enamel apatite (107). With enamel maturation, the amelogenin proteins are completely lost from the tooth, and only low molecular weight forms (2000 to 3000 daltons) of the acidic enamelin protein class persist in the adult tooth (115, 116). Based on a number of criteria (107, 116), we know that the enamelin proteins form a sheath about the growing enamel bioapatite crystallites and cannot be removed from the mineral surfaces until they are completely demineralized (107). The amelogenin proteins are extremely hydrophobic (118), consisting of 30% proline, 25% glutamine, 10% leucine, and 7% histidine, and can be completely removed from enamel tissue without disturbing the mineral phase (107, 109). These amelogenin proteins are found in the intercrystalline spaces of the forming enamel matrix (107, 109, 116). Figures 5 and 6 depict the structure of forming enamel crystals, utilizing this combined biochemical and mor-

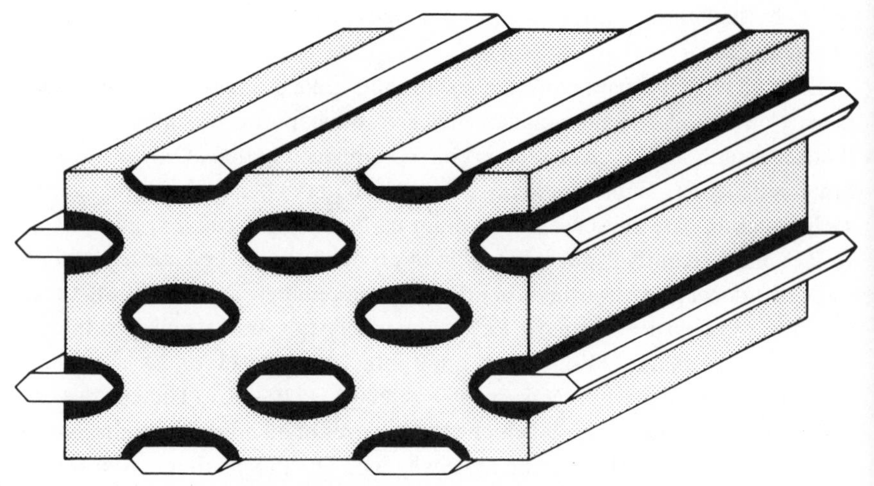

Figure 5 Schematic illustrating the relationship between the growing enamel crystallites (hexagonal prisms), the encapsulating enamelin sheath (dark areas surrounding crystals), and the intercrystalline amelogenin matrix (stippled area) during an early stage of enamel development.

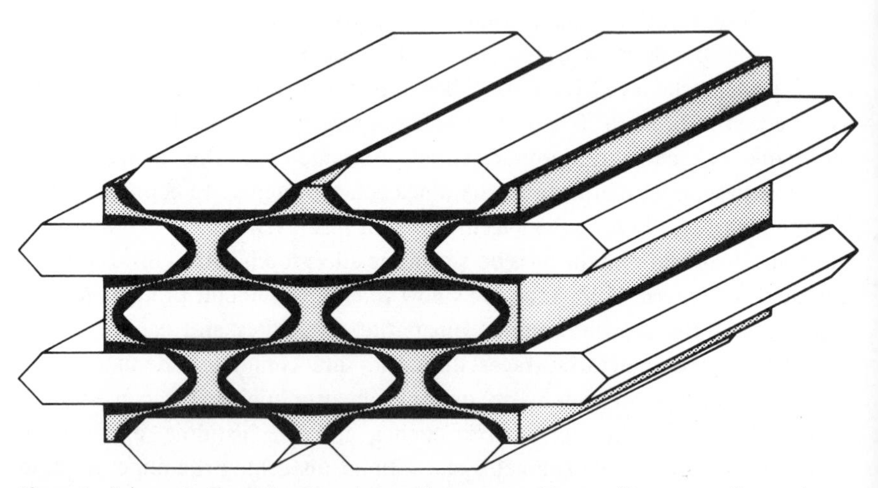

Figure 6 Schematic illustrating the relationship between the crystallites, enamelins, and amelogenins at a later stage of enamel development than shown in Figure 5. As depicted in this illustration, the enamel crystallites have grown larger, at the expense of the enamelin and amelogenin components.

phological information. In these illustrations, the hydrophobic amelogenin matrix (insoluble at physiological pH and temperature) forms a gel through which the developing enamel apatite crystallites and their associated enamelin glycoprotein sheath grow. Further work is needed to determine the exact physicochemical roles that each of these components plays in enamel apatite crystallization. The hydrophobic environment afforded by the amelogenin gel may provide a higher driving force for mineral accretion onto crystal growth sites. The glycoprotein enamelin coat may regulate growth mechanisms at such sites and provide for regulated, directional mineral accretion. Whatever their exact roles, however, these matrix proteins are integral to the growth and packing of the enamel bioapatite. The precise manner in which the enamel-forming cells (ameloblasts) regulate this process has yet to be determined.

5.3 Tooth Dentin Mineralization

The inner, mesenchymal layer of the tooth is a mineralizing tissue similar to bone in that it contains a basic matrix layer consisting of over 85% type I collagen fibrils. Both bone and dentin deposit conventional, small, platy bioapatite crystallites onto an underlying collagenous, extracellular matrix. In both tissues, a final state is reached in which the mineral accounts for approximately 70% of the tissue dry weight and is associated in an intimate manner with the matrix collagen phase. But here, most similarities end. Bone tissue is in a perpetual state of flux, undergoing constant remodeling throughout life. Dentin is never remodeled, physiologically, and in nondeciduous teeth does not contain resorptive cells, only formative ones (odontoblasts). During dentin formation then, only matrix accretion and mineralization occur.

Biochemically, the dentin extracellular matrix contains, as its major noncollagenous component, a protein as yet found no place else in mammalian biology, the dentin phosphoprotein (119–122). From 75 to 80% of the amino acid residues in this protein are either aspartic acid or serine, which exist in equal amounts. Almost half of the serine residues in the molecule are phosphorylated (119–122). The dentin phosphoproteins, called phosphophoryns (123), have differing apparent molecular sizes in rodents and larger animals. Rodent phosphophoryns are ~70,000 daltons, whereas bovine phosphophoryns [and human phosphophoryns, as well (156)] are ~95,000 daltons in size when compared in similar analytical systems (109). The larger phosphophoryns contain greater proportions of nonserine/aspartic acid amino acid residues (especially lysine) than do the rodent molecules.

In vitro, the dentin phosphophoryns display regulatory activities toward both collagen and the calcium phosphates. They inhibit the transformation of amorphous calcium phosphate (124) and octacalcium phosphate (125) precursors to the final apatite mineral phase in spontaneous nucleation experiments and hinder the secondary nucleation of apatite seeds in crystal growth studies using metastable solutions (125). In the latter case, the inhibitory activity of the phosphoproteins was modulated by ionic calcium, perhaps by the induction of conformational changes in the phosphophoryn inhibitor molecules (125). Similarly, the phosphophoryns decrease the rate of assembly of collagen molecules into fibrils in *in vitro* assays (126). This inhibition was dependent on the amount of phosphophoryn added and was significant at physiological levels. Moreover, the hindrance was modulated by ionic calcium; added calcium greatly increased the inhibitor effect (126). These *in vitro* data suggest an essential role for the dentin phosphophoryns in several aspects of dentinogenesis. That such a role may be reasonable is demonstrated by the *in vivo* radioautographic studies of Weinstock and Leblond (127). In forming dentin, the secretory cells (odontoblasts) form a tight layer that remains several microns distant from the mineralization front. Collagen and ground substance molecules appear to be deposited immediately outside the cell surfaces and only become inundated by mineral at a later time, when the actual mineralization front reaches this earlier site of collagen deposition (127). The phosphophoryns, on the other hand, do not accumulate at the immediate periphery of the cell, as does the matrix collagen, but very rapidly migrate to the mineralization front itself, where they presumably exert their influence on the dentin mineralization process (127). It is entirely reasonable to assume that these proteins greatly affect the pattern and rate of dentin mineralization *in vivo*. Whether their influence is entirely physicochemical is uncertain at this point. They may act in concert with other proteins, such as osteonectin (see discussion below), in exerting their normal physiological functions.

5.4 Lamellar Bone Mineralization

Two forms of bone mineralization occur in developing bone tissue. In humans and large animals, such as the cow or pig, early embryonic bone mineralizes via the chondroid (matrix vesicle) format, as described earlier. This type of mineral deposition is characteristic of woven bone, which is found in growing rodents and as an early developmental form in most other species (87, 88). However, in humans and other large animals, this form of bone does not persist for long during fetal growth. It is rapidly replaced by growing lamellar bone, which then becomes the dominant bone form throughout the remainder of fetal development, growth, and adult life. Matrix vesicles are rarely seen in lamellar bone and the dominant mineralization pattern observed is one in which apatite mineral (or its precursors) is deposited directly onto an underlying type I collagen fibrillar matrix (128). Thus, lamellar bone, the major type of bone found in human skeletal tissue, is a collagen-based calcification system.

One of the distinguishing features of collagen calcification in lamellar bone tissue is the periodic arrangement of the apatite crystals along the collagen fiber axis (38, 129) (Fig. 7). This periodicity, which reflects the fundamental 67 nm axial repeat of the fiber, occurs because the so-called hole regions in the fiber structure appear to contain more crystals than other regions within the axial repeat. Such a concentration of mineral in preferential regions of the fiber, most apparent during the early stages of lamellar bone calcification, has led Glimcher (130) to postulate that crystallization is initiated in these regions

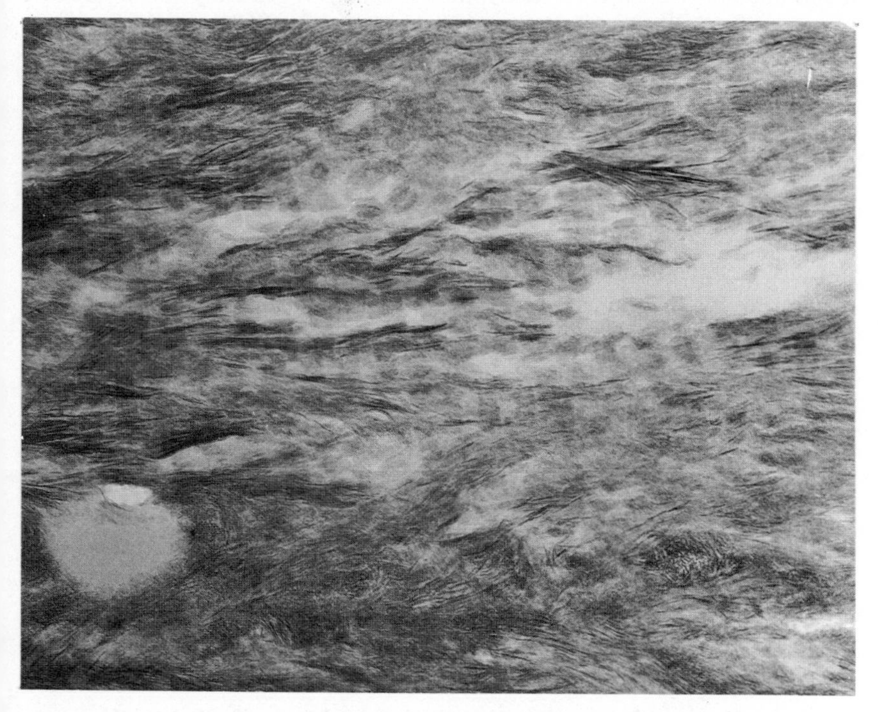

Figure 7 Transmission electron micrograph of a thin section of hydrazine-deproteinated 3-month-old rat tibia (47) showing the periodic arrangement of the apatite crystals.

by a process akin to epitaxy. Studies using x-ray and electron diffraction techniques (131–133) provide support for such a possibility by showing that the length of the apatite crystals generally coincides with the fiber axis. However, little is known about the molecular scale geometry of the holes, other than that they correspond to 40 nm gaps in the end-to-end arrangement of the collagen molecules that comprise the individual strands of the fiber (134). Rapid, almost fluid-like motion of the amino acid side chains that line the hole together with large torsional movements in the backbone chain of the collagen molecules (135, 136) make it difficult to define the charge distribution and surface topology of this region in sufficient detail to assess its potential as a recipient for apatite nucleation. It is possible that the unusual distribution of mineral in the collagenous matrix may simply reflect the added space provided by the holes for deposition.

In addition, only the collagen fibrils found in bone and dentin normally calcify, even though type I collagen molecules identical to those of bone are found in other tissues (137). Also, purified collagens generally do not nucleate mineral in *in vitro* calcification systems (138). Even demineralized bone collagen does not calcify in a normal physiological pattern *in vitro* (139). *In vivo*, implants of demineralized bone matrix also do not calcify, but instead induce a cascade of endochondral bone formation (140, 141). Consequently, it is thought that some other bone factor endows the bone collagen with the ability to calcify, thus mediating the major structural form of bone seen in human skeletal development.

Considerable attention was placed in the past on bone matrix components that might fulfill a mineralization function. These included sialic acid-rich glycoproteins (142), phosphoproteins (143, 144), and serum proteins (145, 146). These proved difficult to characterize in sufficient molecular detail (using the technology of the time) to afford investigations of their specificity of action in skeletal tissue. More recently, attention has focused on a γ-carboxyglutamic acid-containing (M_r = 5,800) component of bone (147, 148) as a mineral-inducing factor. However, animals reared under conditions whereby this low molecular weight protein accumulates in bone to only about 2% of normal display no skeletal defects and have perfectly normal, calcified bone tissue (149).

Recently, technology was developed that provided an enrichment of mineral-specific proteins in dissociative extracts of skeletal tissue (107, 122). Similarly, fractionation procedures were designed to purify skeletal tissue matrix proteins to chemical homogeneity (107, 122). These procedures were applied to subperiosteal fetal calf lamellar bone and led to a thorough characterization of the noncollagenous constituents comprising this tissue (150, 151). The tissue studied was rich in newly synthesized and minimally degraded bone matrix proteins. Three new noncollagenous proteins were purified to homogeneity in this study (151). These included two phosphate-containing glycoproteins of 32,000 and 62,000 daltons, and a 24,000 dalton phosphoprotein containing 2.7% hydroxyproline in its amino acid content (151). All three proteins had a strong affinity for calcium and synthetic hydroxyapatite, and resisted extraction from bone until the tissue was demineralized (151). They all bound, at least to some degree, to insoluble demineralized bone collagen at physiological pH and ionic strength (151). Thus, they were integral bone matrix proteins and excellent candidates for bone-specific functions.

Subsequently, it was found that one of these newly identified lamellar bone matrix proteins had the highest affinity for both apatite mineral and collagen (151, 152). This protein (M_r = 32,000) was called osteonectin, based on its bone origin and its potential to bridge or link the bone collagen and mineral phases (152). *In vitro*, osteonectin mediated the binding of both free calcium ions and synthetic apatite crystallites to type I collagens derived from normally nonmineralized tissues (152). Osteonectin also facilitated the nucleation of calcium phosphate mineral onto insolubilized type I collagen substrates, thus pointing up its potential mineralization role in bone tissue (152). Antibodies to osteonectin localized the protein to bone, where it proved to be an excellent marker for new osteogenic sites (152). A related protein was found (at lower levels) in developing dentin (152). Thus, osteonectin is an excellent candidate for the macromolecular mediator of mineralization to collagenous substrates in developing skeletal tissue. It may afford a new look at the role of the bone matrix in normal metabolism and in disease processes. Future work with this and other bone-specific substances may point to new directions in this area of mineral research.

6 ROLE OF MINERAL IN CALCIUM HOMEOSTASIS

As stated in the introduction to this chapter, the mineral component of skeletal tissues is the major calcium reservoir in the body. Other tissue factors, however, appear to control the utilization of this source for normal calcium metabolism. As Glimcher (128) has developed in detail, active cellular bone formation and resorption, and not passive diffusion of Ca^{2+} to and from the mineral phase, is directly responsible for 95% or more of the movement of this ion into and out of bone tissue. Such remodeling activity is the principal means by which bone tissue contributes to the homeostasis of Ca^{2+} in extracellular fluids. The inability of bone mineral itself to readily exchange Ca^{2+} with the fluid compartment is a consequence of it being well-embedded within a collagenous matrix that serves as an effective barrier to passive diffusion. Nearly all matrix water is firmly bound to the collagen and unavailable for solution-mediated ion movement (153).

Although, as we have seen, there are several active mechanisms for immobilizing Ca^{2+} through mineral deposition, lamellar bone formation appears to be the most important from the viewpoint of homeostatic Ca^{2+} control. Similarly, the release of mineral-bound Ca^{2+} back into circulation appears to occur principally by a single means as well, that of osteoclastic resorption of preexisting bone tisue (154). This process, which is an integral part of bone remodeling, is marked by the complete destruction of entire segments of bone tissue in which both the matrix and mineral components are totally removed. Although the details of the mineral dissolution aspect of this process are not fully understood, the fact that it occurs under the tight control of the bone removing osteoclasts insures that Ca^{2+} movement out of, as well as into bone tissue, is metabolically regulated.

Evidence of a different nature for the absence of direct mineral involvement in Ca^{2+} homeostasis is the fact that, in spite of an enormous mineral/matrix interface [200 m^2/g (155)], extracellular bone fluids are generally supersaturated with respect to the apatite phase (50). As already discussed, to considerable measure the impact of this large surface area is mitigated by the matrix isolation of most of the mineral apatite from the extracellular fluid compartment. However, not all mineral surfaces are protected from fluid contact. In regions occupied by bone cells or cellular extensions, there is generally a narrow fluid-filled osteoid zone separating the cell surface from the adjacent mineral front. The fact that the fluid in this zone does not establish Ca^{2+} equilibrium with accessible apatite surfaces is further support for active cellular control of this ion's movement in bone tissue.

7 CONCLUDING REMARKS

In this chapter, we have presented an overview of vertebrate mineralization in the context of calcium biology. It is evident from research performed during recent decades that a single, all-embracing description of this complex, versatile biological process is not possible. This complexity is brought sharply into focus by the multistep nature of the inorganic precipitation reactions involving calcium, and by the intricate manipulation of these reactions by the physiological processes governing mineral growth and dissolution *in vivo*. One can hope that the unique situations presented by cartilage, dentin, enamel, and bone mineralizations can be fully appreciated as a reflection of nature's diversity, while not losing sight of the fundamental elements of biological calcification common to all vertebrate hard tissues.

REFERENCES

1. W. F. deJong, *Rec. Trav. Chim. Pays-Bas*, **45,** 445 (1926).
2. R. A Young, M. L. Bartlett, S. Spooner, P. E. Mackie, and G. Bonel, *J. Biol. Phys.*, **9,** 1 (1981).
3. J. C. Elliott, *Calcif. Tiss. Res.*, **3,** 293 (1969).
4. D. R. Simpson, *Clin. Orthop. Rel. Res.*, **86,** 260 (1972).
5. J. C. Elliott, *Clin. Orthop. Rel. Res.*, **93,** 313 (1973).
6. R. A. Young, *J. Dent. Res.*, **53,** 193 (1974).
7. R. A. Young, *Clin. Orthop. Rel. Res.*, **113,** 249 (1975).
8. W. E. Brown and L. C. Chow, *Ann. Rev. Mater. Sci.*, **6,** 213 (1976).
9. M. Mehmel, *Z. Kristallogr.*, **75,** 323 (1930).
10. S. Náray-Szabó, *Z. Kristallogr.*, **75,** 387 (1930).
11. A. S. Posner, A. Perloff, and A. F. Diorio, *Acta Cryst.*, **11,** 308 (1958).
12. M. I. Kay, R. A. Young, and A. S. Posner, *Nature,* **204,** 1050 (1964).
13. W. R. Busing and H. A. Levy, *J. Chem. Phys.*, **26,** 563 (1957).
14. I. Zipkin, "The Inorganic Composition of Bones and Teeth," in *Biological Calcification: Cellular and Molecular Aspects*, H. Schraer, Ed., Appleton-Century-Crofts, New York, 1970, p. 69.
15. J. A. Weatherell and C. Robinson, "The Inorganic Composition of Teeth," in *Biological Mineralization*, I. Zipkin, Ed., Wiley-Interscience, New York, 1973, p. 43.
16. G. Bonel and G. Montel, *Compt. Rend. Acad. Sci. Paris*, **258,** 923 (1964).
17. J. C. Elliott, "The Interpretation of the Infra-red Absorption Spectra of some Carbonate-containing apatites," in *Tooth Enamel*, M. V. Stack and R. W. Fearnhead, Eds., Wright, Bristol, 1965, p. 20.
18. D. W. Holcomb and R. A. Young, *Calcif. Tiss. Res.*, **31,** 189 (1980).
19. P. E. Mackie, J. C. Elliott, and R. A. Young, *Acta Cryst.*, **B28,** 1840 (1972).
20. R. Z. LeGeros, *Arch. Oral Biol.*, **20,** 63 (1974).
21. R. A. Young and J. C. Elliott, *Arch. Oral Biol.*, **11,** 699 (1966).
22. A. S. Posner and A. Perloff, *J. Res. Natl. Bur. Stand.*, **58,** 279 (1957).
23. A. S. Posner, J. M. Stutman, and E. R. Lippincott, *Nature*, **188,** 486 (1960).

24. J. M. Stutman, A. S. Posner, and E. R. Lippincott, *Nature*, **193,** 368 (1962).
25. L. Winand, M. J. Dallemagne, and G. Duyckaerts, *Nature*, **190,** 164 (1961).
26. L. Winand and M. J. Dallemagne, *Nature*, **193,** 369 (1962).
27. G. Kuhl and W. H. Nebergall, *Z. Anorg. Allg. Chem.*, **324,** 313 (1963).
28. E. E. Berry, *J. Inorg. Nucl. Chem.*, **29,** 317 (1967).
29. E. E. Berry, *J. Inorg. Nucl. Chem.*, **29,** 1585 (1967).
30. W. E. Brown, J. R. Lehr, J. P. Smith, and A. W. Frazier, *J. Am. Chem. Soc.*, **79,** 5318 (1957).
31. W. E. Brown, J. P. Smith, J. R. Lehr, and A. W. Frazier, *Nature*, **196,** 1050 (1962).
32. W. E. Brown, *Clin. Orthop. Rel. Res.*, **44,** 205 (1966).
33. J. L. Meyer, *Calcif. Tiss. Int.*, **27,** 153 (1979).
34. J. L. Meyer and B. O. Fowler, *Inorg. Chem.*, **21,** 3029 (1982).
35. N. Quinaux, *Bull. Soc. Chim. Biol. Paris*, **46,** 561 (1964).
36. D. J. Greenfield, J. D. Termine, and E. D. Eanes, *Calcif. Tiss. Res.*, **14,** 131 (1974).
37. R. A. Young and S. Spooner, *Arch. Oral Biol.*, **15,** 47 (1969).
38. R. A. Robinson and M. L. Watson, *Ann. NY Acad. Sci.*, **60,** 596 (1955).
39. J. Thyberg, *J. Ultrastruct. Res.*, **46,** 206 (1974).
40. H. Schraer and C. V. Gay, *Calcif. Tiss. Res.*, **23,** 185 (1977).
41. C. V. Gay, *Calcif. Tiss. Res.*, **23,** 215 (1977).
42. R. A. Harper and A. S. Posner, *Proc. Soc. Exp. Biol. Med.*, **122,** 137 (1966).
43. J. D. Termine and A. S. Posner, *Calcif. Tiss. Res.*, **1,** 8 (1967).
44. J. D. Termine and A. S. Posner, *Science*, **153,** 1523 (1966).
45. J. D. Termine, I. Pullman, and A. S. Posner, *Arch. Biochem. Biophys.*, **122,** 318 (1967).
46. J. E. Russell, J. D. Termine, and L. V. Avioli, *J. Clin. Invest.*, **52,** 2848 (1973).
47. J. D. Termine, E. D. Eanes, D. J. Greenfield, M. U. Nylen, and R. A. Harper, *Calcif. Tiss. Res.*, **12,** 73 (1973).
48. S. E. Gruninger, C. Siew, J. J. Hefferren, L. C. Chow, and W. E. Brown, *J. Dent. Res.*, **60A,** 451 (1981).
49. F. C. McLean and M. R. Urist, *Bone, An Introduction to the Physiology of Skeletal Tissue*, University of Chicago Press, Chicago, 1955.
50. W. F. Neuman and M. W. Neuman, *The Chemical Dynamics of Bone Mineral*, University of Chicago Press, Chicago, 1958.
51. K. J. Münzenberg and M. Gebhardt, *Dtsch. Med. Wochenschr.*, **25,** 1 (1969).
52. M. D. Francis and N. C. Webb, *Calcif. Tiss. Res.*, **6,** 335 (1971).
53. A. H. Roufosse, W. J. Landis, W. K. Sabine, and M. J. Glimcher, *J. Ultrastruct. Res.*, **68,** 235 (1979).
54. M. L. Watson and R. A. Robinson, *Am. J. Anat.*, **93,** 25 (1953).
55. B. N. Bachra, O. R. Trautz, and S. L. Simon, *Arch. Biochem. Biophys.*, **103,** 124 (1963).
56. E. D. Eanes, I. H. Gillessen, and A. S. Posner, *Nature*, **208,** 365 (1965).
57. E. D. Eanes and A. S. Posner, *Trans. NY Acad. Sci.*, **28,** 233 (1965).
58. E. D. Eanes, J. D. Termine, and M. U. Nylen, *Calcif. Tiss. Res.*, **12,** 143 (1973).
59. M. U. Nylen, E. D. Eanes, and J. D. Termine, *Calcif. Tiss. Res.*, **9,** 95 (1972).

60. J. L. Meyer and E. D. Eanes, *Calcif. Tiss. Res.*, **25**, 59 (1978).
61. J. L. Meyer, personal communication.
62. E. D. Eanes, L. Powers, and J. L. Costa, *Cell Calcium*, **2**, 251 (1981).
63. E. D. Eanes, *Prog. Cryst. Growth Charact.*, **3**, 3 (1980).
64. J. D. Termine and E. D. Eanes, *Calcif. Tiss. Res.*, **15**, 81 (1974).
65. E. D. Eanes, *Calcif. Tiss. Res.*, **20**, 75 (1976).
66. E. D. Eanes and J. L. Meyer, *Calcif. Tiss. Res.*, **23**, 259 (1977).
67. G. H. Nancollas and B. Tomazic, *J. Phys. Chem.*, **78**, 2218 (1974).
68. B. Tomazic and G. H. Nancollas, *J. Colloid Interface Sci.*, **50**, 451 (1975).
69. E. D. Eanes, *J. Dent. Res.*, **59**, 144 (1980).
70. W. E. Brown, L. W. Schroeder, and J. S. Ferris, *J. Phys. Chem.*, **83**, 1385 (1979).
71. E. D. Eanes and A. S. Posner, *Mat. Res. Bull.*, **5**, 377 (1970).
72. B. N. Bachra, *Ann. NY Acad. Sci.*, **109**, 251 (1963).
73. A. L. Boskey and A. S. Posner, *Mat. Res. Bull.*, **9**, 907 (1974).
74. H. Fleisch, R. G. G. Russell, S. Bisaz, J. D. Termine, and A. S. Posner, *Calcif. Tiss. Res.*, **2**, 49 (1968).
75. J. L. Meyer and A. H. Selinger, *Miner. Electrolyte Metab.*, **3**, 207 (1980).
76. J. D. Termine and K. M. Conn, *Calcif. Tiss. Res.*, **22**, 149 (1976).
77. J. M. Cotmore, G. Nichols, Jr., and R. E. Wuthier, *Science*, **172**, 1339 (1971).
78. R. E. Wuthier and E. D. Eanes, *Calcif. Tiss. Res.*, **19**, 197 (1975).
79. L. A. Cuervo, J. C. Pita, and D. S. Howell, *Calcif. Tiss. Res.*, **13**, 1 (1973).
80. E. D. Eanes and J. L. Meyer, *J. Dent. Res.*, **57**, 617 (1978).
81. H. C. Anderson, *J. Cell. Biol.*, **35**, 81 (1967).
82. H. C. Anderson, *J. Cell. Biol.*, **41**, 59 (1969).
83. E. Bonucci, *J. Ultrastruct. Res.*, **20**, 33 (1967).
84. E. Bonucci, *Z. Zellforsch.*, **104**, 192 (1970).
85. G. W. Bernard, *J. Ultrastruct. Res.*, **41**, 1 (1972).
86. T. Yanagisawa, *Bull. Tokyo Dent. Coll.*, **16**, 109 (1975).
87. G. W. Bernard, *J. Dent. Res.*, **48**, 781 (1969).
88. G. W. Bernard and D. C. Pease, *Am. J. Anat.*, **125**, 271 (1969).
89. R. E. Wuthier, *Calcif. Tiss. Res.*, **23**, 125 (1977).
90. S. Y. Ali, *Fed. Proc.*, **35**, 135 (1976).
91. H. T. Hsu and H. C. Anderson, *Biochim. Biophys. Acta*, **500**, 162 (1977).
92. S. W. Sajdera, S. Franklin, and R. Fortuna, *Fed. Proc.*, **35**, 154 (1976).
93. R. J. Majeska and R. E. Wuthier, *Biochim. Biophys. Acta*, **391**, 51 (1975).
94. H. K. Väänänen, *Calcif. Tiss. Int.*, **30**, 227 (1980).
95. R. E. Wuthier and S. T. Gore, *Calcif. Tiss. Res.*, **24**, 163 (1977).
96. A. Boskey and A. S. Posner, *Calcif. Tiss. Res.*, **19**, 273 (1976).
97. C. A. Tyson, H. V. Zande, and D. E. Green, *J. Biol. Chem.*, **251**, 1326 (1976).
98. C. Serhan, P. Anderson, E. Goodman, P. Dunham, and G. Weissmann, *J. Biol. Chem.*, **256**, 2736 (1981).

99. A. M. Yaari and I. M. Shapiro, *Calcif. Tiss. Int.*, **34,** 43 (1982).
100. C. V. Gay, H. Schraer, and T. E. Hargest, Jr., *Metab. Bone Dis. Rel. Res.*, **1,** 105 (1978).
101. H. S. M. Crabb and A. I. Darling, *The Pattern of Progressive Mineralization in Human Dental Enamel*, Pergamon, Oxford, 1962.
102. J. Deakins, *J. Dent. Res.*, **21,** 429 (1942).
103. M. V. Stack, *J. Bone Jt. Surg.*, **42B,** 853 (1960).
104. M. J. Glimcher, D. Brickley-Parsons, and P. T. Levine, *Calcif. Tiss. Res.*, **24,** 259 (1977).
105. M. Fukae and M. Shimizu, *Arch. Oral. Biol.*, **19,** 381 (1974).
106. C. Robinson, P. Fuchs, and D. Deutsch, *Caries Res.*, **12,** 1 (1978).
107. J. D. Termine, A. B. Belcourt, P. J. Christner, K. M. Conn, and M. U. Nylen, *J. Biol. Chem.*, **255,** 9760 (1980).
108. A. G. Fincham, A. B. Belcourt, and J. D. Termine, *Caries Res.*, **16,** 64 (1982).
109. D. M. Lyaruu, A. B. Belcourt, A. G. Fincham, and J. D. Termine, *Calcif. Tiss. Int.*, **34,** 86 (1982).
110. M. U. Nylen, E. D. Eanes, and K. A. Omnell, *J. Cell. Biol.*, **18,** 109 (1963).
111. P. D. Frazier, *J. Ultrastruct. Res.*, **22,** 1 (1968).
112. J. Arends and W. L. Jongebloed, *J. Biol. Buccale*, **6,** 161 (1979).
113. D. G. Gantt, D. Pilbeam, and G. P. Steward, *Science*, **198,** 1155 (1977).
114. A. G. Fincham, A. B. Belcourt, D. M. Lyaruu, and J. D. Termine, *Calcif. Tiss. Int.*, **34,** 182 (1982).
115. A. B. Belcourt and S. Gillmeth, *Calcif. Tiss. Int.*, **28,** 227 (1979).
116. A. B. Belcourt, A. G. Fincham, and J. D. Termine, *Caries Res.*, **16,** 72 (1982).
117. T. Yanagisawa, M. U. Nylen, and J. D. Termine, *J. Dent. Res.*, **60A,** 995 (1981).
118. A. G. Fincham, A. B. Belcourt, and J. D. Termine, "The Molecular Composition of Bovine Fetal Enamel Matrix," in *The Chemistry and Biology of Mineralized Connective Tissues*, A. Veis, Ed., Elsevier-North Holland, New York, 1981, p. 523.
119. A. Veis and A. Perry, *Biochemistry*, **6,** 2409 (1967).
120. W. T. Butler, J. E. Finch, Jr., and C. V. DeSteno, *Biochim. Biophys. Acta*, **257,** 167 (1972).
121. Y. Kuboki, R. Fujisawa, K. Aoyama, and S. Sasaki, *J. Dent. Res.*, **58,** 1926 (1979).
122. J. D. Termine, A. B. Belcourt, M. S. Miyamoto, and K. M. Conn, *J. Biol. Chem.*, **255,** 9769 (1980).
123. M. T. Dimuzio and A. Veis, *Calcif. Tiss. Res.*, **25,** 169 (1978).
124. J. D. Termine and K. M. Conn, *Calcif. Tiss. Res.*, **22,** 149 (1976).
125. J. D. Termine, E. D. Eanes, and K. M. Conn, *Calcif. Tiss. Int.*, **31,** 247 (1980).
126. R. A. Gelman, K. M. Conn, and J. D. Termine, *Biochim. Biophys. Acta*, **630,** 220 (1980).
127. M. Weinstock and C. P. Leblond, *J. Cell. Biol.*, **56,** 838 (1973).
128. M. J. Glimcher, "On the Form and Function of Bone: From Molecules to Organs. Wolff's Law Revisited, 1981," in *The Chemistry and Biology of Mineralized Tissues*, A. Veis, Ed., Elsevier-North Holland, New York, 1981, p. 617.
129. R. A. Robinson and M. L. Watson, *Anat. Record*, **114,** 383 (1952).
130. M. J. Glimcher, *Rev. Mod. Phys.*, **31,** 359 (1959).
131. A. Engstrom and R. Zetterstrom, *Exp. Cell. Res.*, **2,** 268 (1951).

132. J. B. Finean and A. Engstrom, *Biochim. Biophys. Acta*, **11,** 178 (1953).
133. Z. Molnar, *J. Ultrastruct. Res.*, **3,** 39 (1959).
134. K. A. Piez, "Structure and Function of Collagen," in *Gene Families of Collagen and other Proteins*, D. J. Prockop and P. C. Champe, Eds., Elsevier-North Holland, Amsterdam, 1980, p. 143.
135. L. W. Jelinski and D. A. Torchia, *J. Mol. Biol.*, **133,** 45 (1979).
136. L. W. Jelinski and D. A. Torchia, *J. Mol. Biol.*, **138,** 255 (1980).
137. S. Gay and E. J. Miller, *Collagen in the Physiology and Pathology of Connective Tissue*, Gustav-Fischer Verlag, Stuttgart, 1978.
138. A. S. DeJong, T. J. Hok, and P. Van Duijn, *Conn. Tiss. Res.*, **7,** 73 (1980).
139. B. N. Bachra, *Calcif. Tiss. Res.*, **8,** 287 (1972).
140. M. R. Urist, *Science*, **150,** 893 (1965).
141. A. H. Reddi, *Coll. Res.*, **1,** 209 (1981).
142. G. M. Herring, B. A. Ashton, and A. R. Chipperfield, *Prep. Biochem.*, **4,** 179 (1974).
143. A. Shuttleworth and A. Veis, *Biochim. Biophys. Acta*, **257,** 414 (1972).
144. A. R. Spector and M. J. Glimcher, *Biochim. Biophys. Acta*, **263,** 593 (1973).
145. B. A. Ashton, H. J. Hohlinz, and J. T. Triffitt, *Calcif. Tiss. Res.*, **22,** 27 (1976).
146. J. T. Triffitt and M. O. Owen, *Calcif. Tiss. Res.*, **23,** 303 (1977).
147. P. V. Hauschka, J. B. Lian, and P. M. Gallop, *Proc. Natl. Acad. Sci., USA*, **72,** 3925 (1975).
148. P. A. Price, A. S. Otsuka, J. W. Poser, J. Kristaporis, and N. Ramon, *Proc. Natl. Acad. Sci., USA*, **73,** 1447 (1976).
149. P. A. Price and M. K. Williamson, *J. Biol. Chem.*, **256,** 12754 (1981).
150. J. D. Termine, "Chemical Characterization of Fetal Bone Matrix Constituents," in *The Chemistry and Biology of Mineralized Connective Tissues*, A. Veis, Ed., Elsevier-North Holland, New York, 1981, p. 349.
151. J. D. Termine, A. B. Belcourt, K. M. Conn, and H. K. Kleinman, *J. Biol. Chem.*, **256,** 10403 (1981).
152. J. D. Termine, H. K. Kleinman, S. W. Whitson, K. M. Conn, M. L. McGarvey, and G. R. Martin, *Cell*, **26,** 99 (1981).
153. E. D. Eanes, G. N. Martin, and D. R. Lundy, *Calcif. Tiss. Res.*, **20,** 313 (1976).
154. N. M. Hancox, "The Osteoclast," in *The Biochemistry and Physiology of Bone,* Vol. I, G. H. Bourne, Ed., Academic Press, New York, 1972, p. 45.
155. E. D. Eanes and A. S. Posner, "X-ray Scattering Study of Bone Mineral," in *Small-Angle X-ray Scattering*, H. Brumberger, Ed., Gordon and Breach, New York, 1967, p. 493.
156. J. D. Termine, unpublished data.

CHAPTER 6

Structural Chemistry of Calcium: Lanthanides as Probes

R. BRUCE MARTIN

Chemistry Department
University of Virginia
Charlottesville, Virginia

CONTENTS

1 CALCIUM CHEMISTRY

Calcium ion is one of the four significant alkali and alkaline earth metal ions in living systems. Unlike the other three ions, Na^+, K^+, and Mg^{2+}, Ca^{2+} occurs importantly both in minerals and in solution, often in complexed forms. It also exists plentifully in shells and corals as $CaCO_3$, and occurs as hydroxyapatite, $Ca_{10}(PO_4)_6(OH)_2$, in skeletons of insects and vertebrates. Mineralization is reviewed in another chapter of this volume.

In fluids, Ca^{2+} is involved in muscle contraction, blood clotting, neurotransmitter release, microtubule formation, protein stabilization, intercellular communication, hormonal responses, exocytosis, fertilization, mineralization, cell fusion, adhesion and growth. Many of these Ca^{2+} related activities occur by interactions with proteins, which Ca^{2+} may stabilize, activate, and modulate.

In extracellular fluids, the free or weakly bound Ca^{2+} concentration is about 1 mM. Within many cells the free Ca^{2+} concentration in the cytosol is 0.1 μM, 10^{-4} times less than in extracellular fluids. Cell membranes contain various pumps, such as Ca–ATPases, for maintenance of the high concentration gradient. There is, however, a substantial amount of Ca^{2+} within cells; it is bound tightly to proteins or occurs as phosphate complexes in the mitochondria. In response to a stimulus, the free Ca^{2+} concentration in the cytosol may increase about 10 times. Thus, proteins that participate in these responses possess Ca^{2+} dissociation constants in the μM range. The cytosolic Ca^{2+} concentration change is achieved easily, and free Ca^{2+} serves as a messenger or trigger for other interactions (1).

Measurement of the low free Ca^{2+} concentration within a cell is difficult. The subject has received review (2). Few methods respond to Ca^{2+} concentrations $< 10^{-5} M$. One of these depends upon microinjection of the jellyfish bioluminescent protein aequorin, which emits a blue light in the presence of as little as 10^{-7} M Ca^{2+} (3).

It is anticipated that Ca^{2+} sites in proteins will be composed of negatively charged and neutral oxygen donors; nitrogen donors are unlikely and none have been found. In model amides, there are no examples of Ca^{2+} binding to the amide nitrogen (4). Protein oxygen donors are furnished by carboxylate groups, carbonyl oxygens of the amide backbone, and hydroxy groups of serine and threonine side chains.

The geometry of Ca^{2+} binding to carboxyl and carbonyl groups has been surveyed for 60 crystal structures (5). Some structures show a bidentate carboxylate with both oxygens bound to Ca^{2+}, which is in or near the carboxylate plane. More common is unidentate carboxylate and carbonyl oxygen binding with only one bound carboxylate oxygen. The C-O—Ca^{2+} angle is nonlinear, with favored angles ranging from 120 to 150° almost as if the O atom were trigonally hybridized. Here Ca^{2+} is usually in the carboxylate plane, whereas for the carbonyl group it is generally out of the plane of the trigonal carbon. Various chelate structures also occur, such as between a single carboxylate and an α-hydroxy oxygen. For all structures the Ca^{2+}—O bond lengths are variable, usually from 2.3 to 2.6 Å, with unidentate distances at the short end and bidentate distances at the long end of the scale. Bond distances also increase with coordination number (see Section 3). The frequency of Ca^{2+} coordination numbers decreases in the order $8 > 7 > 6 > 9$ (5).

An unusual amino acid, γ-carboxyglutamate, was discovered in prothrombin and more recently found in other proteins (6). The dicarboxylate side chain has been advanced as a Ca^{2+} binding site. The appropriate model for metal ion binding to the dicarboxylate side chain is not the amino acid itself. The side chain may be viewed as a substituted malonate. For malonate, the stability constant logarithms (7) are Mg^{2+}, 2.1 and Ca^{2+}, 1.5. These low values indicate that the Ca^{2+}-binding capabilities and selectivity against Mg^{2+} of proteins containing γ-carboxylglutamate must be increased due to the juxtaposition of other groups. The amino acid γ-carboxyaspartate has also been found as a component of a bacterial ribosomal proteins (8).

The variability of Ca^{2+} coordination deserves emphasis. Not only is there variability of coordination number and bond lengths among complexes, but also within a single complex in solution. With their penchant for describing coordination geometries in terms of complex polyhedra, such as a square antiprism for 8-coordination, crystallographers imply a static geometry which is misleading. Coordination about Ca^{2+} is basically ionic and spherical. No more than four equivalent groups may be placed on the surface of a sphere (tetrahedron). Even if identical, donor atoms in 7-, 8-, and 9-fold coordination cannot occupy equivalent positions. Thus in solution, even within a single complex, there is variability in bond distances and, in many cases, also in coordination number. Even if the ligand donor atoms remain bound, they are in a dynamic state of flux about the Ca^{2+} with regard to distance and disposition, to the extent allowed by geometric properties of the ligands. Ligands impose their geometry on Ca^{2+}. If one is going to select for Ca^{2+} against other metal ions, a pocket with a specialized number of donor atoms and size is required. To select against Mg^{2+} is relatively easy; to select against lanthanides is difficult.

In its complexes, Ca^{2+} differs in several important ways from Mg^{2+}, which exhibits a strong preference for hexacoordination with oxygen–Mg^{2+} distances restricted to a 2.0 to 2.1 Å range. It also undergoes substitution about 10^3 times slower than Ca^{2+} (9). Thus, both from the static and dynamic viewpoints, Mg^{2+} forms more definite and tighter complexes. With chlorophyll as the supreme example, Mg^{2+} also exhibits a greater tendency to accept nitrogen donors, which Ca^{2+} does unwillingly. It also binds more strongly than Ca^{2+} to iminodiacetate, whereas for oxydiacetate Ca^{2+} binds more strongly (7). We know that Mg^{2+} binding is stronger to small chelate groupings, such as malonate and catecholates; Ca^{2+} forms stronger complexes with large, multidentate ligands, such as nitrilotriacetate, EDTA, and especially EGTA [ethylenebis (oxyethylenenitrilo) tetraacetate], which it binds almost 10^6 times as strongly as Mg^{2+}. To nucleoside phosphates, Mg^{2+} binds only slightly more strongly than Ca^{2+}, but within cells there is insufficient free Ca^{2+} to bind. Within a cell, Mg^{2+} is found to be associated with phosphates; Ca^{2+} is tightly bound by proteins.

It has been suggested that owing to the insolubility of $Ca_3(PO_4)_2$, cells initially extruded Ca^{2+} so they could use phosphate as their basic energy currency (10). Inorganic Ca^{2+} phosphate salts, however, are only about 10 times less soluble than those of Mg^{2+}. Both metal ions form soluble salts with organic phosphates. The strong differentiation occurring in living systems cannot be accounted for by these solubility differences. We have noted that EGTA stability constants are 10^6 times greater for Ca^{2+} than for Mg^{2+}. It is evident that both man and nature can build ligands that complex Ca^{2+} many powers of 10 times more strongly than Mg^{2+}. In living systems, differentiation between Ca^{2+} and Mg^{2+} is provided not by their phosphate insolubilities (which favor Ca^{2+} by only about 10), nor by their relative rates of dehydration (which favor Ca^{2+} by about 10^3), but by their stabilities with both synthetic and natural multidentate oxygen donor ligands (which may favor Ca^{2+} by greater than 10^5). The more likely scenario is that with strong protein complexes able to regulate the Ca^{2+} concentration at μM levels, Ca^{2+} becomes an obvious choice as a second messenger, perhaps the first second messenger.

2 LANTHANIDES (Ln^{3+}) AS PROBES

Electronic transitions of Ca^{2+} cannot be studied by conventional optical absorption and emission spectroscopy, and the absence of unpaired electrons precludes the use of magnetic resonance techniques in probing the chemical and structural nature of Ca^{2+} binding sites. Fortunately, about 12 lanthanide (rare earth) ions (Ln^{3+}) possess qualities that make them excellent probes for Ca^{2+}. (The symbol Ln is a generic symbol for lanthanides as a class; La is the symbol for the first lanthanide, element number 57.) In forming complexes, both Ca^{2+} and Ln^{3+} prefer charged or uncharged oxygen donor groups to nitrogen donor atoms. In aqueous solution, except for some multidentate ligands, hydroxo complex formation almost always occurs before amine nitrogen coordination takes place (11). Both Ca^{2+} and Ln^{3+} display a variable coordination number and a lack of strong directionality in binding donor groups; and Ln^{3+} compete with Ca^{2+} for Ca^{2+} binding sites in many proteins (Section 4).

In cellular and subcellular systems, Ln^{3+} often perform the same roles as Ca^{2+}, occasionally more effectively, and sometimes inhibit selectively processes that require Ca^{2+}. Several examples have been enumerated (12, 13). A more recent review describes Ln^{3+} as probes of Ca^{2+} binding sites on cellular membranes (14). In examples from cellular systems, it is likely that nonspecifically bound Ln^{3+} interact primarily with phospholipids and some proteins, and specifically bound Ln^{3+} have displaced Ca^{2+} from proteins. It is known that Ca^{2+} is specifically bound to a considerable number of proteins that play critical roles in biological processes (15). For these reasons we seek explanations for the profound effects of Ln^{3+} in cellular systems, primarily in reactions of Ln^{3+} with proteins.

3 Ca^{2+} AND Ln^{3+} COMPARED

Effective ionic radii of Ca^{2+} ions increase regularly with coordination number, being 1.00, 1.06, 1.12, and 1.18 Å for 6-, 7-, 8-, and 9-fold coordination, respectively (16). The increase of observed ionic radii with coordination number is a regular feature of metal ion stereochemistry and is often overlooked in comparing radii among metal ions. [Depending upon how observed bond lengths are divided among contributing atoms or ions, absolute values of radii may differ in various compilations. Whatever the basis for the division, however, radii from a single compilation exhibit constant differences when various cations are compared with each other. The ionic radii used here are from a recent, comprehensive analysis (16).]

As shown in Table 1, trivalent lanthanide ions (Ln^{3+}) exhibit effective

Table 1 Effective Ionic Radii (Å)

			Coordination Number			
Element	Atomic Number	Ion	6	7	8	9
Lanthanum	57	La^{3+}	1.03	1.10	1.16	1.22
Cerium	58	Ce^{3+}	1.01	1.07	1.14	1.20
Praseodymium	59	Pr^{3+}	0.99		1.13	1.18
Calcium	20	Ca^{2+}	1.00	1.06	1.12	1.18
Neodymium	60	Nd^{3+}	0.98		1.11	1.16
Samarium	62	Sm^{3+}	0.96	1.02	1.08	1.13
Europium	63	Eu^{3+}	0.95	1.01	1.07	1.12
Gadolinium	64	Gd^{3+}	0.94	1.00	1.05	1.11
Terbium	65	Tb^{3+}	0.92	0.98	1.04	1.10
Dysprosium	66	Dy^{3+}	0.91	0.97	1.03	1.08
Holmium	67	Ho^{3+}	0.90		1.02	1.07
Erbium	68	Er^{3+}	0.89	0.95	1.00	1.06
Thulium	69	Tm^{3+}	0.88		0.99	1.05
Ytterbium	70	Yb^{3+}	0.87	0.93	0.99	1.04
Lutetium	71	Lu^{3+}	0.86		0.98	1.03
Magnesium	12	Mg^{2+}	0.72		0.89	

ionic radii that show a gradual contraction from one end of the series to the other, for example, from 1.16 Å for La^{3+} (atomic number 57) to 0.98 Å for Lu^{3+} (atomic number 71) in 8-fold coordination. These Ln^{3+} commonly engage in 8- and 9-fold coordination, however, for Ca^{2+}, 7- and 8-fold coordination geometries occur most often. For a given coordination number, Table 1 shows that the ionic radii for Ln^{3+} span those for Ca^{2+}, with the values for Pr^{3+} nearly identical to those of Ca^{2+}. However, because of the soft contraction of Ln^{3+} radii with increasing atomic number, it is anticipated that most Ln^{3+} should be able to substitute for Ca^{2+} without causing serious structural modifications in those cases where ion size is of significant importance in determining binding characteristics. The lanthanide Eu^{2+} is the only one with an accessible 2+ state, but its ionic radii are at least 0.12 Å longer than that of Ca^{2+} (16), and it is readily oxidized to Eu^{3+} in protein containing solutions. The unit charge difference between Ca^{2+} and Ln^{3+} is apt to be of secondary importance for substitution of a Ln^{3+} for Ca^{2+}. In minerals, the occurrence of isomorphous replacements is more dependent on radius than on charge. Similarly, in biological systems, Na^{+} and Ca^{2+} of comparable ionic radii are competitive for sites, as are K^{+} and Ba^{2+} (17), which also possess comparable radii. Charge differences may assume more importance in rate phenomena, which may be sensitive to net charge at a binding site. It is suggested below that the effect of Ca^{2+} and Ln^{3+} charge differences on reaction rates may provide a basis for classifying many Ca^{2+} proteins.

It is worthwhile to examine more carefully the combined effects of the lanthanide contraction and coordination number on ionic radii as presented in Table 1. For a given metal ion, each successive increase in coordination number from 6 through 7 and 8 to 9 increases the ionic radii by almost 0.06 Å (frequently only 0.05 Å in going from 8- to 9-fold coordination). For a given coordination number, the decrease in ionic radii on moving from one end of the lanthanide series to the other (0.17 to 0.19 Å) is about equal to the increase in ionic radii on going from 6- to 9-fold coordination. Thus, there are opportunities for fine adjustments upon substitution of heavier Ln^{3+} for Ca^{2+} ions, either by small decreases in ionic radii or by an increase in coordination number. For example, if Ca^{2+} occupies a 7-coordinate site with an average ionic radius of 1.06 Å, substitution of Eu^{3+} or Gd^{3+} would favor a reduction of the ionic radii by 0.05 or 0.06 Å, or an increase in coordination number to 8, where the ionic radii of Eu^{3+} and Gd^{3+} nearly equal that of 7-coordinate Ca^{2+}. In general, Eu^{3+} and Gd^{3+} (and Tb^{3+} nearly so) possess ionic radii that correspond to that of Ca^{2+} in an environment of one less coordination number. Similarly, Er^{3+} possesses ionic radii that correspond to that of Ca^{2+} of two less coordination number. If diagonal lines of constant ionic radii are drawn in Table 1, for each unit increase in coordination number the lines decrease from four to five lanthanide atomic numbers. Thus, we might anticipate small decreases in ionic radii or increases in coordination number when the heavier Ln^{3+} replace Ca^{2+} in proteins.

The replacement of Ca^{2+} by Ln^{3+} has been demonstrated by x-ray diffraction to occur in two proteins, carp parvalbumin (18, 19) and thermolysin (20). In the latter case, the results have been reported with sufficient precision to allow some comparisons of bond distances and coordination numbers between Ca^{2+} and Ln^{3+} containing proteins. Thermolysin is a heat-stable proteolytic enzyme that normally contains four Ca^{2+} and retains enzymatic activity upon their replacement by three Ln^{3+}. Two of the Ca^{2+} are 6-coordinate and occupy a double site only 3.8 Å apart. This pair of Ca^{2+} are triply bridged by carboxylate groups from two glutamate and one aspartyl residues, one oxygen of each of three carboxylate groups being bound to each Ca^{2+}. Only a single Ln^{3+} occupies one of the pair of Ca^{2+} sites, with a probable increase in coordination number by carboxylate side chain movement rendering the other site unattractive. The single Ln^{3+} in this double Ca^{2+} site is firmly bound even under denaturing conditions. The other two Ca^{2+} in thermolysin are bound in single sites. Substitution of Eu^{3+} for 6- and 7-coordinate Ca^{2+} in the single sites appears to result in both cases in increases in coordination numbers. In the 6-coordinate case, the coordination number is increased by slight movement to bring both carboxylate oxygens of an aspartyl side chain into bonding distance to Eu^{3+}. In the 7-coordinate Ca^{2+} enzyme, the coordination number is increased to 8 upon substitution of Eu^{3+} by binding of an additional water molecule. Thus, two different mechanisms, increased carboxylate side chain involvement and additional water binding, are utilized to increase the coordination number upon substitution of Ca^{2+} by Ln^{3+} in thermolysin. By x-ray diffraction criteria, the Ln^{3+} substitutions still exhibit good isomorphism with lesser structural perturbations for the lanthanides of greater atomic number (20). These results are in accord with the principles developed above, wherein Eu^{3+} displays the same ionic radius as Ca^{2+} of one less coordination number and the heavier Ln^{3+} of favored coordination number 8 possess radii (Table 1) nearly identical to that of 6-coordinate Ca^{2+}.

4 LANTHANIDE (Ln^{3+})-PROTEIN INTERACTIONS

Interactions of Ln^{3+} with proteins may be grouped conveniently into four broad classes. Examples of proteins in each class appear in Table 2. In class I are Ca^{2+} (or Mg^{2+}) containing proteins in which Ln^{3+} compete for Ca^{2+} sites, often enhance binding of any substrate, and competitively inhibit the normal activity produced by the presence of Ca^{2+}. The longer class II category contains Ca^{2+} proteins in which Ln^{3+} substitution yields a protein that functions similarly to the Ca^{2+} protein, and sometimes more effectively. The distinction provided by Ln^{3+} substitution between classes I and II Ca^{2+} proteins is worthwhile. Such substitution is likely to be tolerated if the Ca^{2+} is involved only structurally in the protein. Physical and chemical differences between Ln^{3+} and Ca^{2+} are more likely to be important if Ca^{2+} plays a role in the function of the protein. Thus, for class I proteins, Ca^{2+} may participate in part of the mechanism of action or be at the active site. In class II proteins, Ca^{2+} is likely to be more remote from the active site and be involved most often in preserving protein tertiary structure. A role in substrate binding is possible, as seems to be the case with galactosyltransferase (31).

In class III proteins, Ln^{3+} react rather specifically with proteins in which Ca^{2+} is not normally considered to be involved. In my earlier presentation of

Table 2 Lanthanide (Ln^{3+}) Interactions with Proteins

	Reference
I. Ca^{2+} (or Mg^{2+}) proteins inhibited by Ln^{3+} substitution	
Staphylococcal nuclease	21
Blood clotting factor *X*	22
Calcium ATPase	23
Phosphoglycerate kinase (Mg^{2+})	24
Pyruvate kinase (Mg^{2+})	25
Pyrophosphatase (Mg^{2+})	26
II. Ca (or Mg^{2+}) proteins that function similarly upon Ln^{3+} substitution	
Thermolysin	20
α-Amylase	27
Phospholipase A_2	28
Concanavalin A	29
Aequorin luminescence	30
Galactosyltransferase	31
G-actin polymerization	32
Acetylcholine receptor	33
Prothrombin activation	34
Trypsinogen activation	35
Trypsin	36
Chymotrypsinogen	37
Chymotrypsin	38
Elastase	39
Subtilisin	40
Isoleucyl t-RNA synthetase (Mg^{2+})	41
Catechol-O-methyltransferase (Mg^{2+})	42
III. Ln^{3+} interacts with proteins not normally considered to contain Ca^{2+}	
Albumin	43
Lysozyme	44
IV. Ln^{3+} substitution for metal ions other than Ca or Mg^{2+}	
Transferrin, Ovotransferrin (Fe^{3+})	45
Glutamine synthetase (Mn^{2+})	46

this table the serine protease enzymes were listed in Class III (13). Partly from the results of Tb^{3+} luminescence studies, it is now evident that under physiological conditions trypsin, chymotrypsin, elastase, and subtilisin are Ca^{2+} enzymes.

Class IV proteins consist of those in which a Ln^{3+} has been substituted for a metal ion other than Ca^{2+} or Mg^{2+}. Glutamine synthetase is active when Ln^{3+} replace Mn^{2+} (46).

5 MAGNETIC RESONANCE PROBES

Except for La^{3+} and Lu^{3+}, all 12 other available Ln^{3+} contain unpaired f electrons, and both nuclear magnetic resonance (NMR) and electron spin resonance (ESR) methods may be used to investigate their environment. Several Ln^{3+} may be used as chemical shift probes in NMR, and Gd^{3+} with a half-filled $4f$ subshell is a broadening probe in NMR and is also employed in ESR studies (47). Specific application of counting water molecules bound to Gd^{3+} is deferred to Section 6.2.

There are numerous examples of ^{1}H NMR investigations of Ln^{3+} in proteins. Some examples appear in Section 7.4 and others follow: staphylococcal nuclease (48), lysozyme (49), albumin (50), myeloma protein (51), and pyruvate kinase (25).

The ^{1}H nuclear magnetic resonance spectroscopy of Yb^{3+} substituted for Ca^{2+} sites in parvalbumin shows peaks that are shifted over 30 ppm by a dipolar interaction between Yb^{3+} and nearby nuclei (52). The shift magnitude depends upon r^{-3}, where r is the distance between Yb^{3+} and the effected nuclei, and an angular dependent term. Once peak identifications are made, the authors hope to provide a detailed solution structure of the protein.

In one study, Yb^{3+} and Tb^{3+} have been used to shift single carbon peaks in the ^{13}C NMR spectrum of parvalbumin (53). This paper also reports use of ^{160}Tb γ-ray scintillation spectroscopy, in conjuction with a flow dialysis apparatus, to monitor substitution of Tb^{3+} out of parvalbumin by other metal ions.

Despite persistent use of Eu^{3+} in other Ln^{3+} shift studies, it has been recognized for some time that Yb^{3+} possesses the best combination of virtues for dipolar chemical shift studies: significant induced chemical shifts, tolerable line broadening, and a high ratio of dipolar to scalar contributions to the chemical shift (54). The last feature is especially important in ^{13}C and ^{31}P

NMR, where the Ln^{3+} binds to an oxygen only 2 bond lengths away from the observed nucleus. For a nucleus that induces shifts in the opposite direction of Yb^{3+}, and still retains other desirable features, the choice is Pr^{3+}.

Lanthanide ions have been extensively used as shift reagents in lipid bilayer systems (for a review see ref. 55). When placed in a solution containing spherical vesicles, the Ln^{3+} ions shift resonances of outer lipids in ^{1}H, ^{13}C, or ^{31}P NMR spectroscopy so that they are distinguished from resonances of molecules on the inside of the vesicle. In another kind of study, vesicle fusion has been followed by encapsulating Tb $(citrate)_3^{6-}$ in one population of vesicles and dipicolinate (DPA) in another. Formation of luminescent Tb $(DPA)_3^{3-}$ monitors the rate and extent of vesicle fusion (56).

6 LUMINESCENCE PROBES

With the exception of La^{3+} and Lu^{3+}, each of the Ln^{3+} exhibits absorption spectra resulting from intraconfigurational *f-f* transitions that are easily accessible to study by conventional optical absorption techniques. However, the molar absorptivities associated with these transitions are only of the order of unity such that at Ln^{3+} concentrations $<10^{-2}$ M, the low absorbance strains the sensitivity of absorption spectrophotometers. In contrast, luminescence due to intraconfigurational *f-f* transitions in Eu^{3+} and Tb^{3+} ions bound to protein systems remains well above detection limits, even when Eu^{3+} and Tb^{3+} are present at concentrations as low as $\sim 10^{-6}$ M. Thus, a factor of at least 10^4 in sensitivity favors luminescence over absorption spectra.

In order for luminescence to occur, appropriate excited states of the emitting (luminescent) species must be populated. Either direct or indirect methods may be used to excite the Ln^{3+} *f-f* emitting states. In direct excitation, the exciting light is absorbed directly by the Ln^{3+}. Owing to low molar absorptivities, direct excitation requires either relatively high concentrations of Ln^{3+} or an intense excitation source, such as a pulsed laser. Potentially more sensitive is indirect or energy transfer excitation in which Tb^{3+} accepts energy by radiationless energy transfer from an efficient donor, such as an excited aromatic chromophore. In favorable cases, where an aromatic chromophore and Tb^{3+} are closely spaced, the emission from bound Tb^{3+} may become enhanced by a factor of up to 10^5 compared to emission from free aqueous Tb^{3+}.

6.1 Circularly Polarized Luminescence (CPL)

The sensitivity of Tb^{3+} luminescence for revealing fine details of Ca^{2+} binding sites is augmented by the observation of circularly polarized luminescence (CPL) of the emitting Tb^{3+} (40). This CPL is the emission analog of circular dichroism in absorption spectra and is similarly sensitive to details of geometry, but in an excited state rather than the ground state of the emitting species. The term CPL refers to the differential emission of left- and right-circularly polarized light by chiral luminescent systems (57). The sign and magnitude of CPL depends upon the rotatory strength of a transition and may be reported as the luminescence dissymmetry factor, which is the ratio of the CPL intensity to one half the total luminescence (TL) intensity at a specified wavelength. The magnitude of the CPL is generally greater in those systems where chelate rings form. Since oppositely signed CPL spectra are observed for enhanced Tb^{3+} emission in several proteins, all of which are composed of L-amino acid residues, the absolute configuration at the asymmetric α-carbon atom evidently plays only a small role in determining chirality at Tb^{3+} binding sites (40). The CPL of Tb^{3+} emission is likely to result from both chiral disposition of donor groups and a dissymmetric arrangement of chromophores with strong electric dipole transitions, such as the aromatic side chains, in the vicinity of the Tb^{3+} binding site. Since Tb^{3+} at most exterior binding sites contains nonchiral solvent molecules as part of its coordination sphere, CPL should be easier to generate when Tb^{3+} is bound at an interior site. Proteins binding two or more Tb^{3+} may produce a cancellation of oppositely signed CPL components. (Application of CPL to muscle proteins is reviewed in Section 7.)

Appearance of circularly polarized luminescence (CPL) spectra has been invoked to claim tridentate chelation to Tb^{3+} of the ribonucleosides uridine, cytosine, and inosine via the 2′ and 3′ ribose hydroxy groups and a carbonyl oxygen of the nucleic base (58). No CPL was observed below pH 6, and maximum CPL was observed at pH 7 just prior to precipitation of $Tb(OH)_3$. For Tb^{3+} luminescence at 543 nm, a strong excitation peak is reported at 298 nm (58). The reported phenomena occur in the notorious pH 6–7 region, where hydroxo and polymeric complexes form with other ligands (11). No concentration dependent studies accompany the CPL results. The nucleoside concentration was about 15 mM. The reported excitation spectrum maximum (around 298 nm) does not correspond to any absorption band in the ligands and is evidently an artifact resulting from pronounced inner-filtering of incident radiation in the concentrated solutions used (see Section 8). Probably, the observed results are due to polymeric complexes making unnecessary the unlikely, if not even impossible, proposed tridentate chelation.

Since hydroxy complex formation of Tb^{3+} with water occurs at pH ~7 (11), it is plausible that chelation by deprotonation of one or both cis OH groups on sugars also occurs in the same pH region. For deprotonations from water $pK_a = 15.7$ and from ribosides $pK_a \sim 12.5$ (59). Thus, in the same pH region that sugar *cis*-OH groups are chelating, simple and polymeric hydroxo complexes are also forming. Bridges are also possible through deprotonated sugar hydroxy groups.

6.2 Count of Coordinated Water Molecules

Counting of water molecules bound to metal ions in the primary coordination sphere has been beset with difficulties. Proton relaxation enhancement of water protons in NMR spectroscopy induced by Mn^{2+} and Gd^{3+} has been used for many years (60). To interpret the results, several parameters in complex equations must be characterized, often leading to large uncertainties (61). A much simpler and more direct approach is to employ Eu^{3+} or Tb^{3+} luminescence lifetime measurements.

For Eu^{3+} and Tb^{3+}, a radiationless path for deexcitation via coupling with OH vibrational overtones is much more efficient for OH than OD oscillators (62, 63). It has been established that the reciprocal lifetimes for luminescence decay vary linearly with the mole fraction of H_2O in H_2O/D_2O solvent mixtures. Therefore, the observed reciprocal lifetime τ_{obs}^{-1} may be written as the sum of components resulting from all decay processes other than OH induced quenching, τ_o^{-1}, and a term due to deexcitation via OH vibrations, which depends on the mole fraction of H_2O in an H_2O/D_2O mixture.

$$\tau_{obs}^{-1} = \tau_o^{-1} + \tau_H^{-1} X_{H_2O}$$

The reciprocal lifetime for OH induced coupling, τ_H^{-1}, may be found from a plot of τ_{obs}^{-1} versus mole fraction water, X_{H_2O}, in H_2O/D_2O mixtures. (Any small deactivation by OD oscillators is automatically taken into account by this treatment.) It has been demonstrated that in producing radiationless deexcitation the OH oscillators of bound water molecules act independently of each other and of the other ligands bound. Therefore, for the aquo ion we may write $\tau_H^{-1}(\text{aquo}) = cW$, and for a complex, $\tau_H^{-1}(\text{complex}) = cV$, where c is a proportionality constant, W is the number of OH oscillators in the aquo ion, and V is the number of OH oscillators in a complex. The ratio of slopes in a τ_{obs}^{-1} versus X_{H_2O} plot for the complex and aquo ion then yields $r = \tau_H^{-1}$ (complex)/τ_H^{-1} (aquo) $= V/W$. For the aquo ions, the value of W is usually set to 9 H_2O molecules or 18 OH oscillators. The number of H_2O molecules or OH oscillators bound in the complex V, may then be calculated from $V = Wr = 18\ r$. It is advantageous to consider individual OH oscillators in cases where a single OH oscillator may occur, for example, hydroxo complexes, hydroxy groups of serine and threonine side chains, and carbohydrates. Note that the method simply estimates OH oscillators, the coordination number of the complex may be any value and unknown. Several Ca^{2+} binding sites in thermolysin have been characterized by this method (64). The efficiency of the transfer via vibronic OH coupling is greater with Eu^{3+} than Tb^{3+}, giving rise to about a four times greater sensitivity with Eu^{3+}. Since direct excitation of Eu^{3+} and Tb^{3+} must (may) be used, the method is potentially applicable to a variety of systems in addition to proteins, including carbohydrates and lipids.

6.3 Intermetal Energy Transfer

Energy transfer efficiencies from Eu^{3+} or Tb^{3+} to other Ln^{3+} in parvalbumin (65) and thermolysin (63, 64) have been monitored by luminescent lifetimes upon direct excitation of the donor ion by pulsed dye laser radiation. Since the distances between the metal ion sites are known, it is possible to check on the energy transfer mechanism. The results favor a Forster dipole-dipole mechanism with an r^{-6} dependence between donor and acceptor metal ions (66). Energy transfer from Tb^{3+} in a Ca^{2+} site to Co^{2+} in a Zn^{2+} site of thermolysin is consistent with the distance between the two sites found by x-ray diffraction (67).

Distances derived from intermetal energy transfer within a protein are subject to a variety of pitfalls. For Tb^{3+} to Fe^{3+} energy transfer in transferrin (normally a two-Fe^{3+} protein), differences in intermetal distances found in two earlier studies (25 Å and greater than 43 Å) have been resolved by a current distance of 36 Å (68).

6.4 Excitation Spectra (69)

We may presume that Tb^{3+} emission resulting from excitation by energy transfer occurs in four steps, three of which are necessary.

1. A chromophore absorbs radiant energy (in the ultraviolet region in protein and nucleic base systems).
2. The resulting excited state energy may be passed by radiationless energy transfer to other chromophores (70, 71). This step is not necessary.
3. Energy is transferred, most likely by a dipole-dipole mechanism, from an excited chromophore to a Tb^{3+}.
4. Phosphorescence occurs from the Tb^{3+} $^5D_4$ state in the visible region.

By monitoring the intensity of the enhanced green emission band near 545 nm while irradiating the sample, an excitation spectrum is obtained.

A careful study has established that in solutions of low absorbance (Section 8), for the aromatic rings in proteins, the excitation spectrum for enhanced Tb^{3+} emission corresponds closely to the chromophore absorption spectrum (69). (A similar result has been obtained for the nucleic bases. See Section 9.) Thus, the excitation spectrum for Tb^{3+} emission provides a tool for identification of aromatic chromophores positioned near Ca^{2+} binding sites in proteins. In the 240 to 320 nm region of proteins, four aromatic side chain chromophores could be involved in energy transfer: the phenyl group of phenylalanine (phe), the indole group of tryptophan (trp), the phenolic group of tyrosine (tyr), and the ionzed phenolate group of tyrosine (tyr^-). Since these four chromophores exhibit different absorption spectra in the 240 to 320 nm region, it should be possible to differentiate among their excitation spectra. Nevertheless, there has been considerable confusion, uncertainty, and incorrect identifications of groups involved in energy transfer to Tb^{3+}.

In a protein, the excitation spectrum for enhanced Tb^{3+} luminescence depends upon the aromatic side chains present and their proximity to Tb^{3+}. Not only does trp absorb more strongly than the other two amino acid residues, but there is also effective overlap between trp fluorescence from about 320 to 390 nm and the Tb^{3+} absorption band lying at 340 to 380 nm. This overlap makes trp an efficient donor in energy transfer to Tb^{3+}. In more than 20 proteins containing trp in which enhanced Tb^{3+} luminescence occurs, the excitation spectrum corresponds to the trp absorption spectrum (40), even though it is rarely the most prevalent aromatic amino acid residue. Energy transfer from tyr has been observed in the trp free protons, troponin-C, calmodulin (Section 7), and phospholipase A (40). Phenylalanine energy transfer has been observed only in the trp and tyr free parvalbumins. In the presence of trp, some energy transfer from phe and tyr will be difficult to observe. Similarly, in the presence of tyr, energy transfer from phe to Tb^{3+} may not be evident.

Figure 1 displays excitation spectra observed in proteins that may be taken as prototypes for energy transfer from each of the three aromatic side chains. Energy transfer from phe is demonstrated in parvalbumin, from tyr (nonionized) in calmodulin, and from trp in elastase (69). The side chain chromophore engaged in energy transfer to Tb^{3+} in an unknown protein may be deduced by comparison of its excitation spectrum with the prototype spectra of Figure 1. These excitation spectra reflect energy transfer by a through space interaction from the aromatic chromophore to Tb^{3+}; there is no complex formation between the side chain and Tb^{3+}.

A single tryptophan (trp) residue occurs in codfish (III) parvalbumin. If its distance from both Ca^{2+} sites is taken as 11.5 Å, then deductions concerning the efficiency of energy transfer processes may be made (72). Of several Ln(III) investigated, only Eu^{3+} measurably quenches the protein tryptophan fluorescence. The energy transferred to Eu^{3+}, however, dissipates by non-

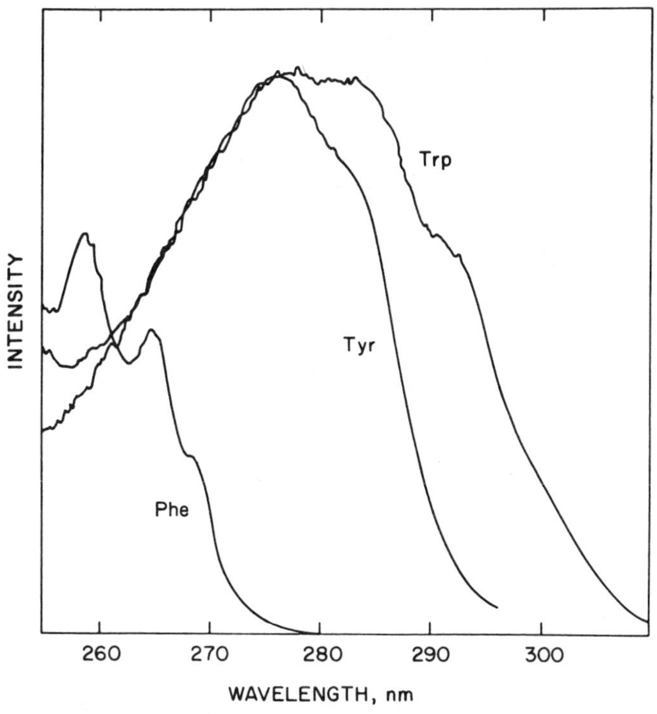

Figure 1 Excitation spectra of Tb^{3+} proteins with emission monitored at 544 nm: carp parvalbumin (Phe), beef calmodulin (Tyr), and porcine elastase (Trp). Intensity scale is arbitrary with no comparison among the three proteins. (From ref. 69.)

radiative processes and does not appear as luminescence. In contrast, energy transfer to Tb^{3+} yields such small quenching of the trp fluorescence that it is not observable. With Tb^{3+}, however, luminescence is highly sensitized and easily detectable. No other Ln^{3+} is so highly sensitized by energy transfer from trp. The distance for 50% energy transfer from tryptophan to Tb^{3+} by the Forster dipole-dipole mechanism is estimated to be $R_o = 3.4$ Å (72). Even though $R_o = 3.4$ Å, and the energy transfer efficiency falls off as r^{-6} (the distance between tryptophan and Tb^{3+}), sensitized Tb^{3+} luminescence is still readily observable when $r = 11.6$ Å. The energy transfer efficiency at 11.6 Å is calculated to be 1600 times less than at 3.4 Å. This high sensitization of Tb^{3+} luminescence is consistent with identification of tryptophan as the energy donor in a variety of proteins (40).

In order to determine the expected excitation spectrum when Tb^{3+} is bound to an ionized phenolate group in a tyrosine side chain, we have employed the model ligand ethylenedinitrilo-*N,N′*-*bis*(2-hydroxyphenyl)-*N,N′*-diacetate or *N,N*-ethylenebis(2- (*o*-hydroxyphenyl)) glycine (EHPG) (73, 74). The neutral ligand contains two carboxylate, two ammonium, and two phenolic groups. The last four groups are titrated from pH 6 to 12 in the free ligand. In the presence of an equimolar amount of Tb^{3+}, the titration is complete by pH 9 and the ligand is hexadentate about Tb^{3+}. In this complex, Tb^{3+} emission at 544 nm is enhanced by a factor of 5×10^5, compared with free Tb^{3+} (measured at pH 9.5 with $\lambda_{ex} = 295$ nm) (69). The enhancement may be the largest yet observed for a Tb^{3+} complex, and is due in large part to energy transfer from two ionized phenolate groups bound to Tb^{3+}. The excitation spectrum for the equimolar hexadentate EHPG:Tb^{3+} complex is shown by the curve labeled 0.035 in Figure 2.

Despite the very high enhancement of Tb^{3+} luminescence in EHPG it is, by some measures, only comparable to that of Tb^{3+} in elastase, where a tryptophan side chain is involved in energy transfer. For elastase the enhancement factor is 2×10^4 for $\lambda_{ex} = 283$ nm (75). When allowance is made for two phenolate groups in the EHPG complex, and intensity more than 10 times stronger in the excitation spectrum of aqueous Tb^{3+} at 283 nm than at 295 nm, the enhancement occurring from a single tryptophan in elastase is comparable to that of one phenolate ring in EHPG (69). The number of Tb^{3+} bound water molecules may differ in the two cases and may provide dissimilar probabilities for radiationless deactivation. The main point, however, is that energy transfer through space from an unbound tryptophan side chain may yield an enhancement of Tb^{3+} emission comparable to that of a bound phenolate.

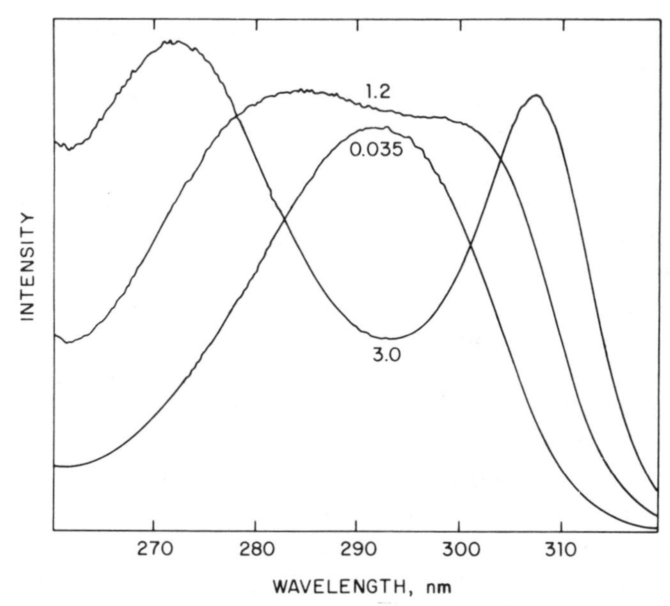

Figure 2 Excitation spectra of equimolar complex of EHPG and Tb^{3+} at pH 9. The intensity scale is arbitrary. Numbers and curves refer to absorbances of the solutions: 0.035, 0.011 m*M* complex in 4 mm cell; 1.2, 0.38 m*M* complex in 4 mm cell; and 3.0, 0.38 m*M* complex in 10 mm cell. At pH 9, the absorption spectrum resembles the curve labeled 0.035 in shape and shows a maximum at 292 nm. (From ref. 69.)

There is strong evidence of several kinds that in transferrin and ovotransferrin (conalbumin) two metal ions, such as Fe^{3+}, Cu^{2+}, and Tb^{3+}, are bound per 76,600 daltons via ionized phenolate groups of tryosyl side chains (45, 68, 73, 76). There is, however, an imperfect match for the excitation spectra for Tb^{3+} emission between the EHPG complex and ovotransferrin (69). A tryptophan side chain appears to make a significant contribution to the ovotransferrin excitation spectrum.

Ionization of the phenolic hydrogen in tyrosine to give the phenolate group is complete by pH 12 (77). Accompanying the phenolic ionization is a pronounced red shift and intensification of the absorption spectrum, with the maximum shifting from 275 to 292 nm. Owing to the strong basicity of the ionized phenolate group, hydroxy complexes of lanthanides occur in aqueous solutions well before phenolate complexation is effective (11). Coordination of ionized tyrosyl phenolates to Tb^{3+} in neutral solutions occurs only in a few special cases, such as transferrin and ovotransferrin. Because of the possibility of hydroxo complex formation, it is desirable to perform experiments with Tb^{3+} at pH <7 (11). At pH <7, there will be few instances in which tyrosyl side chains have undergone deprotonation even in the presence of metal ions. If it is necessary to have pH >7, excess aqueous Tb^{3+} should be avoided.

Absorption spectra for proteins are composites over all aromatic side chains present. The absorption bands are usually shifted to the red, compared to bands of the constituent amino acids in aqueous solutions. Excitation spectra reflect only the one or two side chains engaged in significant energy transfer to Tb^{3+}. Thus, excitation spectra in proteins should not be compared with absorption spectra of individual amino acids or of the whole protein, but rather to representative excitation spectra (as shown in Figs. 1 and 2) because of energy transfer to Tb^{3+} from specified aromatic side chains in proteins.

Comparison of the complete excitation spectrum of an unknown Tb^{3+} protein with the four prototype spectra in Figures 1 and 2 provides the best way to indicate the aromatic side chain engaged in energy transfer to Tb^{3+}. The distinction between tyr and trp depends upon appreciable intensity or even a shoulder with trp at about 292 nm, where there is little intensity with tyr. For all the small ligands and proteins we have investigated, the ratio, R, of excitation spectra intensities at 292 to 276 nm is <0.2 for tyr and from 0.5 to 0.8 for trp. Phenylalanine displays no appreciable excitation spectrum at either wavelength, and Figure 1 shows that its excitation spectrum is unique and easily recognized. For the ionized phenolate side chain of EHPG, $R = 2.0$. We suggest the simple diagnostic guideline that for the ratio of excitation spectrum intensities at 292 to 276 nm, the side chain engaged in energy transfer to Tb^{3+} is indicated as follows: $R < 0.2$, nonionized tyr; $0.5 < R < 1.0$; and $R > 1.8$, ionized tyr^{-}. See, however, the precautions indicated for obtaining reliable excitation spectra in Section 8.

7 PROTEIN EXAMPLES

7.1 Parvalbumin

Parvalbumins are small, highly soluble calcium-binding proteins of the white muscles of vertebrates and mammals. Carp muscle calcium-binding pavalbumin (isotype B or 3) consists of a single polypeptide chain of 108 amino acid residues of known sequence (78) with 10 phenylalanine and no tyrosine or tryptophan residues. Though originally only one Ca^{2+} was found by x-ray crystallography (79), a second Ca^{2+} was located by application of the tangent formula technique (80). The crystal structure has been presented to 1.9 Å resolution (81). The nearly spherical protein is almost half helical with six helical regions of similar length, designated A–F. Two of the loops between helical sections contain Ca^{2+}. One Ca^{2+} located between helical sections C and D, Ca(CD), is 6-coordinated by protein donor atoms. The second Ca^{2+} is positioned between helical sections E and F, Ca(EF), and is 8-coordinated, with one coordination position occupied by a water molecule. Each Ca^{2+} is bound only by oxygen donor atoms, with four negatively charged carboxylate side chains of aspartate and glutamate residues involved in each case. One of the donor atoms for Ca(CD) is the carbonyl oxygen of phe-57. The aromatic side chain of phe-57 is about 6.1 Å from Ca(EF).

Upon addition of Tb^{3+} to a solution containing carp parvalbumin, substitution of Ca^{2+} by Tb^{3+} occurs readily. Along with most other proteins studied, parvalbumin binds Tb^{3+} more strongly than Ca^{2+}. The substitution is indicated by a dramatic enhancement of the characteristic green luminescence of Tb^{3+} from 535 to 555 nm upon irradiation of the protein at 259 nm (82). The Tb^{3+} luminescence is produced by intramolecular energy transfer from the irradiated aromatic ring of phe-57 to the emitting Tb^{3+} bound at the EF site.

A unique feature of the parvalbumin luminescence titration curve in neutral solutions is the intensity maximum at about 1.6 moles Tb^{3+} per mole of parvalbumin (82, 83, 65). Of more than 30 proteins that display enhanced Tb^{3+} emission, only parvalbumin exhibits an intensity maximum (40). The intensity maximum does not occur in the related troponin-C's (84, 85). In parvalbumin, the intensity maximum in the titration curve may be increased 40% by adding 4 equivs Tb^{3+} and dialyzing (83). Dialysis removes excess Tb^{3+} and displaced Ca^{2+}. Experiments in Tb^{3+} atomic absorption establish that the most highly emitting protein contains two bound Tb^{3+} (86, 87). These results unequivocally rule out Tb(CD) quenching of Tb(EF) luminescence as the cause of the maximum in the parvalbumin luminescence titration curve.

The maximum in the parvalbumin luminescence titration curve may be accounted for by weaker Tb^{3+} coordination at a third site. Two third sites have been suggested. Requirements for the third site are that energy transfer occurs from Tb(EF) to Tb^{3+} at the third site and that the latter Tb^{3+} bears water molecules. Bound water molecules provide a pathway for radiationless energy dissipation in which H_2O is many times more effective than D_2O (62). That water is involved in the Tb^{3+}–parvalbumin quenching process is supported by greater emission in D_2O (83, 88). Quenching of Tb(EF) luminescence by Tb(CD) is specifically excluded (83). The ion Ca(EF) in parvalbumin bears a water molecule and Tb(EF) presumably binds at least one water molecule. One proposal for the third interaction site to produce quenching is weak coordination of aqueous Tb^{3+} at the water molecule(s) bound to Tb(EF) (83). The second suggestion is that four aspartate residues bind the third Tb^{3+} in a site near Tb(CD) (88). Three of the four aspartate residues bind to both Tb^{3+} in the suggested third site and Tb(CD).

Existence of a weaker, metal ion binding site near the EF site is supported by two NMR studies. Characteristics of the third site have been deduced from ^{113}Cd NMR (89) and proton relaxation enhancement of water protons by Gd^{3+} and Mn^{2+} (90). The third site offers only weak binding, with the metal ions remaining extensively hydrated. In contrast, up to 4.5 equivs Yb^{3+} produced no shifts in the ^1H NMR spectrum attributable to a third site (52).

Some sort of third metal ion interaction center in parvalbumin is required to account for the intensity maximum in the Tb^{3+} luminescence titration curve. This interaction center need not be a site that one necessarily expects to find by static methods such as x-ray diffraction or NMR chemical shifts. All that is required to account for the luminescence intensity maximum is an additional pathway for radiationless deactivation of Tb^{3+} bound in the EF site ascribable to excess Tb^{3+} not bound in the CD site. Weak, momentary association of unbound Tb^{3+} at the water molecule bound to Tb(EF) is one way to provide this additional pathway.

Crystallographers have contributed confusion to the site specificity and quenching of Tb^{3+} luminescence in parvalbumin. Preparation of Tb^{3+}-substituted parvalbumin from solutions containing 2 equivs Tb^{3+} at pH 7 produced first Tb^{3+} only in the EF site (18) and second Tb^{3+} in both the EF and CD sites (19). From the experimental descriptions, the major difference between the two preparations appears to be the use of phosphate buffer in the former (18), which would withdraw Tb^{3+} from solution. In the second study (19), difference Fourier maps calculated at 4 and 8 equivs Tb^{3+} failed to show greater occupancy of CD and EF sites than did 1 equiv, which is inadequate to saturate 2 sites. Despite inconsistencies within the crystal structure papers, evidently in neutral solutions of 2.9 M $(NH_4)_2SO_4$, the equilibrium constant for Tb^{3+} substitution is greater for Ca(EF) than for Ca(CD).

Based on their inability to locate a Tb^{3+} at a parvalbumin third site, the crystallographers objected to Tb^{3+} binding at a third site, and reintroduced Tb(CD) quenching of Tb(EF) luminescence as the mechanism to account for the maximum in the Tb^{3+} luminescence titration curve (19). As reviewed above, significant quenching by this mechanism had been specifically excluded. Moreover, conditions in the solutions used to prepare crystals make arguments from crystal structure irrelevant to commenting on the quenching mechanism in the Tb^{3+} titration curve. Only a weak interaction is proposed for Tb^{3+} binding at a third site. Crystals were grown in the presence of 2.9 M $(NH_4)_2SO_4$. Because of ion-pair formation with SO_4^{2-}, less than 2% of the total Tb^{3+} exists as the free aqueous ion. This small amount of free aqueous Tb^{3+} precludes binding at a weak third site. There is no SO_4^{2-} in the spectroscopic experiments (87).

7.2 Troponin-C (TN-C)

Regulation by Ca^{2+} of contraction in vertebrate skeletal muscles results from Ca^{2+} binding to the protein complex troponin. Troponin consists of three different subunits, only the lightest of which, TN-C, binds up to four Ca^{2+}. The second component, TN-I, inhibits actomyosin ATP-ase activity. The third subunit, TN-T, interacts with the protein tropomyosin.

Rabbit skeletal muscle TN-C consists of a single polypeptide chain of 159 amino acid residues that is divided into four internally homologous regions designated I–IV from the amino terminus. Each region contains a 10-residue Ca^{2+} binding loop, with carboxylate side chains, flanked on both sides by α-helical segments. Regions I and II each bind Ca^{2+} specifically, with a dissociation constant of about 2 μM; regions III and IV each bind Ca^{2+} strongly, with a constant of 0.05 μM, and Mg^{2+} weakly, with a constant of 0.2 mM (91, 92). In bovine cardiac, TN-C amino acid substitutions in Region I have destroyed its Ca^{2+} binding capability (93) and the protein binds only three Ca^{2+}, two of high affinity and one of low affinity (94).

Upon addition of Tb^{3+} to solutions containing either rabbit skeletal muscle TN-C (84) or bovine cardiac TN-C (85), and irradiation in the ultraviolet region, the green Tb^{3+} luminescence increases and reaches more than half maximum intensity after addition of 1 equiv Tb^{3+}. Since both proteins behave similarly, and bovine cardiac TN-C lacks a Ca^{2+} binding region I, that region must not be involved in the favored substitution of Ca^{2+} by Tb^{3+}.

The Tb^{3+} luminescence from troponin-C and parvalbumin is partially circularly polarized (Section 6.1). Luminescence dissymmetry factors (ratio of CPL intensity to one-half total luminescence (TL) intensity) for parvalbumin and troponin components are compared in Table 3. Table 3 shows that CPL

Table 3 Luminescence Dissymmetry Factors for Tb^{3+}

Protein	545 nm	550 nm
Carp parvalbumin	−0.050	+0.058
Rabbit skeletal TN-C	−0.058	+0.054
Rabbit skeletal troponin	0	0
Bovine cardiac TN-C	−0.060	+0.048
TN-C + TN-1	0	0
TN-C + TN-T	0	0
TN-C + TN-1 + TN-T	0	0

does not appear in the whole troponin trimer from rabbit skeletal muscle, or in any combination of TN-C with other troponin components of bovine cardiac muscle. The other troponin components also partially quench the TL of Tb^{3+}-substituted TN-C. Thus, CPL and TL spectroscopies provide probes into the microenvironment of Tb^{3+} in TN-C undergoing subunit interactions with other troponin components. The probes are so sensitive that they may reveal structural alterations not apparent at the level of resolution achieved in x-ray diffraction studies of proteins.

For Tb^{3+}-substituted rabbit skeletal TN-C and bovine cardiac TN-C, the wavelengths and dissymmetry factors at the CPL extrema are virtually identical, as shown in Table 3. In addition, excitation spectra for both proteins identifies a tyrosyl side chain as the donor group in the energy transfer process leading to Tb^{3+} luminescence. Thus, in both troponins excitation spectra indicate a tyrosine side chain overlies a Tb^{3+} site, and similar CPL spectra suggest that the chirality about the site is nearly identical. Two tyrosyl residues occur in rabbit skeletal TN-C and three in bovine cardiac TN-C. The amino acid sequences of both proteins are known, and the only homology involving tyrosyl residues occurs at 109 for rabbit skeletal TN-C and at 111 for bovine cardiac TN-C (93). This single homologous tyrosyl residue is thereby strongly implicated as the one providing energy transfer to Tb^{3+} in the emission experiments.

Further support for this identification comes from the CPL result for Tb^{3+}-substituted carp parvalbumin (Table 3), which is closely similar to those of the two TN-C molecules. For parvalbumin, however, the excitation spectrum displays a maximum at 259 nm, identifying a phenylalanyl side chain as the group participating in energy transfer to Tb^{3+} (82). In the alignment of amino acid sequences, phe-57 of parvalbumin matches with tyr-109 of rabbit skeletal muscle TN-C (95). The crystal structure of carp parvalbumin shows the aromatic side chain of phe-57 near Tb^{3+} in the Ca(EF) site (81). Thus, a persistent feature of a Ca^{2+}-binding site of the three multiple Ca^{2+} proteins from different species is an aromatic side chain near the most easily Tb^{3+}-substituted Ca^{2+} in a site of closely similar chiralities.

The two homologous tyrosine side chains involved in energy transfer to Tb^{3+} occur in region III of the two TN-C's. Troponin-C regions III and IV are thus analogous to parvalbumin CD and EF regions, respectively. On the basis of the Tb^{3+} luminescence results, in TN-C the region III tyr 109 (111) overlies the region IV Ca^{2+}, which is also the first to be substituted by Tb^{3+}. And Tb^{3+} substitutes for the region III TN-C Ca^{2+} almost as readily (96).

7.3 Calmodulin

Calmodulin, a protein of 148 amino acid residues (beef brain) found in nonmuscle tissue, activates several enzymes only when Ca^{2+} is present (97). Like rabbit skeletal TN-C, calmodulin also possesses four Ca^{2+}-binding regions (98). Several groups of investigators report a variety of Ca^{2+}-binding strengths. The most recent reported that Ca^{2+} dissociation constants are 3 to 20 μM (99). Thus, calmodulin does not appear to exhibit the very strong binding of the two high affinity sites of TN-C.

An even stronger contrast with TN-C is observed when a Tb^{3+} luminescence titration is performed on beef brain calmodulin. As mentioned above, with TN-C virtually all the luminescence intensity is generated by addition of 1 to 2 equivs Tb^{3+} (84, 85, 96). In contrast, the calmodulin, both with (69) and without (100) Ca^{2+}, significant luminescence enhancement appears only after addition of 2 to 3 equivs Tb^{3+} (69, 100). The excitation spectrum of calmodulin identifies a tyrosine as the energy donor to Tb^{3+} (69). In calmodulin tyr-99 is homologous with tyr-109 in rabbit skeletal TN-C. These tyrosines occur in region III of both proteins and may, in calmodulin, be similarly responsible for energy transfer to Tb^{3+} in region IV. Hence, the Tb^{3+} titration results indicate that substitution for Ca^{2+} in regions III and IV, which takes place early in the TN-C titrations, occurs later and only with some difficulty in calmodulin. In calmodulin, regions I and II add Tb^{3+} and undergo Ca^{2+} substitution by Tb^{3+} before regions III and IV.

7.4 Serine Proteases

Except for the role of Ca^{2+} in the activation of trypsinogen and in the resistance of the latter to autolysis, metal ions have usually not been considered with the serine proteases. The Ln^{3+} ions are more effective than Ca^{2+} in accelerating conversion of trypsinogen to trypsin, presumably because they are more effective in charge neutralization at the amino terminus of the zymogen substrate (35). Nuclear magnetic resonance methods show that Gd^{3+} binds competitively with Ca^{2+} to trypsin (101) and elastase (39). We have observed enhanced Tb^{3+} luminescence from all the serine proteases we examined: trypsinogen, trypsin, chymotrypsinogen, chymotrypsin, and elastase; and also in subtilisin Carlsberg and subtilisin BPN (40). In all cases, tryptophan is the energy donor. The Tb^{3+} luminescence from porcine elastase is unusually strong (40, 75) and follow-up studies have been made on this enzyme. Inhibitors change the Tb^{3+} luminescent intensities, providing a probe of protein conformational changes (75, 102), and permitting evaluation of inhibitor binding constants (75). Recent independent investigations also support Ca^{2+} and/or Ln^{3+} binding in trypsin (36, 37, 103, 104), chymotrypsin (37, 38), chymotrypsinogen (105), and the subtilisins (40, 106). The Ln^{3+} studies played a part in pointing out the Ca^{2+} binding capabilities of these enzymes.

For all the enzymes mentioned, the references above report stability constants for Ln^{3+} and Ca^{2+} binding. Unfortunately, there is a less than satisfying agreement among the several studies for each enzyme. It seems evident, however, that under physiological conditions all the enzymes mentioned are Ca^{2+} proteins. In a crystal structure determination, a Ca^{2+} binding loop has been located in residues 70 to 80 of bovine trypsin (107). Tryptophan-141 is near enough to the Ca^{2+} site to serve as the energy donor for enhanced Tb^{3+} luminescence. A similar loop occurs in all the serine proteases. Substitution of negatively charged glu-80 in trypsin with neutral isoleucine in chymotrypsin is consistent with significantly weaker Tb^{3+} and Ca^{2+} binding in the latter enzyme (37).

8 INNER FILTER EFFECTS (69)

Since only one, or a few, of many absorbing aromatic chromophores in proteins or nucleic acids are actually involved in energy transfer to Tb^{3+}, a large background absorption occurs. In a typical luminescence experiment, emitted light is measured at right angles to the incident light. The observation window for emitted light is usually not as wide as the optical cell containing the solution. In a highly absorbing solution, appreciable fractions of incident light are absorbed before the observation window for emitted light is reached. As a consequence, less than the optimum number of Tb^{3+} molecules achieve excited states and the resulting excitation spectrum is distorted. The distortions are greatest in the wavelength regions of highest absorption of incident light. It may be shown (69) that the luminescence intensity L is given by

$$L = K(10^{-A(1-f)/2} - 10^{-A(1+f)/2}) \tag{1}$$

where K is a proportionality constant, A is the absorbance of the solution, and f is the fraction of the entire optical cell length utilized by a cell-centered emission window. When the emission window becomes as wide as the solution in the cell, $f = 1$, and $L = K(1 - 10^{-A}) \simeq A \ln 10 - (A \ln 10)^2/2! + \ldots$. It is on the basis of the first term in the expansion that luminescence intensity is often assumed to be proportional to the absorbance. The proportionality $L \sim A$ is valid, however, only in the limit of low absorbances, $A < 0.1$. At absorbances $A > 0.1$, increasing amounts of inner-filtering occur. For values of $f \sim 0.5$, when $A \sim 1.0$, spurious maxima begin to appear in the excitation spectrum.

Figure 3 shows theoretical curves of excitation spectra versus frequency. The curves were constructed by assuming an absorption spectrum gaussian in

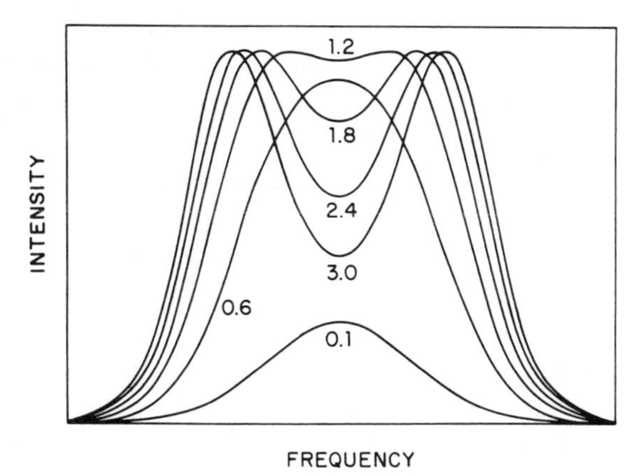

Figure 3 Excitation spectra intensity versus frequency at indicated values of the maximum absorbance in an absorption band. (From ref. 69.)

frequency and applying eq. 1 across the absorption band. All the curves refer to an emission window to whole solution width $f = 0.5$; similar results are obtained at other f values except $f = 1.0$, where only flattening appears. Numbers on the curves in Figure 3 indicate the maximum absorbance. Maxima appear in the center of the excitation spectra band at all $A > 0.9$, giving rise to two spurious maxima on both sides of the true maximum. Increasing the absorbance burns deeper holes at the maximum in the true excitation spectrum. To avoid the appearance of spurious maxima, excitation spectra should never be taken in solutions with absorbances greater than 0.8. Only two curves, with $A = 0.1$ and 0.6, appear normal in Figure 3. Actually, the maximum in the $A = 0.6$ curve is appreciably flattened, and occurs at only 3.4 instead of 6.0 times the intensity of the $A = 0.1$ curve (69).

Experimental excitation spectra similar to the double peaked theoretical curves in Figure 3 are easily realizable, and inadvertent examples appear in the literature. The theoretical curve with $A = 3.0$ in Figure 3 is plotted because the maximum absorbance corresponds to that of the solution of the Tb^{3+} complex of EHPG, the excitation spectrum of which is the pronounced double peaked curve, also labeled 3.0 in Figure 2. Numbers on curves in Figure 2 represent absorbances of the solutions. The similarities between the theoretical and experimental curves are evident: the spurious maximum at 307.5 nm in Figure 2 has the characteristics anticipated from the $A = 3.0$ curve in Figure 3. The shorter wavelength region of the double peaked curve in Figure 2 is distorted by an additional absorption band at still shorter wavelengths. The flattest excitation spectrum in Figure 2 corresponds closely to the theoretical curve for $A = 1.2$ in Figure 3. The remaining simple curve in Figure 2 is the excitation spectrum for a solution with absorbance of only 0.035, and offers the most reliable excitation spectrum for Tb^{3+} and EHPG. It corresponds in shape to the theoretical curve for $A = 0.1$ in Figure 3. Thus, each of the experimental curves in Figure 2 finds its close theoretical counterpart in Figure 3.

Two excitation spectra maxima for Tb^{3+} luminescence have been reported at 295 and near 260 nm for a solution containing 0.1 mM porcine trypsin (103). The maximum absorbance for this solution is estimated to be 3.5. At this high absorbance we predict a pair of spurious maxima at all f values <0.9994. The excitation spectrum for Tb^{3+} luminescence from porcine trypsin obtained in our research appears in reference 37 and is similar to that labeled trp for elastase in Figure 1. The true excitation spectrum of porcine trypsin shows a maximum at 277 nm with a tryptophan shoulder at 292 nm.

As anticipated from Figure 3, the true excitation spectrum maximum at 277 nm is midway between the pair of reported spurious maxima at 260 and 295 nm.

9 NUCLEIC ACIDS

In addition to inner filter effects, procuring reliable excitation spectra of enhanced Tb^{3+} emission due to energy transfer from nucleic bases is made difficult by the short ultraviolet absorption maxima near 260 nm, where there is a pronounced fall off of xenon lamp intensity. Virtually all published excitation spectra for Tb^{3+} and Eu^{3+} enhanced emission from nucleic bases suffer from one or both of these limitations. Figure 4 shows a corrected (by a ratiometric method) excitation spectrum for enhanced Tb^{3+} emission from 5′-GMP and its near correspondence with the absorption spectrum. Both curves exhibit similar shapes and show a maximum near 252 nm (108). In contrast, the excitation spectrum uncorrected for the wavelength dependence of the xenon lamp intensity displays little intensity near 252 nm and a spurious maximum 35 nm to the red at about 287 nm. Polyguanylic acid gives results similar to GMP (108). The uncorrected curve in Figure 4 appears several times in the lit-

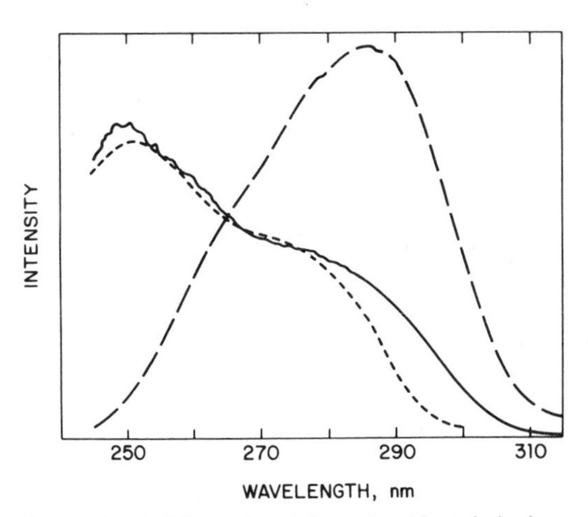

Figure 4 Corrected excitation (solid curve) and absorption (short dashed curve) spectra for a solution containing 42 μM 5′-GMP and 167 μM Tb^{3+} at pH 6.0. The excitation spectrum uncorrected for the wavelength dependence of the xenon lamp intensity (long dashed curve) exhibits a maximum near 287 nm. (From ref. 108.)

erature. It might be mistaken for the excitation spectrum for tryptophan energy transfer to Tb^{3+}.

Figure 5 shows excitation spectra for enhanced Tb^{3+} emission from polyguanylic acid (poly G) and polycytidylic acid (poly C). The maximum of the latter excitation spectrum lies at 270 nm, identical to that of the poly C absorption spectrum (108). Thus, for Tb^{3+} enhanced emission by energy transfer from aromatic chromophores in proteins and nucleic bases, the excitation spectrum at >250 nm is virtually identical to the absorption spectrum of the chromophore engaged in energy transfer.

Figure 5 also shows the excitation spectrum for enhanced Tb^{3+} emission at 545 nm from unfractionated *E. coli t*-RNA. The excitation spectra maxima in Figures 4 and 5 observed at 251 to 252 nm for GMP and poly G, and at about 253 to 254 nm for both *E. coli* and yeast (not shown) *t*-RNA, are markedly shorter than reported maxima at wavelengths of 287 to 290 nm (109–112). In comparison with the curve labeled trp for tryptophan in Figure 5, these spurious reported maxima lie even to the long wavelength side of the excitation spectra maxima in proteins. The probable cause for spurious maxima near 290 nm is the lack of correction for decrease of xenon lamp intensity at short wavelengths.

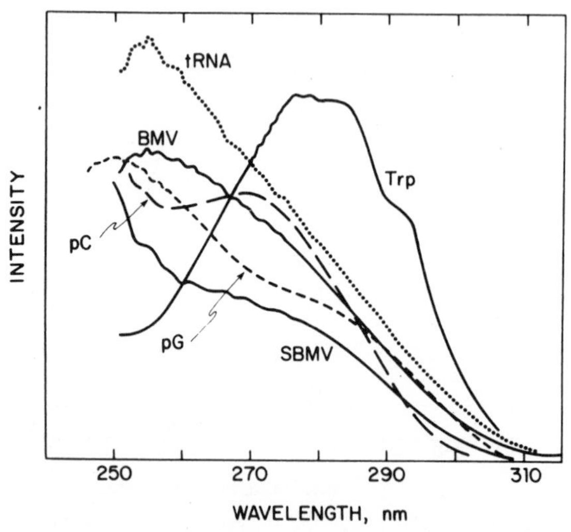

Figure 5 Excitation spectra for Tb^{3+} emission at 544 nm in solutions containing poly G, poly C, unfractionated *E. coli t*-RNA, bromegrass mosaic virus (BMV), southern bean mosaic virus (SBMV), and elastase (Trp). Intensities are not normalized. (From ref. 108.)

Comparison of the excitation spectrum for *t*-RNA with those for the homopolynucleotides in Figure 5 suggests that guanine bases are primarily responsible for enhanced Tb^{3+} luminescence in the 250 to 310 nm region for both unfractionated *t*-RNA samples. A contribution from cytosine accounts for the relatively featureless *t*-RNA excitation spectrum at >260 nm. At wavelengths longer than 310 nm in unfractionated *E. coli t*-RNA, a new maximum appears at 343 nm due to energy transfer from 4-thiouridine (110, 113).

We conclude that excitation spectra for enhanced Tb^{3+} emission owing to energy transfer from the nucleic bases strongly resemble the absorption spectra for the nucleic bases that serve as donors in the energy transfer process. Samples of RNA show excitation spectra maxima in the 250 to 260 nm region that are primarily the result of energy transfer to Tb^{3+} from guanine bases.

10 VIRUSES

Recent studies have shown the importance of Ca^{2+} in virus assembly and disassembly. It has been suggested that a low Ca^{2+} concentration within plant cells leads to virus disassembly and RNA release (114). We know that Ca^{2+} stabilizes the small icosahedral viruses, bromegrass mosaic virus (BMV) (114), and southern bean mosaic virus (SBMV) (115). This stabilization has been ascribed to cation bridges between carboxylate groups in protein subunits with a contribution from RNA phosphate (116).

Excitation spectra for Tb^{3+} enhanced emission for the two icosahedral viruses BMV and SBMV also appear in Figure 5. The spectra were obtained with about a fourfold excess of Tb^{3+} per nucleotide residue. Figure 5 shows strong similarities among the excitation spectra for poly G, *t*-RNA, BMV, and SBMV. Whatever differences exist among them (108), all of the excitation spectra contrast sharply with that for the trp side chain in proteins, also shown in Figure 5. Tryptophan has been found to be the predominant side chain for energy transfer to Tb^{3+} in proteins (40). Guanine is the major chromophore engaged in energy transfer to Tb^{3+} in BMV and SBMV, with some contributions from other bases. A small contribution from trp in the intact virus cannot be completely ruled out, but is not supported by the negligible enhancement observed for BMV capsids (108).

We conclude that guanine bases closely overlie putative Ca^{2+} binding sites in BMV and SBMV viruses. The simplest explanation would have Tb^{3+} (Ca^{2+}) bound to nucleic acid phosphate near a guanine base. The Tb^{3+} is not bound to the guanine because energy transfer from a nucleic base is anticipated to be due to a through space interaction. Both Ca^{2+} and Ln^{3+} are bound to the phosphate and not the base portion of nucleotides (59). It is possible, as has been suggested (116), for Ca^{2+} and Tb^{3+} to be bound in sites formed by RNA-phosphates and protein carboxylates at a RNA-protein interface. A guanine base, but not a trp side chain, is near this interfacial metal ion binding site in the virus.

REFERENCES

1. H. Rasmussen, P. Jensen, W. Lake, N. Friedmann, and D. B. P. Goodman, *Adv. Cyclic Nucl. Res.*, **5,** 375 (1975).
2. R. H. Kretsinger and D. J. Nelson, *Coord. Chem. Rev.*, **18,** 29 (1976).
3. J. R. Blinks, F. G. Prendergast, and D. G. Allen, *Pharmacol. Rev.*, **28,** 1 (1976).
4. H. Sigel and R. B. Martin, *Chem. Rev.*, **82,** 385 (1982).
5. H. Einspahr and C. E. Bugg, in *Calcium-Binding Proteins and Calcium Function*, R. H. Wasserman et al., Eds., Elsevier North-Holland, New York, 1977, p. 13.
6. J. Stenflo, P. Fernlund, W. Egan, and P. Roepstorff, *Proc. Natl. Acad. Sci.*, **71,** 2730 (1974). G. L. Nelsestuen, T. H. Zytkovicz, and J. B. Howard, *J. Biol. Chem.*, **249,** 6347 (1974). S. Magnusson, L. Sottrup-Jensen, T. E. Peterson, H. R. Morris, and A. Dell, *FEBS Lett.*, **44,** 189 (1974). P. V. Hauschka and S. A. Carr, *Biochemistry*, **21,** 2538 (1982).
7. A. E. Martell and R. M. Smith, *Critical Stability Constants*, Plenum Press, London, 1974.
8. M. R. Christy, R. M. Barkley, T. H. Koch, J. J. Van Buskirk, and W. M. Kirsch, *J. Am. Chem. Soc.*, **103,** 3935 (1981).
9. C. M. Frey and J. Stuehr, *Met. Ions Biol. Syst.*, **1,** 51 (1974).
10. R. M. Kretsinger, in *Calcium-Binding Proteins and Calcium Function*, R. H. Wasserman et al., Eds., Elsevier North-Holland, New York, 1977, p. 63.
11. R. Prados, L. G. Statdherr, H. Donato, Jr., and R. B. Martin, *J. Inorg. Nucl. Chem.*, **36,** 689 (1974).
12. E. Nieboer, *Struct. Bond.*, **22,** 1 (1975).
13. R. B. Martin and F. S. Richardson, *Quart. Rev. Biophys.*, **12,** 181 (1979).
14. C. G. dos Remedios, *Cell Calcium*, **2,** 29 (1981).
15. R. H. Kretsinger, *Ann. Rev. Biochem.*, **45,** 239 (1976).
16. R. D. Shannon, *Acta Crystallogr.*, **A32,** 751 (1976).
17. N. Sperelakis, R. Valle, C. Orozco, A. Martinez-Palomo, and R. Rubio, *Am. J. Physiol.*, **225,** 793 (1970). K. Hermsmeyer and N. Sperelakis, *Am. J. Physiol.*, **219,** 1108 (1970).
18. P. C. Moews and R. H. Kretsinger, *J. Mol. Biol.*, **91,** 229 (1975).
19. J. Sowadski, G. Cornick, and R. H. Kretsinger, *J. Mol. Biol.*, **124,** 123 (1978).

20. B. W. Matthews and L. H. Weaver, *Biochemistry*, **13,** 1719 (1974).
21. B. Furie, A. Eastlake, A. N. Schechter, and C. B. Anfinsen, *J. Biol. Chem.*, **248,** 5821 (1973).
22. B. C. Furie and B. Furie, *J. Biol. Chem.*, **250,** 601 (1975).
23. G. Meissner, *Biochim. Biophys. Acta*, **298,** 906 (1973). S. Yamada and Y. Tonomura, *J. Biochem.*, **72,** 417 (1972).
24. P. Tanswell, E. W. Westhead, and R. J. P. Williams, *FEBS Lett.*, **48,** 60 (1974); *Europ. J. Biochem.*, **63,** 249 (1976).
25. K. M. Valentine and G. L. Cottam, *Arch. Biochem. Biophys.*, **158,** 346 (1973). G. L. Cottam, K. M. Valentine, B. C. Thompson, and A. D. Sherry, *Biochemistry*, **13,** 3532 (1974).
26. B. S. Cooperman and N. Y. Chiu, *Biochemistry*, **12,** 1670 (1973). L. G. Butler and J. W. Sperow, *Bioinorg. Chem.*, **7,** 141 (1977).
27. G. E. Smolka, E. R. Birnbaum, and D. W. Darnall, *Biochemistry*, **10,** 4556 (1971). D. W. Darnall and E. R. Birnbaum, *Biochemistry*, **12,** 3489 (1973).
28. R. D. Hershberg, G. H. Reed, A. J. Slotboom, and G. H. deHaas, *Biochemistry*, **15,** 2268 (1976).
29. A. D. Sherry, A. D. Newman, and C. G. Gutz, *Biochemistry*, **14,** 2191 (1975).
30. K. T. Izutsu, S. P. Felton, I. A. Siegal, W. T. Yoda, and A. C. N. Chen, *Biochem. Biophys. Res. Commun.*, **49,** 1034 (1972). O. Shimomura and F. H. Johnson, *Biochem. Biophys. Res. Commun.*, **53,** 490 (1973).
31. E. T. O'Keefe, R. L. Hill, and J. E. Bell, *Biochemistry*, **19,** 4954 (1980).
32. J. A. Barden and C. G. Dos Remedios, *Biochim. Biophys. Acta*, **624,** 163 (1980); *Biochem. Biophys. Res. Commun.*, **86,** 529 (1979). A. Ferri and E. Grazi, *Biochemistry*, **20,** 6362 (1981).
33. H. Rubsamen, G. P. Hess, A. T. Eldefrawi, and M. E. Eldefrawi, *Biochem. Biophys. Res. Commun.*, **68,** 56 (1976). H. Rubsamen, M. Montgomery, G. P. Hess, A. T. Eldefrawi, and M. E. Eldefrawi, *Biochem. Biophys. Res. Commun.*, **70,** 1020 (1976).
34. B. C. Furie, K. G. Mann, and B. Furie, *J. Biol. Chem.*, **251,** 3235 (1976).
35. J. E. Gomez, E. R. Birnbaum, and D. W. Darnall, *Biochemistry*, **13,** 3745 (1974). D. W. Darnall and E. R. Birnbaum, *Met. Ions Biol. Sys.*, **6,** 251 (1976).
36. M. Epstein, A. Levitzki, and J. Reuben, *Biochemistry*, **13,** 1777 (1974).
37. J. de Jersey, R. S. Lahue, and R. B. Martin, *Arch. Biochem. Biophys.*, **205,** 536 (1980).
38. E. R. Birnbaum, F. Abbott, J. E. Gomez, and D. W. Darnall, *Arch. Biochem. Biophys.*, **179,** 469 (1977).
39. J. L. Dimicoli and J. Bieth, *Biochemistry*, **16,** 5532 (1977).
40. H. G. Brittain, F. S. Richardson, and R. B. Martin, *J. Am. Chem. Soc.*, **98,** 8255 (1976).
41. M. S. Kayne and M. Cohn, *Biochem. Biophys. Res. Commun.*, **46,** 1285 (1972).
42. D. R. Quiram and R. M. Weinshilboum, *Biochem. Pharmacol.*, **25,** 1727 (1976).
43. E. R. Birnbaum, J. E. Gomez, and D. W. Darnall, *J. Am. Chem. Soc.*, **92,** 5287 (1970).
44. K. Kurachi, L. C. Sieker, and L. H. Jensen, *J. Biol. Chem.*, **250,** 7663 (1975).
45. A. Gafni and I. Z. Steinberg, *Biochemistry*, **13,** 800 (1974).
46. F. C. Wedler and V. D'Aurora, *Biochim. Biophys. Acta*, **371,** 432 (1974).
47. E. M. Stephens and C. M. Grisham, *Biochemistry*, **18,** 4876 (1979).

48. B. Furie, J. H. Griffen, R. J. Feldman, E. A. Sokoloski, and A. N. Schechter, *Proc. Natl. Acad. Sci.*, **71,** 2833 (1974).

49. R. Jones, R. A. Dwek, and S. Forsen, *Eur. J. Biochem.*, **27,** 548 (1974). I. D. Campbell, C. M. Dobson, and R. J. P. Williams, *Proc. Roy. Soc.*, **A345,** 41 (1975).

50. J. Reuben, *Biochemistry*, **10,** 2834 (1971); *J. Phys. Chem.*, **75,** 3164 (1971).

51. R. A. Dwek, J. C. A. Knott, D. Marsh, A. C. McLaughlin, E. M. Press, N. C. Price, and A. I. White, *Eur. J. Biochem.*, **53,** 25 (1975).

52. L. Lee and B. D. Sykes, *Biochemistry*, **20,** 1156 (1981); **19,** 3208 (1980).

53. D. J. Nelson, A. D. Theoharides, A. C. Nieburgs, R. K. Murray, F. Gonzalez-Fernandez, and D. S. Brenner, *Int. J. Quantum Chem.*, **16,** 159 (1979).

54. J. Reuben, *J. Mag. Reson.*, **11,** 103 (1973).

55. W. C. Hutton, P. L. Yeagle, and R. B. Martin, *Chem. Phys. Lipids*, **19,** 255 (1977).

56. J. Wilschut, N. Duzgunes, R. Fraley, and D. Papahadjopoulos, *Biochemistry*, **19,** 6011 (1980).

57. F. S. Richardson and J. P. Riehl, *Chem. Rev.*, **77,** 773 (1977).

58. S. A. Davis and F. S. Richardson, *J. Inorg. Nucl. Chem.*, **42,** 1793 (1980).

59. R. B. Martin and Y. H. Mariam, *Met. Ions Biol. Syst.*, **8,** 57 (1979).

60. R. A. Dwek, *Nuclear Magnetic Resonance in Biochemistry*, Clarendon Press, Oxford, 1973.

61. D. R. Burton, S. Forsen, R. A. Dwek, A. C. McLaughlin, and S. Wain-Hobson, *Eur. J. Biochem.*, **71,** 519 (1976).

62. W. D. Horrocks, Jr. and D. R. Sudnick, *J. Am. Chem. Soc.*, **101,** 334 (1979).

63. W. D. Horrocks, Jr. and D. R. Sudnick, *Acc. Chem. Res.*, **14,** 384 (1981).

64. A. P. Snyder, D. R. Sudnick, V. K. Arkle, and W. D. Horrocks, Jr., *Biochemistry*, **20,** 3334 (1981).

65. M. Rhee, D. R. Sudnick, V. K. Arkle, and W. D. Horrocks, Jr., *Biochemistry*, **20,** 3328 (1981).

66. T. Forster, *Disc. Farad. Soc.*, **27,** 7 (1959).

67. V. G. Berner, D. W. Darnall, and E. R. Birnbaum, *Biochem. Biophys. Res. Commun.*, **66,** 763 (1975). W. D. Horrocks, Jr., B. Holmquist, and B. L. Vallee, *Proc. Natl. Acad. Sci. USA*, **72,** 4764 (1975).

68. P. O'Hara, S. M. Yeh, C. F. Meares, and R. Bersohn, *Biochemistry*, **20,** 4704 (1981).

69. J. de Jersey, P. J. Morley, and R. B. Martin, *Biophys. Chem.*, **13,** 233 (1981).

70. J. W. Longworth, in *Excited States of Proteins and Nucleic Acids*, R. F. Steiner and I. Weinryb, Eds., Plenum Press, New York, 1971.

71. E. A. Burstein, E. A. Permyakov, V. I. Emelyanenko, T. L. Busheva, and J. F. Pechere, *Biochem. Biophys. Acta*, **400,** 1 (1975).

72. W. D. Horrocks, Jr. and W. E. Collier, *J. Am. Chem. Soc.*, **103,** 2856 (1981).

73. R. Prados, R. K. Boggess, R. B. Martin, and R. C. Woodworth, *Bioinorg. Chem.*, **4,** 135 (1975).

74. R. K. Boggess and R. B. Martin, *J. Am. Chem. Soc.*, **97,** 3076 (1975).

75. J. de Jersey and R. B. Martin, *Biochemistry*, **19,** 1127 (1980).

76. R. C. Warner and I. Weber, *J. Am. Chem. Soc.*, **75,** 5094 (1953).

77. R. B. Martin, J. T. Edsall, D. B. Wetlaufer, and B. R. Hollingworth, *J. Biol. Chem.*, **233,** 1429 (1958).
78. C. J. Coffee and R. A. Bradshaw, *J. Biol. Chem.*, **248,** 3305 (1973).
79. C. E. Nockolds, R. H. Kretsinger, C. J. Coffee, and R. A. Bradshaw, *Proc. Natl. Acad. Sci. USA*, **69,** 581 (1972).
80. W. A. Hendrickson and J. Karle, *J. Biol. Chem.*, **248,** 3327 (1973).
81. R. H. Kretsinger and C. E. Nockolds, *J. Biol. Chem.*, **248,** 3313 (1973). P. C. Moews and R. H. Kretsinger, *J. Mol. Biol.*, **91,** 201 (1975).
82. H. Donato, Jr. and R. B. Martin, *Biochemistry*, **13,** 4575 (1974).
83. D. J. Nelson, T. L. Miller, and R. B. Martin, *Bionorg. Chem.*, **7,** 325 (1977).
84. T. L. Miller, D. J. Nelson, H. G. Brittain, F. S. Richardson, R. B. Martin, and C. M. Kay, *FEBS Lett.*, **58,** 262 (1975).
85. H. G. Brittain, F. S. Richardson, R. B. Martin, L. D. Burtnick, and C. M. Kay, *Biochem. Biophys. Res. Commun.*, **68,** 1013 (1976).
86. T. L. Miller, personal communication, 1977.
87. T. L. Miller, R. M. Cook, D. J. Nelson, and A. D. Theoharides, *J. Mol. Biol.*, **141,** 223 (1980).
88. M-J. Rhee, D. R. Sudnick, V. K. Arkle, and W. D. Horrocks, Jr., *Biochemistry*, **20,** 3328 (1981).
89. A. Cave, J. Parello, T. Drakenberg, E. Thulin, and B. Lindman, *FEBS Lett.*, **100,** 148 (1979).
90. A. Cave, M-F. Daures, J. Parello, A. Saint-Yves, and R. Sempere, *Biochimie*, **61,** 755 (1979).
91. J. D. Potter and J. Gergely, *J. Biol. Chem.*, **250,** 4628 (1975).
92. I. L. Sin, R. Fernandes, and D. Mercola, *Biochem. Biophys. Res. Commun.*, **82,** 1132 (1978).
93. J-P. van Eerd and K. Takahashi, *Biochem. Biophys. Res. Commun.*, **64,** 122 (1975).
94. P. C. Leavis and E. L. Kraft, *Arch. Biochem. Biophys.*, **186,** 411 (1978).
95. J. H. Collins, *Biochem. Biophys. Res. Commun.*, **58,** 301 (1974).
96. C-L. A. Wang, P. C. Leavis, W. D. Horrocks, Jr., and J. Gergely, *Biochemistry*, **20,** 2439 (1981).
97. W. Y. Cheung, *Science*, **207,** 19 (1980). C. B. Klee, T. H. Crouch, and P. G. Richman, *Ann. Rev. Biochem.*, **49,** 489 (1980).
98. D. M. Watterson, F. Sharief, and T. C. Vanaman, *J. Biol. Chem.*, **255,** 962 (1980).
99. T. H. Crouch and C. B. Klee, *Biochemistry*, **19,** 3692 (1980). J. Haiech, C. B. Klee, and J. G. Demaille, *Biochemistry*, **20,** 3890 (1981).
100. M-C. Kilhoffer, J. G. Demaille, and D. Gerard, *FEBS Lett.*, **116,** 269 (1980). M. C. Kilhoffer, D. Gerard, and J. G. Demaille, *FEBS Lett.*, **120,** 99 (1980).
101. F. Abbott, D. W. Darnall, and E. R. Birnbaum, *Biochem. Biophys. Res. Commun.*, **65,** 241 (1975). F. Abbott, J. E. Gomez, E. R. Birnbaum, and D. W. Darnall, *Biochemistry*, **14,** 4935 (1975).
102. G. Duportail, J-F. Lefevre, P. Lestienne, J-L. Dimicoli, and J. G. Bieth, *Biochemistry*, **19,** 1377 (1980).
103. M. Epstein, J. Reuben, and A. Levitzki, *Biochemistry*, **16,** 2449 (1977).

104. D. W. Darnall, F. Abbott, J. E. Gomez, and E. R. Birnbaum, *Biochemistry*, **15,** 5017 (1976).
105. J. Osborne, A. Lunasin, and R. F. Steiner, *Biochem. Biophys. Res. Commun.*, **49,** 923 (1972).
106. G. Voordouw, C. Milo, and R. S. Roche, *Biochemistry*, **15,** 3716 (1976).
107. W. Bode and P. Schwager, *J. Mol. Biol.*, **98,** 693 (1975); *FEBS Lett.*, **56,** 139 (1975).
108. P. J. Morley, R. B. Martin, and S. Boatman, *Biochem. Biophys. Res. Commun.*, **101,** 1123 (1981).
109. C. Formoso, *Biochem. Biophys. Res. Commun.*, **53,** 1084 (1973).
110. M. S. Kayne and M. Cohn, *Biochemistry*, **13,** 4159 (1974).
111. M. D. Topal and J. R. Fresco, *Biochemistry*, **19,** 5531 (1980).
112. D. P. Ringer, S. Burchett, and D. E. Kizer, *Biochemistry*, **17,** 4818 (1978).
113. T. M. Wolfson and D. R. Kearns, *Biochemistry*, **14,** 1436 (1975).
114. A. C. H. Durham, D. A. Henry, and M. B. von Wechmar, *Virology*, **77,** 524 (1977).
115. I. Rayment, J. E. Johnson, and M. G. Rossmann, *J. Biol. Chem.*, **254,** 5243 (1979).
116. P. Pfeiffer and A. C. H. Durham, *Virology*, **81,** 419 (1977).

Index